Applied Ecology

EDWARD I. NEWMAN
School of Biological Sciences, University of Bristol, England

b
Blackwell
Science

Editorial Offices:
Osney Mead, Oxford OX2 0EL
25 John Street, London WC1N 2BL
23 Ainslie Place, Edinburgh EH3 6AJ
238 Main Street, Cambridge
Massachusetts 02142, USA
54 University Street, Carlton
Victoria 3053, Australia

Other Editorial Offices:
Arnette Blackwell SA
224, Boulevard Saint Germain
75007 Paris, France

Blackwell Wissenschafts-Verlag GmbH
Kurfürstendamm 57
10707 Berlin, Germany

Zehetnergasse 6
A-1140 Wien, Austria

First published 1993
Reprinted 1994, 1995, 1996

Set by Excel Typesetters, Hong Kong
Printed and bound in Great Britain
at the University Press, Cambridge

DISTRIBUTORS
Marston Book Services Ltd
PO Box 269
Abingdon
Oxon OX14 4YN
(*Orders:* Tel: 01235 465500
Fax: 01235 465555)

USA
Blackwell Science, Inc.
238 Main Street
Cambridge, MA 02142
(*Orders:* Tel: 800 215-1000
617 876-7000
Fax: 617 492-5263)

Canada
Copp Clark, Ltd
2775 Matheson Blvd East
Mississauga, Ontario
Canada, L4W 4P7
(*Orders:* Tel: 800 263-4374
905 238-6074)

Australia
Blackwell Science Pty Ltd
54 University Street
Carlton, Victoria 3053
(*Orders:* Tel: 03 9347 0300
Fax: 03 9349 3016)

A catalogue record for this title
is available from the British Library

ISBN 0-632-03657-5

Library of Congress
Cataloging-in-Publication Data

Newman, E. I.
Applied ecology / Edward I. Newman.
p. cm.
Includes bibliographical references and index.
ISBN 0-632-03657-5
1. Environmental sciences. 2. Ecology.
I. Title.
GE105.N48 1993
333.95—dc20

Contents

Preface

The world faces very serious environmental problems. This book is about what science—and especially biological science—can do to help. The book deals with a wide range of topics which are usually covered in separate books by different experts, who may well have been trained in different university departments—biological science, environmental science, forestry, agriculture, range science, fisheries and wildlife, marine science and others. Here the topics are covered in a single book written by one person. I have written such a book because I believe the world needs people who have studied a wide range of environmental problems, who understand how they relate to each other and how they are based on underlying principles of ecological science.

How did I come to be interested in all these aspects of applied ecology? Part of the credit must go to two places where I lived during formative periods of my childhood: Keene Valley, among the forests and lakes of the Adirondack Mountains in northern New York state; and Comberton, Cambridgeshire, in the fertile farmlands of eastern England. In Keene Valley many people were concerned about keeping the Adirondacks 'forever wild'. But in Comberton I lived among people whose main aim was to make a living by growing as much food as possible. This book is about exploitation of resources—including how to obtain timber, fish and meat on a sustainable basis, how to maintain the fertility of farm soils, how to control pests and diseases in crops—as well as about how to preserve wild species and prevent damage by polluting chemicals.

In 1992 much of the Adirondack region is still dense forest, in spite of acid rain and Olympic skiing. The forest still grows close to the house in Keene Valley where I lived as a child, chipmunks still play, the stream still flows clear. In contrast, the countryside around Comberton has changed greatly; Dutch elm disease killed off many of the trees, but the primary cause of the changes has been new farming practices. This book does not assume that change is always bad (or always good). It does not assume that modern farming methods are all bad, or that cutting down trees in ancient forest is always wrong, or that we should cure all pollution by stopping the production of all the chemicals that cause it. The real world is not so simple.

Sometimes science can *suggest solutions* to ecological problems: for example, ways of controlling diseases or minimizing effects of pollution. Sometimes it can *answer practical questions*, such as how many fish we can take from an ocean this year without reducing the catch in future years. Sometimes it can help with the *resolution of conflicts*, for example over alternative demands on land. This book is concerned with each of those aspects of applied ecology. The actual resolving of conflicts is likely to involve politics, but this book steers clear of politics: it aims to provide the scientific background to help people make wise decisions.

To find out more on what this book is about, I suggest you look at the 'Questions' section at the start of each of the main chapters (Chapters 2–8). Then you could start reading the Introduction (Chapter 1), which will lead you into the heart of the book. You do not hear any more about my life, past or present: science must be based on sound evidence, not on anecdotes.

I hope there will be a second edition of this book, in due course. If you have comments or suggestions about how to make the second edition even better, please write to me. I promise to reply.

Edward I. Newman

Chapter 1: Introduction

This chapter explains what the book is about, and who it is intended for. It also says something about how the book is organized.

This book is about how ecology can be applied to important practical problems. Some people may think that ecology and concern for the environment are luxuries for the rich, but this book is about fundamentals that affect everyone, such as food, fuel, timber, harmful chemicals and climate change. It is also about things that affect the quality of almost everyone's life, including landscape, clean rivers and lakes, diseases of animals and plants, and the preservation of wild species.

Why things cannot all stay unchanged

Table 1.1 and Fig. 1.1 present the basic ecological problem facing humankind: we live in a world of fixed size, but the human population is increasing. During the 40 yrs from 1950 to 1990 the population of the world doubled (Fig. 1.1(a)). Part (b) of Fig. 1.1 shows how the population changed in the continents that had the fastest increase (Latin America) and the slowest (Europe). This population growth underlies problems that feature in every chapter of this book: it increases the world demand for food, timber and other biological products, which in turn increases the pressure to convert wildlife habitats to other uses. As the human population increases it is almost inevitable that there is more production of many industrial chemicals, some of which are harmful pollutants; and also increased use of fossil fuels, producing more carbon dioxide which contributes to climate change. I have not extrapolated the curves in Fig. 1.1 into the future; but it is virtually certain that the world's human population will continue to increase for at least several decades. Therefore our aim as ecologists cannot be simply to keep things the same as they are now, or as they were at some supposedly perfect time in the past: that is not an option. Our ultimate aim should be to make things better, for all people of the world and for the other species in it too. But often this book will be about how to maintain things as good as they are now, the difficult task of preventing them from getting worse. And sometimes we must accept that harmful changes will occur, for example amounts of some pollutants will increase and areas of natural ecosystems will be destroyed; the most useful thing ecologists can do then may be to give advice on how to minimize the harm.

'Sustainable', a key word

The word *sustainable* occurs often in this book. 'Sustainable' does not mean the same as 'renewable'. For example, rice is a renewable resource, since after we harvest the grain and eat it we can grow some

Table 1.1 Area of water, land and main land uses on the earth. Land use data are for 1989, from *FAO Production Yearbook 1990*

	Area (million km^2)	Percentage of whole earth	Percentage of land area
Whole earth	510		
Oceans	376	74	
Fresh water	3	<1	
Land	131	26	
Crops	14.8		11
Permanent grazing land	33.0		25
Forest and woodland	40.9		31
Other land*	42.1		32

* Includes ice, tundra, desert, towns.

more. But it is only sustainable if the farm can continue to produce as high a yield of rice year after year indefinitely. If the structure of the soil or its nutrient status becomes less favourable, or soil is lost by erosion, or if insects or fungi harmful to the rice plant increase, then rice production may not be sustainable long-term. So we need to consider the whole system, not just a single species. This book assumes that 'Is the system sustainable long-term?' is an important question. Some people might, on the contrary, adopt the attitude 'as long as things keep going all right during my life-time, what happens after that does not concern me'. That is not a point of view taken in this book.

Pure science and applied science

This book is not just basic ecology with applications tacked on to the end of each chapter; applied ecology is the heart of it. One way to express the difference between pure and applied science is 'pure science is problem-orientated, applied science is solution-orientated'. Pure scientists are happy if they can end a scientific paper with the words 'this research has asked more questions than it answered': that leaves plenty of opportunity for more research, which is what they like doing. I am sympathetic to that attitude: I have spent much of my working life doing research that was motivated by intellectual curiosity. But when there are important practical problems to be solved the statement 'we cannot answer that question until we have done more research' may well be unacceptable. We probably need an answer now, or anyway soon. Whether the problem is climate change, declining fish catches or a new insect pest spreading into a forest, waiting even a few years may well make the problem more difficult to solve. So the challenge in applied ecology is often to reach the best decisions possible on the basis of present information. This book is much concerned with existing information and evidence, and what we can conclude from it. It does not hestitate to say when more research is needed, but it tries to make the most of what knowledge we have.

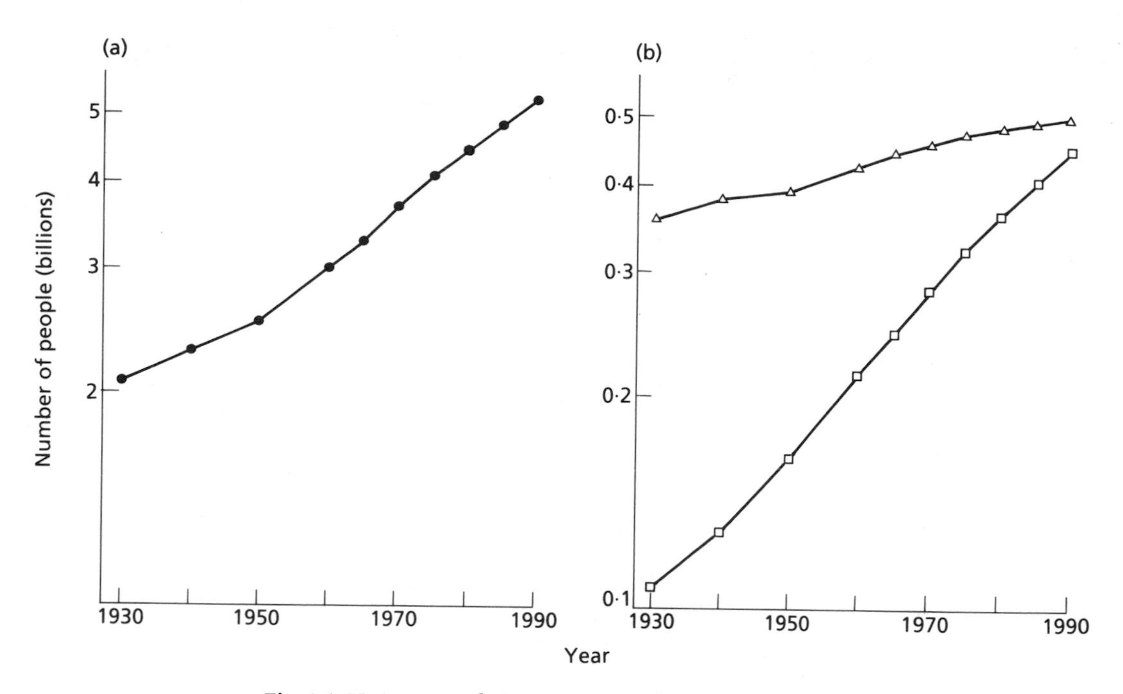

Fig. 1.1 Human population 1930–90. The vertical axes are on log scales. (a) Whole world. (b) △, Europe, excluding U.S.S.R.; □, South and Central America. Data from *UN Statistical Yearbooks* and *FAO Production Yearbook 1990.*

Applied ecology approached at a fundamental level

Table 1.2 shows, for some of the main groups of animals, plants and microbes, approximately how many thousands of species there are in the world. This serves to illustrate why, although this book is about applied problems, it approaches them at a fundamental level. We wish to preserve as many as possible of those species, but we do not have the time or resources to do research on every one of them; therefore their conservation and management must be based substantially on fundamental scientific understanding. So the chapter on conservation considers questions such as 'How can management promote high species diversity?'; 'If a large area of natural vegetation is reduced to small, separated fragments, how will this affect the plant and animal species living in it?'; 'In some ecosystems is there a *keystone* species, whose extinction would have a major effect on the whole ecosystem?'. Another example is biological control of pests and diseases. Some books deal with this case by case, describing in turn each pest species and its successful control. Here, Chapter 6 instead considers basic questions such as 'Can we tell in advance whether a particular species is likely to be a successful biological control agent?'; 'Are species that initially provide effective control of a pest likely to evolve to become less effective?' In these and other chapters the questions are answered with the aid of examples—particular ecosystems, particular species, particular pollutant chemicals—examples chosen to illuminate the

Table 1.2 Approximate number of species in major groups. Based on Table 1.1 of Holdgate (1991)

	Thousands
Mammals	4
Birds	9
Reptiles	6
Amphibians	4
Teleost fishes	19
Crustaceans	40
Insects	750*
Molluscs	50
Angiosperms	250
Gymnosperms	1
Bryophytes	17
Fungi	50
Cyanobacteria	2
Bacteria	4

* Probably several million other species not yet described.

fundamental question, to provide scientific evidence, but never aiming to be a complete list of all those that have been studied.

Within the book I have tried to maintain some balances. It shows how ecology at different levels—ecosystem, population, physiological, biochemical, genetic and evolutionary—can all contribute to applied ecology. The book says much about plants and animals; and micro-organisms have not been forgotten. Oceans, fresh water and land all feature. On land there are examples from temperate and tropical regions, and from different continents, although you may detect a preponderance of examples from western Europe and North America. These balances are maintained within the whole book, not necessarily within each chapter—inevitably the chapter on fisheries is mainly about animals, and the chapter on timber production mainly about plants. In order to cover subjects with adequate scientific rigour I have had to be selective. Topics that could have formed whole chapters do not appear, and within each chapter there are major subject areas not covered. If I have missed out something you particularly wanted to read about, I apologize; but a book twice as long would not necessarily have been more useful to you.

Who is the book for?

This book has been written primarily for undergraduates studying biological science. It should also be useful to students of environmental science and geography, and to many other people who want to find out about the scientific background to current ecological problems, provided they accept the book's strong biological emphasis. For example, the section on global climate change passes rather rapidly over the difficulties of predicting how increases in greenhouse gases will affect future climate, and pays much more attention to how plants and animals will respond to changes in climate and carbon dioxide concentration. The chapter on pollution skims briefly over how pollutant chemicals are produced and dispersed, and is mainly concerned with their effects on living things, how to predict these effects and how to minimize them. That chapter says virtually nothing about rules and regulations controlling release of pollutants. The book as a whole touches only occasionally on social systems, laws, politics or economics. Applied ecology interacts with all of these, and it is important that some ecologists should be involved with them, to take an active part in the making of decisions and in their application. But this book sticks to the scientific side.

The book is aimed primarily at people who want to find out about how ecology can contribute to solution of a wide range of practical problems. It may be that you are only interested in one aspect. It should be possible to benefit from reading any chapter of this book on its own; there are many cross-connections between chapters, but each chapter should be comprehensible without the others. But if you are really only interested in pollution, or fisheries management, or pest control you may well find the relevant section in this book too short, and you would probably do better to look for a whole book on your chosen subject. A suitable book will probably be suggested in 'Further Reading' at the end of one of the chapters in this book.

What I expect you to know already

What do you need to know to understand this book? I have assumed some knowledge of basic ecology and other aspects of biology, such as would occur in an introductory biology course at university. You also need some knowledge of basic physics and chemistry, such as any biology or environmental science student at university should have. What about mathematics? Ecology is a quantitative subject. This chapter and every other chapter in the book contain graphs and tables of numbers which are essential to the subject matter. But the mathematics in the book is sparse and simple. Mathematical models are important in ecology, and feature in several chapters, but they are usually presented by words and graphs rather than by equations. There are a few exceptions: the densest mass of equations is in Box 6.2, so if you can cope with that the maths elsewhere in the book should be no problem for you. You also need to know a little about statistics, enough to understand what a correlation coefficient shows and what is meant by 'this difference is statistically significant ($P < 0.01$)'.

I hope this book will be useful to people in many countries. Therefore I have not assumed that you, Dear Reader, have personal experience of any particular country. If you are American you must excuse my explaining where Yellowstone National Park is, just as

British readers must put up with being told what the Norfolk Broads are.

How the book is organized

Following this short introductory chapter, there are seven main chapters. Of these, five have each a clear subject area to cover: Chapter 3, food from farming; 4, fish from the sea; 5, timber from forests; 7, pollution; 8, conservation of wild species. The other two each include several topics, and deserve a little more explanation. Chapters 3–5 could be said to have the common theme, how to make useful organic materials using energy from the sun, and Chapter 2 starts by providing some background science for that. It then considers the use of non-renewable energy sources by people, and whether biological sources could replace them. And the final section of that chapter is about a product of fossil fuels, CO_2, its likely effect on global climate and how living things may respond to climate change. So this last part can be seen as a foretaste of the Pollution chapter, if you view CO_2 as a pollutant.

The themes of Chapters 2 and 6

Much of Chapter 6 (Invaders and Pests) is about diseases and herbivorous insects, and ways of controlling them. However, the early part of the chapter is about the biologically related topic of invaders, i.e. about animals, plants or microbes that arrive naturally or are introduced to a new area, whether they establish and if so how fast they spread. We may want them to establish (e.g. biological control agents) or we may not. Release of genetically engineered organisms poses special risks and receives special attention in that chapter.

Each chapter after this one starts with a section headed 'Questions', which indicates the main applied problems around which the chapter is constructed. Below that, under 'Background science' are listed more fundamental topics that are covered in the chapter, to allow a more thorough understanding of how the applied problems can be tackled. In each chapter the background science is interwoven with its application, but separation of the 'Questions' and 'Background science' at the start will, I hope, make clearer the aims of each chapter. So these two lists may not show you the structure of the chapter. To find your way about within the chapter, you need to use the headings. You will find that headings within the text are fairly sparse, since I do not want to divide the text up unnecessarily. I have preferred to use many side headings, in the margin, to show you what is going on and how the subject matter is developing. The book has an index, of course, which provides another route to a particular topic you want to read about.

This is a side heading

The book does not have summaries. At the end of each chapter there are 'Conclusions'. They are not all the conclusions from the chapter, just a selection of them; and some of them are a dangerous simplification of what was said earlier in the text. So if you read the Conclusions and nothing else you will miss a lot.

The glossary

There is a glossary near the end of the book. It gives the meanings of technical and specialist words, and of strings of initials. However, you are expected to know the more basic scientific terms. For example, none of the groups named in Table 1.2 is defined or described in the glossary. In the text I call a species by its English name, if it has one that is widely used and precise enough. If not, the Latin name is used; this applies to most invertebrate animals and most microbial species. If the Latin name is used the glossary may tell you what major group the species belongs to. If the English name on its own has been used in the text the glossary will give the Latin name.

I have enjoyed writing this book. I hope you will enjoy reading it.

Further reading

Ecology texts:
- Begon *et al.* (1990)
- Ricklefs (1990)
- Chapman & Reiss (1992)

How people have interacted with, and changed, their environment:
- Simmons (1989)

Economics of natural resources:
- Common (1988)

Chapter 2: Energy, Carbon Balance and Climate Change

Questions

- Are present rates of food and wood production near the upper limit set by incoming solar radiation?
- Could fuels from plants replace fossil fuels?
- How fast is CO_2 concentration in the atmosphere increasing? Where does it come from?
- What effect will increases in CO_2 and other gases have on world climate?
- Could we stop the CO_2 increase by planting more trees?
- What effect will increasing CO_2 and temperature have on crop production? And on wild plants and animals?

Background science

- Solar energy reaching the earth and what happens to it.
- Energy used by people: how much and where it comes from.
- Photosynthesis and primary productivity.
  Its efficiency in capture of solar energy.
  What limits that efficiency.
  Ways of measuring productivity in the field, including use of satellites.
- To grow biomass containing 1 GJ of energy:
  How much land area is needed?
  How much energy is needed?
- The carbon cycle of the earth: processes, amounts, rates.
- The greenhouse effect: how it operates, what gases contribute to it.
- How climate changed in the past.
  How plants and animals responded.

All life depends on energy. Nearly all of that energy comes ultimately from the sun: chlorophyll-containing plants and microorganisms capture solar energy by photosynthesis, and almost all of the remaining living things obtain energy from them, along food chains. Many people also use energy from fossil fuels—coal, oil and gas—and sometimes from other sources such as nuclear reactors. Consideration of fossil fuels leads on, in this chapter, to the increasing carbon dioxide concentration of the world's atmosphere and what effects it may have. Carbon dioxide can be regarded as a pollutant, but it is dealt with here, rather than in the chapter called 'Pollution', because it is so closely related to energy use.

Solar radiation and primary production

Net primary production

Box 2.1 summarizes what happens to the energy in the solar radiation that reaches the earth. Only a tiny proportion of it is used in photosynthesis, but that is what concerns us here. On the ecological scale this is measured as *net primary production* (or net primary productivity, meaning *rate* of production). *Primary* means production by photosynthetic organisms, as opposed to secondary production by non-photosynthetic (heterotrophic) organisms; *net* means excluding organic matter used by the green plants for respiration, so the net production is potentially available to heterotrophs. The net primary production over a year is rarely all still present as extra standing biomass at the end of the year: plants or parts of them are eaten by herbivorous animals, attacked by parasites or die and are attacked by decomposer

Box 2.1 Radiation from the sun and what happens to it.

Radiation emitted by the sun (*solar radiation*) mostly has wavelengths within the range *0.2–3 µm*. This is called *short-wave radiation.*

Fate of the solar radiation reaching the top of the earth's atmosphere:

- Reflected by clouds,
- Absorbed by gases, especially ozone, carbon dioxide and water vapour, or
- Reaches the earth's surface.

Fate of short-wave radiation hitting plants:

- Reflected, passes through or
- Absorbed. Fate of absorbed energy:
  - Radiated, as *long-wave radiation* (wavelength >3 µm),
  - Used in transpiration,
  - Used in photosynthesis (primary production), or
  - Warms plants and surrounding soil and air.

Of the short-wave solar radiation reaching the earth's surface about half is *photosynthetically active radiation*, i.e. within the wavelength range 0.4–0.7 µm which can be absorbed by photosynthetic pigments.

Further information: Nobel (1991a).

organisms. In a true climax ecosystem we should expect that on average the biomass present now is the same as that a year ago: reproduction and growth of some individuals is on average equalled by death and decomposition of others.

Net primary production can be expressed in terms of the dry weight of the plant biomass produced, or if we take account of the energy content of the plant material, in energy terms. The energy content of most plant materials, when dry, differs little: it is usually within the range 17–21 $kJ\,g^{-1}$ (FAO 1979, Lawson *et al.* 1984), though a few storage tissues such as oil-rich seeds give higher values. The net primary production of the whole earth, land plus sea, is probably within the range $30–50 \times 10^{20}\,J\,yr^{-1}$. This is about 0.1% of the incoming short-wave radiation (Table 2.1). The energy content of the food consumed by the world's population is about 0.5% of the total net primary production. That might suggest that there is no need to worry about whether we can feed the expanding human population: why not just channel more of the primary production into human food? But things are not as simple as that. Table 1.1 shows that at present only about one-tenth of the world's land area (i.e. about 3% of the total surface area of the globe, land and water) is used for producing crops. Human food can also be obtained from the oceans (three-quarters of the world's surface) and from grazing lands, but fish and meat are much less efficient ways of converting solar energy to human food energy. Increasing crop production will require either: (1) expanding the area under crops on to land at present under other uses, a topic that recurs in several chapters of this book; or (2) increasing crop production per hectare. Chapters 3 and 4 discuss in detail food production from land and from the oceans, how it can be maintained and whether it can be increased.

How much energy is used by people?

Other forms of energy used by people, most of it from fossil fuels, amount to more than 10 times as much as food consumption (Table 2.1), but to only about one-tenth of net primary production and only 1/10 000 of incoming short-wave radiation. Wood provides only a minor contribution to energy in most developed countries, but in some other countries it is extremely important, especially for domestic cooking. Worldwide, the total energy content of the fuelwood used each year is about equal to that of food.

We should consider how the data in Table 2.1 were obtained and how accurate they are likely to be. You will notice that all but one example has been quoted to the accuracy of only one significant figure. Many of the data come from United Nations yearbooks, which sound authoritative, but they are compiled from data supplied by individual countries, which are likely to vary in accuracy. If a country has been in the throes of a civil war for several years, the official government may not know accurately even how many people there are in the country, let alone how much firewood they collect. Coal, oil and natural gas

Table 2.1 Basic energy data for the world. Values are accurate to only one significant figure, except for fossil fuels

	Total energy per year ($\times 10^{20}$ J)	Source of data
Incoming short-wave radiation reaching surfaces of oceans or land cover	30 000	1
Net primary production	30–50	2
Human food consumption	0.2	3
Human energy use		
Fossil fuels	3.0	3
Fuelwood	0.2	3
Others*	0.1	3
Total	3	

* Includes nuclear and hydro-electricity.
Sources of data: 1, Harte (1985); 2, values within this range given by Whittaker (1975), Vitousek *et al.* (1986); 3, data for 1987–9, from *UN Statistics Yearbook 1987*, *UN Energy Statistics Yearbook 1989*, *FAO Production Yearbook 1990.*

are extracted by large organizations, so the amounts involved are known and reported with reasonable accuracy, but food consumption in countries practising subsistence agriculture is bound to involve estimates.

Measuring net primary productivity

Special difficulties apply to estimating global net primary production. For example, to determine the productivity of an area of forest requires measurement of production of ephemeral parts such as leaves, fruits and fine roots, including those eaten by animals; and of new woody material both below and above ground. In most parts of the oceans, phytoplankton carry out all the photosynthesis, so small volumes of ocean water with their plankton can be enclosed and the photosynthesis rate measured, e.g. by ^{14}C uptake or oxygen production. Each of these measurements of productivity, on land and at sea, is made at a particular site, and the productivity figure for the world is calculated by scaling up from them. A major aim of the International Biological Program, which operated during the late 1960s and early 1970s, was to obtain measurements of net primary productivity for a representative set of sites throughout the world. In the event, the sites studied were not evenly spread across the world; not surprisingly, inaccessible regions, for example in tropical and boreal forests, were scarcely sampled. Although there have been subsequent measurements of productivity at other sites, information about some major vegetation types, such as tropical forest, is still inadequate. This is why I have given a range rather than a single figure for net primary productivity in Table 2.1, and even this range is open to some doubt.

There is a different technique which may in future be able to provide much more extensive measurements of primary productivity:

Box 2.2 Information from satellite remote sensing that can be useful to ecologists.

Artificial satellites now in orbit provide images of the earth's surface, at various resolutions from 20 m upwards. Most record in more than one waveband, and comparison of the intensity of different wavebands provides much of the information of interest to ecologists. Some satellites can measure in microwave wavelengths, which can pass through cloud, but many can only record ground images when not obscured by cloud.

Well-established methods are already in use for:

- Distinguishing different types of vegetation, hence measuring areas of different types, and changes (e.g. deforestation);
- Recording fires, their frequency and area;
- Measuring surface temperature of vegetation and exposed soil;
- Measuring ocean surface temperature, which can indicate regions of upwelling and hence high productivity (see Chapter 4);
- Determining a 'greenness index' of vegetation, which is related to green leaf area and amount of chlorophyll (see below).

Methods are under research or development for measuring:

- Vegetation water content;
- Transpiration rate;
- Concentration of constituents in plants, e.g. nitrogen, cellulose, lignin;
- Net primary productivity (see text).

An example of the use of two wavebands

A 'greenness index' is given by

$(IR - R)/(IR + R)$

where IR and R are the amounts of radiation measured by the satellite in the near infra-red and the red wavebands, respectively. This method is based on the fact that chlorophyll absorbs much of the red radiation that falls on it, but little of the near infra-red (about 1 μm wavelength), i.e. it reflects more near infra-red than red. The difference between the amount reflected in these two wavebands is much greater for green leaves than it is for dead leaves, stems or soil, so the greenness index given above is related to the amount of chlorophyll or green leaf biomass per unit ground area.

Further information: Steven & Clark (1990), Matson & Ustin (1991).

remote sensing from satellites. Remote sensing can provide various sorts of information useful to ecologists, and these are summarized in Box 2.2. Measurements of amount of green foliage per unit land area, and also of amount of chlorophyll (i.e. phytoplankton) in oceans, are accurate enough to be useful. These are sometimes used to estimate the primary productivity, on the assumption that photosynthesis rate will be proportional to amount of chlorophyll. Box *et al.* (1989) tested the accuracy of this method for measuring net primary productivity of a wide range of terrestrial vegetation types, in various parts of the world. After excluding mountainous and very wet areas, where they

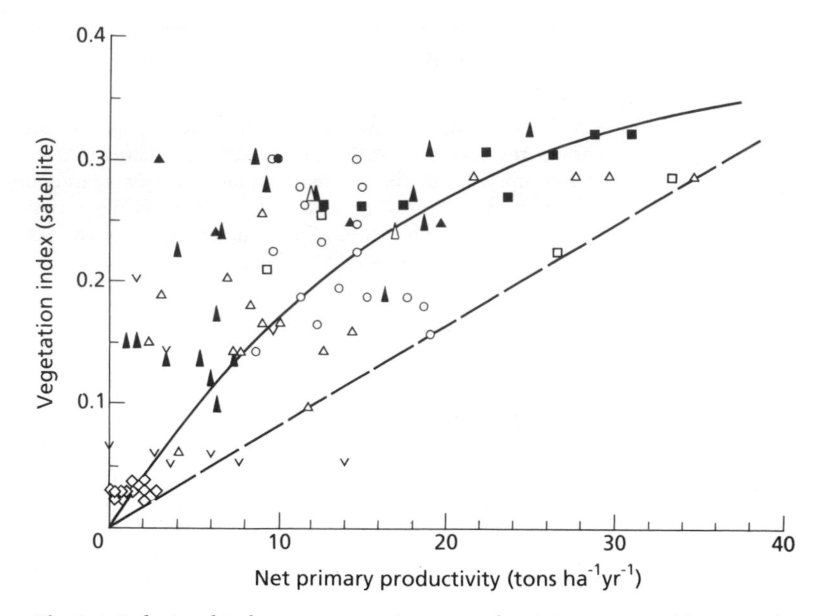

Fig. 2.1 Relationship between net primary productivity measured by ground studies and a vegetation index measured by a satellite. Each symbol represents a different type of vegetation. The curved line is a best fit provided by the authors. The dashed straight line is an approximate lower boundary line drawn by eye, by me. From Box *et al.* (1989).

thought remote sensing would run into special problems, there were data from 95 sites where net primary productivity had been adequately determined by conventional ground studies. Figure 2.1 shows how the measured productivity related to a vegetation index calculated from amount of red and infra-red reflected, i.e. similar to the 'greenness index' of Box 2.2. Clearly there is some relationship between the two measures: the curve shown is a highly significant fit statistically. However, the scatter of points is large; for example, if the satellite vegetation index was 0.2 then measured productivity could be anywhere from 2 to 27 tons $ha^{-1} yr^{-1}$. So the greenness index evidently has only a limited ability to predict productivity of individual sites. An alternative way to interpret Fig. 2.1 is that nearly all of the points lie above a boundary line, shown by the straight dashed line. The vegetation index would then indicate a maximum possible productivity, e.g. if the vegetation index is 0.15 the productivity could be anything from 18 tons $ha^{-1} yr^{-1}$ down to zero. This makes sense: the vegetation index is measuring amount of foliage or chlorophyll, which determines the *potential* rate of photosynthesis at the site, but the actual photosynthesis could be reduced below that rate by unfavourable conditions of weather or soil.

Thus measurement of productivity from satellites looks potentially very useful but its accuracy needs improving. Active research on this is being carried out. Chlorophyll fluoresces at a particular wavelength, and it may be possible to use this as a measure of how much of the solar radiation absorbed by chlorophyll is actually used in carbon assimilation.

The low efficiency of photosynthesis

Table 2.1 shows that only about 0.1% of the incoming short-wave radiation that reaches the earth's surface is captured by photosynthesis and used for synthesis of organic compounds. This may seem a remarkably low efficiency. In the next three chapters should we accept this low productivity as inevitable, or expect that scientists will soon be able to increase it? The figure of 0.1% is an average for the whole world, including the oceans. Much of the radiation is not absorbed by chlorophyll-containing tissue: in the oceans much is absorbed by water, and on land some hits soil or ice. When the radiation does hit green tissue, both the radiation-capture and CO_2-capture mechanisms are usually working below 100% efficiency. When leaves are directly exposed to bright light photons arrive too fast for all of them to be processed by the photosynthetic apparatus, so a substantial proportion may be 'wasted'. And for plants with the C_3 photosynthetic carbon cycle (which is by far the most widespread) the concentration of CO_2 in the atmosphere, about $350\,\mu l\,l^{-1}$, is well below optimum. These plants will photosynthesize faster in an atmosphere with two or three times this concentration of CO_2. This is partly because the enzyme responsible for the initial trapping of CO_2, ribulose bisphosphate carboxylase/oxygenase (Rubisco), can react with oxygen as well as CO_2, resulting in *photorespiration* which releases some CO_2; in other words the net amount of CO_2 incorporated into organic chemicals is reduced. Plants with C_4 photosynthesis or Crassulacean acid metabolism (CAM) have a preliminary CO_2-concentrating mechanism which largely overcomes this limitation of Rubisco.

In theory there is scope for major increase in plant productivity by increasing the efficiency of the photosynthetic process. But in practice, in spite of extensive fundamental research on photosynthesis, most of the increases in crop productivity during the 20th century have not arisen from changes in the photosynthetic mechanism itself, but from other things, including increased supply of mineral nutrients, reduced damage by pests and pathogens, increasing the amount of incoming radiation absorbed (by faster canopy development and longer growing season), and by breeding plants that put a greater proportion of their growth into the useful parts such as seeds. The next chapter gives more information about factors that influence crop yields.

Productivity: some measured rates

Table 2.2 gives examples of productivities that have actually occurred, in crops, forestry plantations and natural vegetation. The values in part (a) have been chosen to show the high end of the known range. The grass in the Amazon floodplain was growing partly submerged in nutrient-rich water, in a region warm throughout the year, so environmental conditions were very favourable. This is the highest measured productivity of natural vegetation, as far as I know. The percentage

Table 2.2 Some measured rates of net primary productivity

	Parts of plant included	Productivity tons $ha^{-1} yr^{-1}$	Productivity TJ $ha^{-1} yr^{-1}$	Percentage efficiency*	Source of data
(a) *Some of the highest recorded values*					
Highly productive natural vegetation					
Tropical rainforest	All	10–35	0.2–0.7†		1
Echinochloa polystachya (a grass), in Amazon floodplain	All	99	2.0†	4.6†	2
Crops, exceptionally favourable conditions	Above-ground	26–70	0.5–1.4†		3
				1.6–4.2†	4
(b) *Typical or average values*					
Phytoplankton in oceans	All	0.5–10	0.01–0.2†		Chapter 4
Maize, U.S.A., mean 1986–90	Grain	7	0.1		Table 3.1
Forestry plantations	Stem wood	1–29	0.02–0.6†		Table 5.6

TJ = 10^{12} J.
* [(Energy content of plant material)/(Energy of intercepted short-wave radiation)] × 100.
† Assumes energy content 20 kJ g^{-1}.
Sources of data: 1, Lieth & Whittaker (1975); 2, Piedade *et al.* (1991); 3, Nobel (1991b); 4, Cannell *et al.* (1987).

efficiency column shows the percentage of short-wave radiation *intercepted by the plants* that was converted to chemical energy (if radiation 'wasted' because it hit the ground were included the efficiency would be less). Calculated in this way the efficiency of use of solar radiation can be several per cent, but values above 5% have not been reported, as far as I know. Part (b) of the table shows more typical values for productivity; their percentage efficiency of radiation use is mostly less than 1%. These figures for average and exceptionally high productivities will form a useful basis for more detailed study of food and timber production in the next three chapters.

Energy sources now and for the future

How long will the reserves of fossil fuels last?

The life-style of the world's richer countries is much dependent on fossil fuels (see Table 2.1). Nowadays it seems obvious that the world's resources of fossil fuels cannot last for ever, if we continue to use them at our present rate or anything near it. Predicting *when* the supplies will be exhausted is more difficult: more reserves will probably be discovered, methods of extracting the fuels and of using them are likely to change. A very simple approach is to compare the world's present rate of use with the known or estimated reserves; this is done for oil and coal in Table 2.3. The right-hand column of figures shows how long the reserves will last if (1) we go on using oil and coal at our present rate, and (2) all the presently known or guessed reserves are used, but (3) no others are discovered. None of these three 'ifs' is likely

Table 2.3 World oil and coal reserves and rate of consumption. Values quoted to two significant figures

	Consumption (Gtons yr^{-1})	Reserves (Gtons)	Ratio $\frac{\text{reserves}}{\text{consumption}}$
Oil	3.0	120*	40
Coal	4.7	(a) 3200	680
		(b) 11 000	2300
Data for year	1989	1987	

* In addition oil shale + bituminous sands, reserves 50.
(a) Known, (b) known + additional estimated.
Data From *UN Energy Statistics Yearbook 1989.*

to prove correct, but the table nevertheless gives a first guess at a time-scale: oil is likely to last for decades rather than centuries, but coal for centuries. During the 1970s there was great concern about exhaustion of fossil fuel supplies. During the 1980s public attention switched more to consequences of use of fossil fuels, especially increase in atmospheric carbon dioxide and its possible effects, a topic considered later in this chapter. Both of these are reasons for considering alternative energy sources to fossil fuels. Biologists can help by deciding whether using plants as energy sources, *biomass energy*, could replace a substantial proportion of fossil fuels.

Biomass as an energy source

Could biomass be a major energy source?

Until a few hundred years ago plants, especially wood, provided most human fuel; a little energy was provided by animals, wind and water. Today wood is still a very important fuel in many less-developed countries, but as Table 2.1 shows, it forms less than 10% of total world energy consumption. Table 2.1 also shows that even to replace the whole of our present fossil fuel use by biomass would require only about 10% of the world's net primary productivity. To suggest that we use wood instead of coal and oil may sound like abandoning the modern industrial world and returning to a simple life off the land, but the proposal is worth examining.

Biomass energy: Sweden and U.K. as examples

Prospects for biomass energy have often been assessed on a national or regional scale. To provide an example of how the potential for biomass energy can differ between countries with similar levels of economic development, I compare Sweden and the United Kingdom (U.K.). Table 2.4 shows that their energy use per person is almost identical. However, they differ greatly in their native fuel resources:

Table 2.4 Land area, human population and energy use in Sweden and U.K. in 1989

	Sweden	U.K.
Area ($\times 10^3\,km^2$), excluding fresh water	412	242
Population		
Total (millions)	8.5	57.4
People per km^2	21	238
Energy consumption		
Total ($\times 10^{18}\,J\,yr^{-1}$)	1.25	8.45
Per person ($GJ\,yr^{-1}$)	149	147

Data from *UN Energy Statistics Yearbook 1989, FAO Production Yearbook 1990.*

U.K. has oil, natural gas and substantial reserves of coal; in contrast, Sweden has no oil or gas and very little coal, but it does have uranium. U.K. has since the start of the industrial revolution made much use of fossil fuel, though it has at present also some nuclear power stations. Sweden developed a nuclear power programme, but during the 1970s attitudes among Swedish people became less favourable to nuclear power, thus turning their attention to other possibilities. Lönnroth *et al.* (1980) presented a report from a committee that considered energy options for Sweden, including whether biomass could provide a major contribution. Their 'solar Sweden' alternative suggested that about half of Sweden's energy could come from biomass, mainly 'energy forests' of fast-growing tree species. Let us assume that harvestable growth in these plantations could average 5 tons dry matter $ha^{-1}\,yr^{-1}$, which is about the middle of the range of yields of forestry plantations in temperate regions given in Table 5.6; to expect much higher values from large areas of forestry in northern temperate climates is probably unrealistic. Since tree stem material has an energy content of about $20\,kJ\,g^{-1}$, 5 tons would contain 0.1 TJ. Then to provide 50% of its present energy use on a continuous basis, Sweden would need 60 000 km^2 of biomass plantations, or 15% of its land area. Assuming the same yield per hectare in U.K., over 400 000 km^2 of energy plantations would be needed to provide half of U.K.'s present energy use, which is more than the country's total land area (about 70% more). Thus Sweden *might* be able to get a substantial proportion of its energy from biomass, U.K. clearly cannot. A study by Price and Mitchell (1985) of the potential of wood as an energy source in Britain, which took into account variations in productivity across the country and also other competing uses for land, estimated that only about 1% of Britain's energy could realistically come from biomass. The difference between Sweden and U.K. arises because U.K. has a much larger population than Sweden but a smaller land area (Table 2.4).

Box 2.3 Fuels that can be made from biomass.

Gas

Methane gas can be made from cellulose and hemicellulose. The process involves three separate bacterial stages:

Cellulose → sugars → acetate → CH_4 + CO_2

Hence control of the process in fermenters is complex (Klass 1984). Research, including selection and genetic modification of the microorganisms, can be expected to improve it. So far methane production has been mainly from waste, e.g. cattle dung, sewage sludge and domestic refuse, but there have been tests on gas production from harvested plant material. Plants low in lignin, such as grasses or floating aquatic plants, seem more promising than woody plants, since lignin is more slowly decomposed than cellulose.

Liquid

The best-known liquid plant product that can be a petrol substitute is ethanol, which can be produced by yeasts and some bacteria. The traditional starting point for ethanol production has been starch or sugar. In the largest biomass ethanol-for-fuel programme, in Brazil, sugarcane has been the principal biomass crop. Processes involving cellulose as a starting product tend to be slow, but research on how to speed them up could well be successful. There are also chemical processes, not involving microbes, for producing liquid fuels from plant materials; the best-known one makes methanol (Gates 1985).

Solid

If a solid fuel is required, for direct combustion, wood has the advantage of being more compact and hence more easily transported than, for example, dried grass, though it still has less energy per cubic metre than oil or coal. An alternative to conventional timber trees is short-rotation coppice, for example of willow or poplar (Evans 1984). *Coppicing* means that the trees are cut back to the stump, and new shoots then grow from it. After a few years each tree is likely to comprise many stems several metres tall and a few centimetres in diameter. They can then be cut and harvested by a machine and the stumps left to grow new stems. The plantation is thus rather like a scaled-up hay-meadow. See Chapter 5 for more details.

If we are planning to use biomass as an energy source, a key question is what sort of fuel do we want: gas, liquid or solid? Box 2.3 summarizes how gas, liquid or solid fuel can be obtained from biomass. The choice may depend on what type of land is available, since each type of fuel is likely to come from a different type of plant and thus to be in competition with a different existing land use: gas production is most likely to compete with grazing land, liquid fuel with arable crops and solid fuel with timber land.

How much net *energy can be obtained from biomass?*

We should now consider how much energy needs to be invested to produce biomass fuels. The *net energy* obtained from biomass is the energy obtained from using the fuel, less the energy expended in growing the biomass, harvesting it, converting it to the fuel and trans-

Table 2.5 Energy balance of ethanol production from sugarcane. Units of energy are GJ ha^{-1} yr^{-1}.

	Brazil		Louisiana, U.S.A.		
Inputs					
On farm					
Machinery*	2.7		6.6		
Fuel	8.3		17.0		
N fertilizer	3.8		8.3		
Other fertilizer	1.0		1.0		
Seed, pesticides	1.1		2.4		
Human labour	0.5		0.1		
Total on farm	17.4		35.4		
In ethanol distillation					
Machinery*	?		1.7		
Total up to here		17.4		37.1	= *U*1
Fuel	45.2		43.5		
Total inputs		62.6		80.6	= *U*2
Produce					
Ethanol	78.4		76.9		= *E*
Waste	73.2		71.9		
Net energy gain					
(a) Not using waste (=*E* − *U*2)		15.8		−3.7	
(b) Using waste to power distillation (=*E* − *U*1)		61.0		39.8	
Energy return on investment					
(a) (=*E*/*U*2)		1.25		0.95	
(b) (=*E*/*U*1)		4.5		2.1	

* Energy used for construction and maintenance of machinery.
Data from da Silva *et al.* (1978) and Hopkinson & Day (1980).

porting it to the point of use. Table 2.5 shows the estimated energy costs of producing ethanol from sugarcane in Brazil and in southern U.S.A. The amount of ethanol produced per hectare was similar in the two regions. The energy costs on the farm include not just the fuel used by the tractors and other machinery, but also the energy used to make that machinery, to make ammonium fertilizer and pesticides chemically, to mine and transport other fertilizers. The energy in the food eaten by the farm workers is also included, but is a small proportion of the total. In Brazil the total energy used on the farm was nearly a quarter of the energy content of the ethanol produced, and in Louisiana it was nearly a half. The difference between the two areas was partly due to greater use of fertilizers and pesticides in Louisiana, and partly due to more use of machines there and less of human labour. Another major energy cost arises because yeasts and other microorganisms can tolerate only a limited concentration of ethanol: ethanol production by yeast is much slowed down when the concentration (by volume) reaches 5–7% and ceases at 13–18%. This dilute alcohol solution is not suitable as a fuel until most of the water has been removed by distillation. The energy required for this is more than half the energy content of the pure ethanol. The result is that in Louisiana the total energy expended to obtain the ethanol is more than the ethanol contains, so no energy is actually gained; the amount of net energy is a negative number. In Brazil the net energy is a positive amount, 15.8 GJ ha^{-1} yr^{-1}, but this is only about one-fifth of the total energy in the ethanol. A way of expressing the efficiency of the system is the Energy Return on Investment (E_R).

Energy return on investment

$$E_R = \frac{\text{energy in fuel}}{\text{energy used to obtain fuel}}$$

For ethanol the energy return on investment, E_R, was 1.25 for Brazil and 0.95 for Louisiana. If E_R is 1 or less no net energy is gained from the system. A higher E_R can be regarded as indicating more efficient use of resources. However, the sugarcane system need not be as inefficient as these calculations make it appear. The 'waste', i.e. non-sugar material from the cane plants, has an energy content greater than the fuel requirement for distillation; so if the distillery can be entirely fuelled by this waste, fuel for distillation effectively ceases to be an input. Both countries' ethanol then becomes a substantial net energy provider.

Zavitkovski (1979) gave figures for the energy inputs required for growing forest in Wisconsin and harvesting it at 10 yrs of age. He compared poplar grown with irrigation and fertilizer to a naturally regenerating forest with no irrigation or fertilizer. Table 2.6 shows

Table 2.6 Energy inputs and outputs (GJ ha^{-1} yr^{-1}) for two forests in Wisconsin, U.S.A., averaged over the 10-yr growth period. Inputs include manufacture and operation of machinery, manufacture and application of fertilizers, harvesting and drying of trees; they do not include solar energy falling on the forest. Outputs (B) are the energy contents of all above-ground parts present at harvest

	Poplar plantation, irrigated and fertilized*	'Natural' forest regrowth
Energy inputs ($=U$)	75	14
Energy in biomass ($=B$)	323	67
Net energy ($=B - U$)	248	53
Energy return on investment ($=B/U$)	4.3	4.7

* *Populus tristis × balsamifera.*
Data from Zavitkovski (1979).

that, as we should expect, the intensively managed poplar required more energy input, but it also provided more biomass. A substantial proportion of the energy input was used in drying the wood after harvest. The two stands gave a similar energy return on investment, but the intensively managed stand gave far more net energy per hectare.

It is interesting to compare these figures for energy return on investment for biomass fuels with corresponding figures for fossil fuels. Hall *et al.* (1986) gave values of about 10 for oil and about 30 for coal, higher than any of the values calculated for biomass fuels. Energy costs are not necessarily proportional to money costs, but these comparisons indicate one reason why fossil fuels are often preferred to biomass: they give a higher return on investment. Calculating net energy is often difficult. It is, for example, not easy to estimate how much energy is used to make an oil rig or even a tractor. And none of the calculations included the energy used in transporting the fuel to the place where it is used.

Comparing biomass and wind as energy sources

It seems, then, that it is technically possible to produce fuel by growing biomass for energy. But are there better alternative energy sources? To ask a more restricted question, are plants the best way of capturing solar energy? Wind provides energy that is derived from the sun, and the technology for converting it to electricity is already well developed. Grubb (1988) reviewed the potential for 'wind farms', i.e. arrays of large turbines (propellers) on tall towers, to provide electricity in Britain. His calculations indicate that in Britain the wind energy captured by an array of turbines, per unit area of ground, would average about 1.5 TJ ha^{-1} yr^{-1}. If the wind blew steadily (which it would not in practice), this would be equivalent to about 50 kW ha^{-1}. The electricity actually generated would be less, perhaps about half as much, due to various steps having less than 100% efficiency. The figure of 1.5 TJ ha^{-1} yr^{-1} can be compared with plant productivities listed in Table 2.2: it is near the top end of the highest productivities, which occur under extremely favourable conditions, and well above the typical or average values. The productivity I took as a realistic average for biomass forests in Britain was 0.1 TJ ha^{-1} yr^{-1}, a full order of magnitude lower than the energy from the wind turbines. So wind turbines can, at least in some areas, generate far more energy per hectare than plants; and plants can still be grown on much of the land between the wind towers. On the other hand, wind is a very variable energy source, and storage of large amounts of electricity presents problems. Modern wind turbines can alter the appearance of the landscape: they are often 50–100 m tall and are usually sited in exposed situations, where they are clearly visible.

The conclusion is that biomass energy can in some parts of the world make a contribution to reducing our dependence on fossil fuels, but the contribution is likely always to be a limited one. There are other possible alternative energy sources, including wind, water flow, waves, tides, photovoltaic cells, heat deep in the earth's crust, nuclear fission, maybe even nuclear fusion. Long-term solutions to the world's energy problems are likely to depend much more on physical scientists and engineers than on biologists.

Carbon dioxide and global climate change

The world's carbon balance

Burning fuels releases carbon dioxide. Since 1958 the concentration of CO_2 in the atmosphere has been accurately measured on Mount Mauna Loa in Hawaii, so it has been known since the 1960s that the CO_2 concentration is increasing, and how fast. It is also known that the CO_2 concentration of the atmosphere influences the world's climate, though to predict with any precision how the climate of the world will change in the future is difficult. Because this book is mainly about biological aspects, I shall in this chapter make much use of the conclusions of an Intergovernmental Panel on Climate Change (IPCC) which was set up in 1988 by world governments to assess the available evidence on changes in the atmosphere and their likely effects, and to suggest 'realistic response strategies'. One of the outcomes was a book (Houghton *et al.* 1990) giving the scientific conclusions. There were also three Policymakers' Summaries by working groups; the first (which I call here IPCC1) was on predictions of atmospheric gas changes and resultant climate change. That 26-page document was further distilled into a two-page Executive Summary, which is perhaps all the politicians were expected to read. Research on this very important subject has continued actively since 1990, and I mention a few later publications. But my basic approach in this section is to accept the IPCC's predictions of climate change over the next century, without a critical discussion of the evidence, so that I can move on to consider in more detail how such climate change, if it occurs, will affect (1) food production and (2) wild plants and animals.

Changing CO_2 concentrations, past and future

Figure 2.2 shows the concentration of CO_2 in the world's atmosphere since 1700, and predictions for the future. The values from 1958 to the present are from direct measurements of air samples, at first from Mount Mauna Loa in Hawaii, later also from a few other sites. Samples of older air are preserved in bubbles in Antarctic ice, which allow the composition back to AD 1700 to be measured. Between the 18th century and 1990 the CO_2 concentration increased by one-quarter, from about 280 to about $350\,\mu l\,l^{-1}$ ($\mu l\,l^{-1}$ is the same as parts per million by volume; $350\,\mu l\,l^{-1}$ equals 0.035% of the total gases, by volume). CO_2 is at present increasing by $1.8\,\mu l\,l^{-1}\,yr^{-1}$. The prediction up to the year 2100 shown in Fig. 2.2 assumes what IPCC called 'business-as-usual', meaning what is expected to happen if no steps are taken in future to limit CO_2 emissions. Use of fossil fuels per year would be expected to

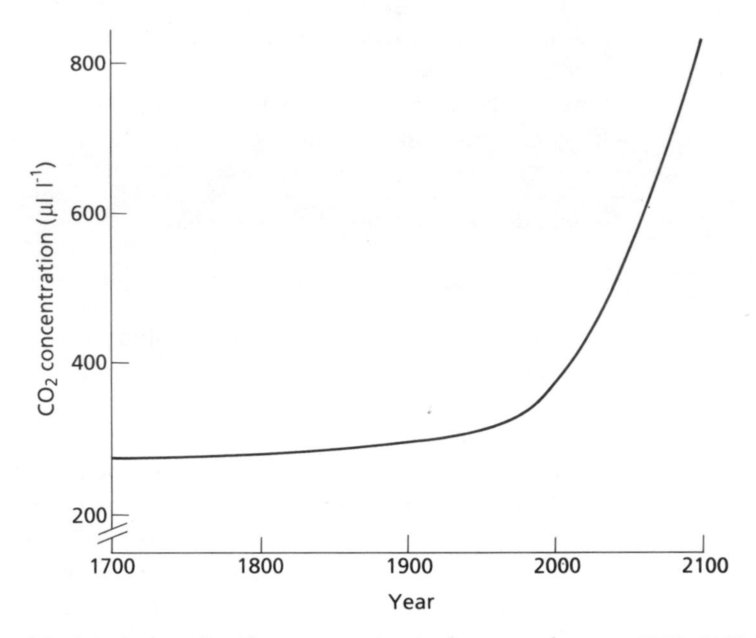

Fig. 2.2 Carbon dioxide concentration in the atmosphere, AD 1700–2100. Basis: 1700–1958, measurements on air bubbles trapped in ice (Pearman *et al.* 1986); 1958–88, measurements on atmospheric air samples (Krupa & Kickert 1989); 1990–2100, IPCC1 (1990) prediction on 'business-as-usual' scenario.

increase, because of population increase and also increased use of fuel per person. If CO_2 production per year is held at the 1990 rate the atmospheric concentration is predicted to reach 520 $\mu l\,l^{-1}$ by AD 2100, as against 825 $\mu l\,l^{-1}$ for 'business-as-usual'. Even 520 $\mu l\,l^{-1}$ is nearly double the pre-industrial level.

Carbon cycling on a world scale

Figure 2.3 shows estimates of the amount of carbon held in various pools, and rates of transfer between pools. The numbers should be regarded as being accurate only to one significant figure, and some of them are less certain than that. The world has very large pools of carbon in rocks, 75 million Gtons (75×10^{15} tons) according to Berner & Lasaga (1989); much of this is in limestone. Some exchange of this with the atmospheric CO_2 occurs by weathering and volcanic eruptions. This is likely to be significant on a geological time-scale, but not within decades or a few centuries (Berner & Lasaga 1989). Compared with the amount in rocks, the total weight of C in the world's atmosphere is modest; it is the same order of magnitude as the C in the world's biomass (which is nearly all on land), and as the C in the dead organic matter on land and dissolved in the oceans. There is estimated to be more than ten times as much C in reserves of fossil fuels as there is in the atmosphere, giving the possibility for humans to greatly increase atmospheric CO_2 by burning fossil fuel.

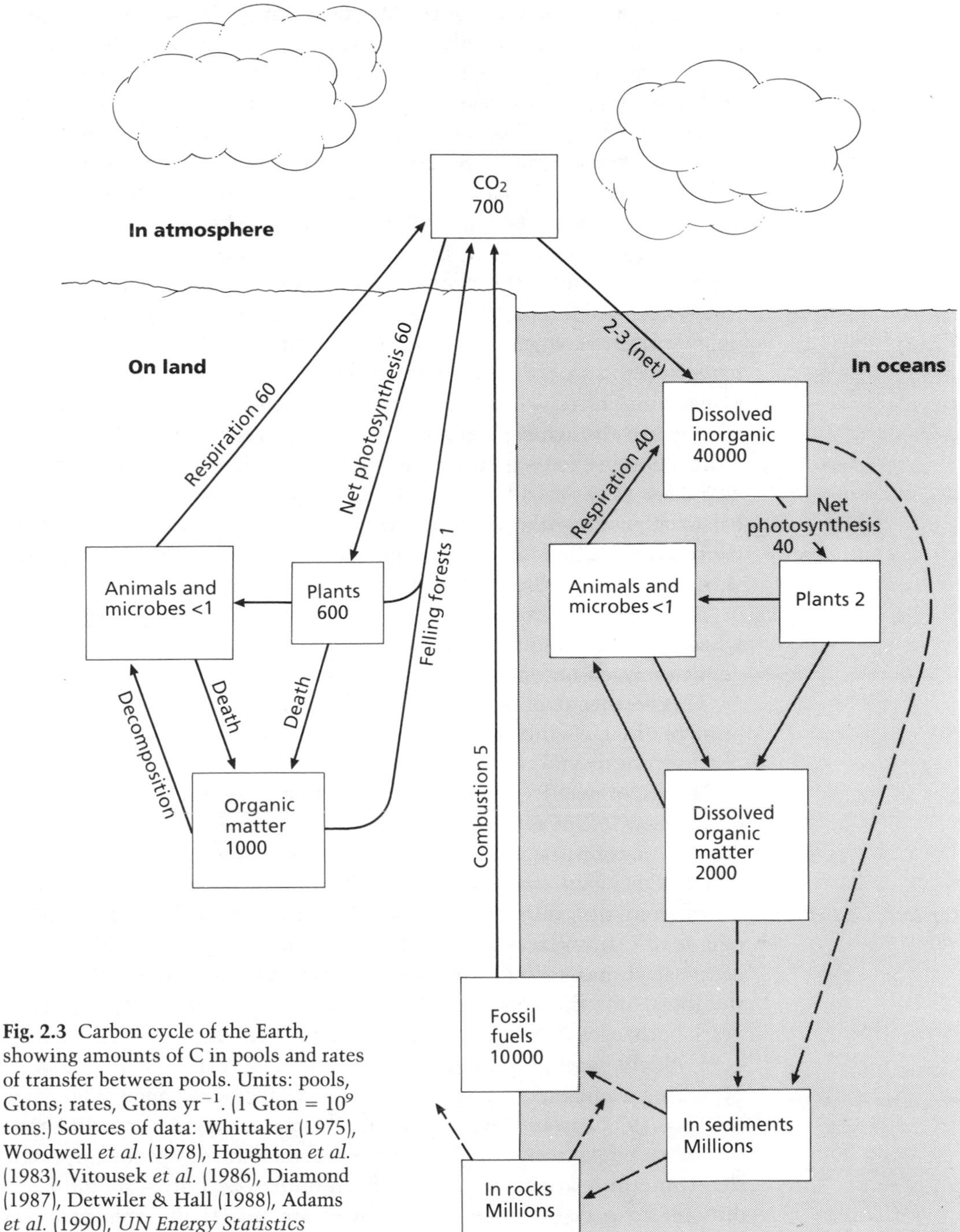

Fig. 2.3 Carbon cycle of the Earth, showing amounts of C in pools and rates of transfer between pools. Units: pools, Gtons; rates, Gtons yr^{-1}. (1 Gton = 10^9 tons.) Sources of data: Whittaker (1975), Woodwell *et al.* (1978), Houghton *et al.* (1983), Vitousek *et al.* (1986), Diamond (1987), Detwiler & Hall (1988), Adams *et al.* (1990), *UN Energy Statistics Yearbook 1988.*

Determining the rates of transfer between pools presents considerable difficulties, and the only rate that is reliably known to be accurate even to one significant figure is the C release from fossil fuels, 5 Gtons yr^{-1}. The rate of increase of C as CO_2 in the atmosphere is also reliably known to be 3 Gtons yr^{-1}, so some of the fossil fuel C evidently goes elsewhere. Figure 2.3 shows photosynthesis and respiration on land being exactly in balance, and this also being true in the oceans. This surprises some people, who have been taught that plants take up CO_2 and give out oxygen. But ecologists should think of the whole ecosystem, not just the plants: viewed over a large area and a long time-scale, stable ecosystems are not net absorbers of CO_2. A wheat field, while the wheat is growing, is absorbing CO_2; but when it is harvested the stubble and roots are burnt or left to rot and the grain is made into bread which is eaten and respired by people, so the CO_2 is returned to the atmosphere. In a hectare of climax tropical rainforest, although many trees increase the stored C in their woody parts over a year, a few trees die and decay, and on average respiration by heterotrophs balances photosynthesis by autotrophs. The same balance occurs for oxygen production and consumption. If anyone tells you that the world depends on tropical forests to produce the oxygen we breathe, he or she is wrong: climax forests are not net producers of oxygen. The wheat plants that produced the bread you ate today also produced enough oxygen for you to respire the bread.

Climax ecosystems are not C sinks

This balance of photosynthesis and respiration is a sort of ecological principle that is embodied in Fig. 2.3; we cannot measure the world's photosynthesis and respiration rates accurately enough to prove it. An imbalance of just 1 Gton yr^{-1}, e.g. if net photosynthesis on land were 60 Gtons yr^{-1} but respiration 59, could have a significant effect over decades or centuries on the CO_2 pool in the atmosphere. One way in which the C gain and loss of living things is at present not in balance arises from the felling of forests. We are considering here a one-off release of C from trees that are cut down and then burnt or allowed to decay. If climax forest is replaced by grassland (for example), neither the forest nor the grassland is a net source or sink for C, year after year; both ecosystems should be in long-term balance. There have been widely varying estimates of the amount of C released worldwide by felling of forests: Houghton *et al.* (1983) listed papers published between 1977 and 1981 which contain estimates ranging from 20 Gtons of C released per year to 2 Gtons yr^{-1} absorbed. It may seem surprising that something arising so directly from human activities should be so difficult to measure. The uncertainties are in: (1) the area of forest felled each year, (2) the biomass of the forests, (3) how much C is released from soil humus as a result of felling, and (4) what happens to the area after felling. The last of these is relevant because if the area is subsequently maintained as grassland or arable farmland the stored C is greatly reduced, but if forest is allowed to regrow then there may be

C released by felling forests

little long-term effect on stored C (though this does depend on how well the forest regenerates). More recent calculations have in general considered the higher previous estimates of C release to be exaggerated. Detwiler & Hall (1988) estimated the loss of C from forests cleared in 1980 to be somewhere between 0.4 and 1.6 Gtons. Nearly all of this was from tropical forests. Clearance of forests in North America and Europe has contributed to C release in the past, but in recent decades these continents have been approximately in balance for C (Houghton *et al.* 1983).

Where does the extra CO_2 go?

Thus in Fig. 2.3 there are two significant net sources of CO_2, 5 Gtons yr^{-1} from fossil fuels and 1 Gton yr^{-1} from clearance of forests. It is known that the increase in the atmosphere totals 3 Gtons yr^{-1}, so half of the CO_2 produced each year must be taken up by one or more sinks. It is assumed that the principal sink is the oceans; as the concentration in the atmosphere rises the amount dissolved in the oceans tends to rise too. As Fig. 2.3 shows, the amount of C as CO_2, carbonate and bicarbonate dissolved in the oceans is vastly larger than in the atmosphere. The rate at which CO_2 passes into the oceans has not been measured directly. Estimates by oceanographers tend to put the rate at 2 rather than 3 Gtons yr^{-1}, which seems to leave 1 Gton yr^{-1} unaccounted for. Maybe each of the relevant figures is wrong by a fraction of 1 (Detwiler & Hall 1988), or maybe there are several small sinks that together have a significant effect. For example, the amount of organic matter in temperate soils may be increasing.

Climate change

The greenhouse effect

As explained in Box 2.1, radiation from the sun is short-wave (wavelength less than 3 μm), whereas radiation from plants and any other object at a temperature that occurs on earth is long-wave (greater than 3 μm). Short-wave radiation mostly passes through the glass of a greenhouse. Inside, much of it is absorbed by the plants, benches, floor and other objects, which reradiate some of it as long-wave. The glass is less transparent to long-wave than to short-wave, so it absorbs some of the out-going long-wave and reradiates some of it back inwards. This keeps the greenhouse warmer than the outside air during daylight hours. There are gases in the atmosphere whose molecules act in a similar way to the glass of a greenhouse, letting much short-wave radiation pass through but absorbing more out-going long-wave and radiating it back again. These are known as *greenhouse gases* (see Box 2.4). The principal natural greenhouse gases are water vapour, carbon dioxide, methane, nitrous oxide and ozone. If all these were removed from the atmosphere the temperature near the ground would quickly become about 33°C colder than it is at present. So the greenhouse effect is undoubtedly a Good Thing for human beings and for life on earth. What we are concerned about here is a potential *change* in the

Box 2.4 The principal greenhouse gases.

	Main sources of origin
Water vapour	Evaporation from water surfaces. Transpiration by plants.
Carbon dioxide	See Fig. 2.3.
Methane	Produced by microorganisms in natural wetlands, rice paddy fields, rumens of sheep and cattle. Fossil natural gas, leaking from gas wells, oil wells and coal mines.
Nitrous oxide	Produced by microorganisms in soil (denitrifiers). From oxidation of nitrogen-containing compounds in plant materials and fossil fuels when they are burnt.
Ozone	Photochemical reactions between other gases.
Chlorofluorocarbons (CFCs)	No natural sources. Manufactured for use in refrigerators and aerosols.

Further information: Houghton *et al.* (1990), UNEP (1991).

greenhouse effect: if the concentration of greenhouse gases increases we should expect the world to get warmer. In addition to the increase in CO_2 (Fig. 2.2), methane and nitrous oxide are increasing. Ozone is decreasing in some parts of the upper atmosphere but increasing in the lower atmosphere. These gases do not have the same amount of greenhouse effect, molecule-for-molecule: a methane molecule has 21 times as much effect as a CO_2 molecule, and a molecule of CFC-11 about 12 000 times as much. From the increase in amount of each gas and its effect per molecule, one can calculate how much each contributed to the increase in greenhouse effect between 1980 and 1990 (IPCC1): CO_2 contributed 55%, methane 15%, N_2O 6% and CFCs 24%.

Predicting future climate change

Predicting how climate will be affected by increases in greenhouse gases is difficult. Among the problems are that changes in temperature would be expected to affect amounts of water vapour, sea ice and clouds and to alter ocean currents, all of which would probably in turn affect the climate; in other words there will be feedbacks. The Intergovernmental Panel on Climate Change used 'General Circulation Models' to predict how temperature will change up to the year AD 2100 if greenhouse gases increase according to their 'business-as-usual' predictions. Because of the uncertainties involved in the predictions, their report (IPCC1) provided a high estimate and a low estimate as well as a central 'best estimate', and these are shown in Fig. 2.4. All of these are based on the same predicted rate of increase of greenhouse gases, which assumes that no active steps are taken to slow down the increases: if the gases increase more slowly then the temperature

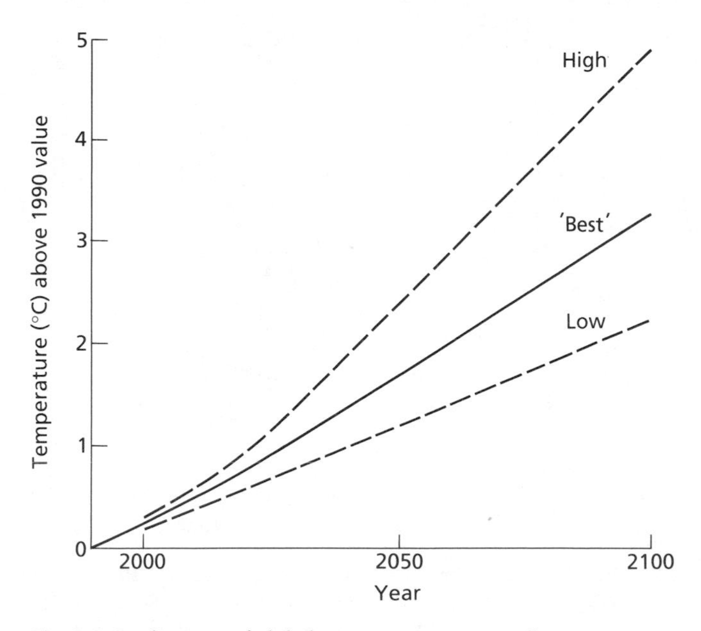

Fig. 2.4 Predictions of global mean temperature, relative to temperature in 1990. The three lines show the high, 'best' and low estimates. From IPCC1 (1990).

increase will be slower, too. Each of the three lines in Fig. 2.4 is nearly straight, i.e. the predicted rate of temperature rise is fairly constant. For the 'best estimate' it is about 0.3°C per decade, leading to a temperature in AD 2100 some 3°C higher than at the end of the 20th century. Predictions have also been made of how rainfall will change. IPCC1 predicted that global average rainfall will increase by only a few per cent up to AD 2030.

Can the greenhouse effect explain past temperature change?

One way to test the accuracy of this model is to use it to project backwards. We know the CO_2 concentrations during the last few hundred years (Fig. 2.2) and methane and nitrous oxide concentrations also. IPCC1 used these to predict with its model the expected temperatures back to 1850. Figure 2.5 shows their 'best estimate' backward projection compared with means of actual measured air temperatures. Considered over the whole period of a century, the agreement is not bad, but evidently the real world experienced a major temperature fluctuation not predicted by the model, a faster-than-expected rise from 1880 to 1940, followed by a fall for 30 yrs. The actual temperatures suggest why, although the annual increase in atmospheric CO_2 had been known since the 1960s, and its potential for causing increased warming was known to scientists, the governments of the world showed no response until they suddenly became interested in the mid-1980s. While the temperature of the world was falling it was reasonable for governments not to worry about global warming. It is unlikely that the temperature rise during the next hundred years will be a nearly straight line as in Fig. 2.4: fluctuations as large as that in Fig. 2.5 may well be

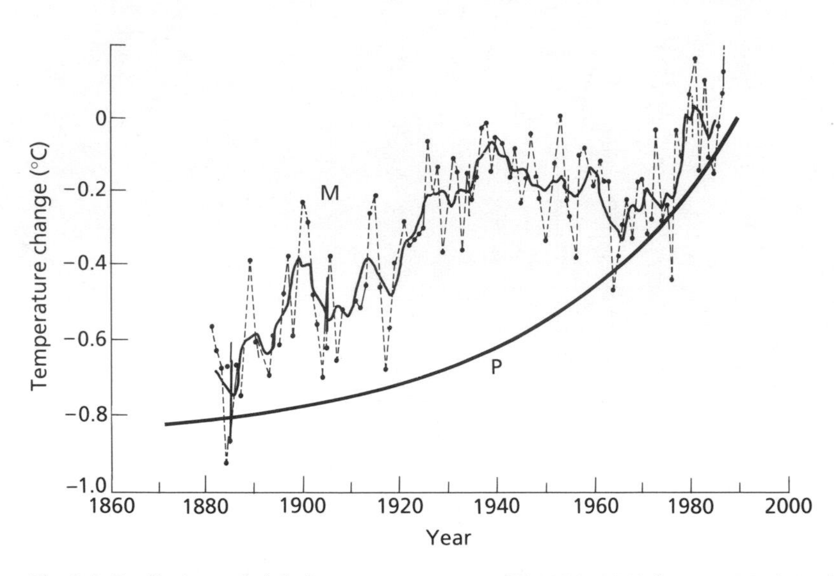

Fig. 2.5 Predictions of global mean temperature (P) 1870–1990 from IPCC1 (1990). Global mean measured temperatures (M) from Schneider (1989). Dotted line joins annual means, continuous line shows 5-yr running means.

superimposed. If these include decades of actual cooling, will the governments of the world stop worrying about the greenhouse effect?

Future rise in sea levels

If the world gets warmer it is likely that sea levels will rise, because of melting of ice and expansion of the water in the oceans. IPCC1 estimated that if greenhouse gases increase at the 'business-as-usual' rate, the sea level rise will average between 3 and 10 mm yr^{-1} up to AD 2100. However, not all scientists are agreed that sea levels will rise at all (Schneider 1992): it has been suggested that in polar regions greater winter snow fall might balance greater summer melting. Because of the uncertainties about rates of sea level rise I am not discussing its effects in this chapter. If the upper end of the IPCC range of estimates (10 mm yr^{-1}) is correct, the rise of 1 m in a century could be very serious for low-lying coastal regions.

It is unlikely that the climate changes will be uniform across the world. Clearly people in individual countries want to know whether their climate will warm up faster or slower than the average global rate, whether it will warm more in winter or in summer, whether the rainfall will increase or decrease. Although predictions of such things have been made for different regions, we cannot at the moment place much confidence in them. We need better understanding of the greenhouse effect and better resulting models.

Can biologists help to reduce global warming?

All these predictions of future climate are for 'business-as-usual', assuming that no steps are taken to limit production of greenhouse gases, so the amounts produced per year increase as world population and industrial activity continue to expand. In fact reductions in use of CFCs are already being actively organized. But use of fossil fuels is so closely linked with world economic activity that reducing the production of CO_2 is much more difficult. Possibilities to be considered for reducing atmospheric CO_2 are:

1 Reduce CO_2 production,
  (a) replace fossil fuels by other energy sources,
  (b) reduce felling of forests.
2 Remove some CO_2 from the atmosphere.

I consider each of these briefly in turn, concentrating on possible biological contributions. Replacement of fossil fuels by biomass fuels would reduce the increase in atmospheric CO_2. Although burning biomass produces CO_2, this is equal in amount to CO_2 that was removed from the atmosphere by the growing plants only years or decades earlier, so on that time-scale it is in carbon balance. Referring back to Fig. 2.3, we would in effect be generating energy for people to use from organic matter that would otherwise be substrate for decomposer organisms. Biomass as a source of fuel was discussed earlier in this chapter, ending with the conclusion that, although it may have some contribution to make, there are other and more efficient ways of generating energy without producing CO_2.

Suggestion 1(b), that we reduce the felling of forests, cannot be discussed here because it requires information from future chapters. It relates to the forest chapter (Chapter 5), where I show that, at least from a biological point of view, it should be possible to obtain all the world's wood requirement on a sustainable basis, without any decrease in forested area. Much of the decrease in forested area that is occurring is because of demands for land for other uses, particularly food production, and this is discussed in Chapter 3.

Could we reduce CO_2 by planting more forests?

Possibility 2 was to remove some of the CO_2 from the atmosphere. It has been suggested that this could be done by planting new forests, which would lock C up in their biomass. Vitousek (1991) has discussed whether this could provide a significant contribution to controlling CO_2 increase. The key point to emphasize is that steady-state forests are not net absorbers of C. If C is to be removed from the atmosphere year by year on a long-term basis, we should need to establish 'C-sink' plantations, harvest them while they are still growing actively, replant the site, and store the harvested wood permanently so that it does not rot (since that would return CO_2 to the atmosphere). This storage would be a formidable activity. It has been estimated (Vitousek 1991) that the total amount of C stored in all cut timber, worldwide, in use in houses, furniture, fences, etc. plus wood products such as paper, is 4–5 Gtons. This is about equal to a year's release of C from burning fossil fuels (Fig. 2.3). Therefore we would need to add each year to the world's C-sink store an amount of wood about equal to the present total; and this wood has to be stored *for ever*. The area required for growing these C-sink forests would also be formidable. If we assume 5 tons $ha^{-1} yr^{-1}$ as an average productivity (see earlier), this would absorb 2 tons C $ha^{-1} yr^{-1}$ (since plant dry matter is about 40% C), so to absorb 5 Gtons C yr^{-1} would require 25 million km^2 to be permanently dedicated to C-sink forests; this is 19% of the world's land surface. These calculations are based on the aim of absorbing all the CO_2 generated from burning fossil fuels each year, but they serve to show that even if we had a more limited aim, C-sink forests are not a realistic way of helping to solve the problem.

Effects of increased temperature and CO_2 on living things

Increases in temperature and in CO_2 concentration are both likely to affect plants, sometimes in the opposite direction (see Box 2.5). Here I consider how crop plants are likely to respond during coming decades, and I then go on to consider how wild plants and animals may be affected.

Effects of CO_2 concentration on photosynthesis

Temperature and CO_2 concentration can both be controlled in controlled-environment plant growth cabinets, and these have been used to investigate the response of a range of crop and wild species. Limitations in such experiments include: (1) the intensity of radiation in cabinets is usually much below natural; (2) temperature is usually varied step-wise, e.g. constant for 12 h then suddenly switching to a lower 'night' temperature; (3) rooting volume available to the plants (e.g. in pots) is often very small. Experiments have been conducted on plants outdoors in open-topped chambers in which elevated CO_2 concentrations can be maintained. These give more natural radiation and rooting conditions, but the air temperature cannot be controlled. It is important to conduct long-term experiments, since some experiments have shown that an initial large response of photosynthesis to increased CO_2 may diminish with continued exposure over many weeks (Eamus & Jarvis 1989). In spite of these technical problems there is no doubt that increasing atmospheric CO_2 concentration above the present ambient does often increase rates of photosynthesis and growth. Results of numerous experiments on crop plants and on tree species have been summarized by Kimball (1983) and Eamus & Jarvis (1989). Although a few species showed no significant response to increased CO_2, in most the rates of photosynthesis and growth were increased. The average response to a doubling of CO_2 concentration was for growth rate to increase by about one-third, though there was a lot of variation and in a few species the growth rate doubled.

One reason that photosynthesis often does not increase fully in proportion to CO_2 concentration is that increased CO_2 causes stomata to close partially. As well as reducing the rate of CO_2 entry to the leaf, this will reduce transpiration rate. That could be harmful in some very

Box 2.5 Summary of effects of increased atmospheric CO_2 concentration and temperature above present ambient.

+ means an increase in the process (++ greater increase than +),
− means a decrease (−− greater decrease than −),
blank indicates response slight or variable.

	CO_2 increase	*Temperature increase*
Solubility of CO_2 in water		−−
Solubility of O_2 in water		−
Specificity of Rubisco (CO_2/O_2)		−
Rate of photorespiration	−	+
Rate of net photosynthesis:		
C_3 plants	++	*
C_4 plants	+	†
Stomatal aperture	−	
Transpiration rate	−	
Water use efficiency $= \frac{\text{dry wt increase}}{\text{transpiration}}$	+	
Rate of canopy expansion		+

* Fastest about 20–35°C; † fastest about 30–45°C.

Further information: Lawlor & Mitchell (1991), Nobel (1991a), Taiz & Zeiger (1991).

hot climates, since it reduces transpirational cooling and so makes the leaf hotter; but where water supply to the roots is limited it may well be beneficial. A plant's transpiration rate usually rises and falls in parallel with the rate of photosynthesis, because both are strongly affected by intensity of incoming radiation and by stomatal aperture. The ratio (dry weight increase)/(weight of water transpired) for any period is known as the *water use efficiency*. Expressed as mmol CO_2 per mol H_2O lost, over 24 h periods it is commonly in the range 0.5–1.5 for C_3 plants, 1–2 for C_4 plants and 4–10 for CAM plants, at present ambient CO_2 concentrations (Nobel 1991b). Increased CO_2 will increase water use efficiency (Box 2.5), and this would be beneficial in any area where water supply commonly limits crop production. It would allow, for example, increased production from a given amount of irrigation.

Water use efficiency

Effects of temperature on photosynthesis and growth

Temperature has complex effects on photosynthesis and growth (Box 2.5). Photorespiration increases with increasing temperature, partly because the solubility of CO_2 decreases faster than that of oxygen, partly because the greater affinity of Rubisco for CO_2 than for oxygen is reduced. This is one reason why there is an optimum temperature for photosynthesis (which differs between species); in cool climates temperature rise will increase photosynthesis per unit of leaf area, but above the optimum further warming will reduce it. In warmer conditions the foliage will develop faster, and this at first sight would indicate faster photosynthesis per plant or per square metre of ground. However, in annuals it may also hasten senescence at the end of the growth period. Monteith (1981) concluded that for cereals in Britain higher temperatures, by hastening senescence, reduce the growing season so much that grain yields are reduced; controlled-environment experiments support this. Correlations between past yields of wheat and summer temperature indicate a yield reduction of 5–6% per 1°C rise in temperature. Because this reduction is caused by end-of-season senescence, it does not occur in perennial crops such as grassland.

Predicting effects on crop production

Thus increased CO_2 is likely to increase crop growth, but temperature rise could either increase or decrease it. That might lead us to expect that the combined effect of increased CO_2 and the warming caused by it would be increased yields. That will often be so, but not necessarily always. Predictions for some crops in some areas (for example in Goudriaan *et al.* 1990) have shown yield little changed, or even reduced. In order for farmers to benefit from warming and extra CO_2 they may need to sow different crop varieties. It may be necessary to breed new cereal varieties that can make use of longer growing seasons. Or it may be better to change species; C_4 crops such as maize may be grown in regions, e.g. in Britain and Canada, where up to now wheat (which is C_3) has predominated. Parry & Carter (1990) discuss changes of farming practice that may be appropriate, for a climate caused by CO_2 doubling, in Japan, Saskatchewan and several parts of northern Europe. For example, they predict that in the region of Russia around Moscow it will be beneficial to reduce the area sown to barley, oats and potatoes and to increase maize, winter wheat and vegetables.

There are other possible effects of climate change that could be important but are at present difficult to predict; for example, responses of pests and diseases. But on the basis of the evidence I have summarized, the main conclusion is that the combined effects of increased CO_2 and temperature will not on a world scale reduce food production, and should allow increased production in many areas. There may be problem areas, for example where rainfall decreases, but we are not yet able to predict with confidence where these will be.

Effects on wild species

In the previous section evidence about how crop plants will respond to future climate change came mainly from experiments. But the key questions about how wild species will respond are different: will they change genetically to adapt to climate change? or will they migrate—can they migrate fast enough? We have to base the answers to such questions mainly on study of how plants and animals responded to climate changes of the past. A considerable amount of information about this is now available. Box 2.6 summarizes some of the methods that have been used to obtain it.

Temperature changes in the past

Table 2.7 gives a very simplified summary of temperature changes during the last 20 000 yrs. For about 8000 yrs, from 15 000 to 7000 yrs BP (before present), there was a long-term warming trend. During the warm plateau from about 7000 to 5000 yrs BP the temperature was

Box 2.6 Techniques for studying living things and climate of the past 100 000 years.

Distribution of animals and plants

The past distribution of vertebrate animals is indicated by bones, and of arthropods by exoskeletons. Macroscopic remains of plants that can provide information are seeds, leaves and wood. Pollen grains have the advantage of very large numbers, allowing quantitative assessment of changes in abundance, but the disadvantage that they can be widely dispersed, hence the area of catchment is not well defined. Very useful sources of these remains are sediments at the bottom of lakes, and peat in growing bogs. These combine: (1) little physical damage; (2) anaerobic conditions, so slow microbial decomposition; and (3) continued accumulation of the surrounding medium, so the remains are in a vertical time sequence. Remains are also sometimes well preserved in caves; cave-dwelling animals may have left faeces, pellets (e.g. from owls) or remains of collected food or bedding, in which animal or plant remains can be identified.

Dating the remains of living things

The age of organic materials up to tens of thousands of years old can be measured by *radiocarbon dating*. Among the CO_2 molecules in the atmosphere a small proportion contain the natural radioactive isotope ^{14}C, which has a half-life of 5730 yrs. The isotope is incorporated into living plants by photosynthesis and from them down food chains. The amount of ^{14}C remaining in dead plant or animal material provides a measure of when the plant it originated from was alive. There are technical problems arising from the fact that the $^{14}C/^{12}C$ ratio in the atmosphere has not remained constant throughout the last 100 000 yrs.

Past temperatures

Temperatures that occurred in the past can be estimated using the stable isotopes ^{2}H (deuterium) and ^{18}O. These both occur naturally in a small proportion of water molecules. Because they alter the molecular mass, they alter slightly the rate at which the molecules evaporate, condense or freeze. Hence the *isotope ratios*, $^{2}H/^{1}H$ and $^{18}O/^{16}O$ in ice indicate the temperature at which snow formed in the air overhead. Oxygen isotope ratios in $CaCO_3$ in skeletons of ocean animals can be used in a similar way to indicate the temperature of the water at the time they were formed. This has been applied particularly to cores from Antarctic and Arctic ice and to ocean sediments containing foraminiferan shells, since both these sources provide long vertical time sequences.

Further information: Lamb (1977), Moore & Chapman (1986).

Table 2.7 Summary of temperature changes during the last 20 000 yrs

Thousand yrs BP	Temperature		Period
0 (=present)			
0.4–0.2	Cold period	'Little Ice Age'	Post-glacial (=Holocene)
	Getting cooler		
7–5	Warmest	'Climatic optimum'	
10			
	Getting warmer		Late-glacial
15			
	Much colder than present		Full-glacial

BP, before present.
For further information see Pennington (1974), Lamb (1977).

Table 2.8 Periods of warming

Period	Place	Length (yrs)	Temperature rise (°C)	Rate of rise (°C per century)	Method of temperature measurement
AD 2000–2100	World (mean)	100	3	3	Predicted
AD 1885–1940	World (mean)	55	0.6	1.1	Thermometers
AD 1690–1730	Central England	40	1.8	4	Thermometers
AD 1127–1176	Northern Sweden	50	2.4	5	Tree rings
15 000–9000 BP	Antarctic	6000	9	0.15	$^{2}H:^{1}H$
15 000–8000 BP	Southern Indian Ocean	7000	6	0.1	$^{18}O:^{16}O$
13 000–7000 BP	Tropical Indian Ocean	6000	2.5	0.04	$^{18}O:^{16}O$

Data from: Lamb (1977), Jouzel *et al.* (1987), Schneider (1989), Briffa *et al.* (1990), IPCC1 (1990), van Campo *et al.* (1990).

about 2°C warmer than it is now. This period used to be called the 'climatic optimum' on the assumption that warmer means better; nowadays people do not seem to be so sure. Since 5000 yrs BP the long-term trend has been a slow cooling; but there have been shorter-term fluctuations superimposed on that, one of which, the 'Little Ice Age', I have shown in Table 2.7. Atmospheric CO_2 concentrations were lower than at present during most of the glacial but rose towards the end (Shackleton *et al.* 1983). So there are similarities between past events and the predictions for the future. But the average rate of warming after the full-glacial, over several thousand years, was far slower than is predicted for the next hundred years (Table 2.8), in fact less than one-tenth of the rate. There have been periods of a few decades when temperatures rose at about the 0.3°C per decade predicted for the next century (Table 2.8), and even during the earlier part of the 20th century the warming was at about one-third of that rate. We may also be able to learn from the response of species to the cooling during the last 5000 yrs.

Species migrated in response to past climate changes

Information from the past can help to answer the question: will species adapt to changed conditions or will they migrate? The answer is that in the past they have often migrated, presumably because they did not adapt. Figure 6.3 shows the spread of two tree groups in North America after the full-glacial, and Table 6.4 summarizes rates of spread of tree genera in North America and Europe. These results are discussed in more detail in Chapter 6, including how species with large, heavy seeds managed to achieve such high rates of spread. For now, note that the species did not all spread at the same rate, or even in the same direction. Plants have also changed their ranges in response to cooling since the 'climatic optimum'. For example, remains of tree leaves recovered from lake mud and bogs show that 5000–9000 yrs BP tree species grew 300–400 m higher in mountains in New England than they do today (Davis *et al.* 1980). The present lapse rate (decrease of temperature with altitude) is 0.6°C per 100 m, which would agree with the temperature being about 2°C higher then.

Migration of animals after the last Ice Age cannot be mapped as precisely as plants, because remains are less frequent, but we can at least compare known sites of occurrence in the past with present ranges. These show that not only mammals and birds, but some invertebrates, have moved their ranges by several thousand kilometres since the full-glacial. For example, several beetle species are known to have lived near the Great Lakes 13 000–10 000 yrs ago that are now found only in northern Canada or Alaska (Morgan *et al.* in Porter 1983). Graham (1986) provides maps for North America, Eurasia and Australia showing sites where individual small mammal species were towards the end of the full-glacial and in the late-glacial, and their ranges today. The conclusion is the same as for higher plants, only more strikingly so: as the climate changed, the animal species did not all move at the same rate or even in the same direction.

A species that has failed to adapt to temperature change

An example of a species failing to adapt to climate change (though to cooling, not warming) is provided by the native lime tree of England, *Tilia cordata*. The pollen record shows that it invaded southern England between 8000 and 7000 yrs BP and at first spread rapidly northwards (Birks 1989). However, its spread slowed after 6500 yrs BP and stopped about 5500 yrs BP, in northern England. Its northern limit as a native plant is still there, in the Lake District and nearby. In a detailed study Pigott & Huntley (1978, 1980, 1981) found that nearly all the lime trees in the Lake District were very old, and set seed only occasionally in unusually hot summers. The most northerly sites where it regularly sets seed have a mean temperature about 2°C higher than the Lake District. In the Lake District viable pollen and ovules are produced in many years, but the pollen fails to fertilize the ovules, perhaps because of inadequate pollen tube growth. Thus it appears that this species reached its present northern limit at the end of the 'climatic optimum', when the mean temperatures were about 2°C warmer than nowadays; as the climate subsequently cooled it has been able to set viable seed less and less frequently. Since many of the surviving trees are of the order of 1000 yrs old, they may have established during the period AD 1150–1300, when mean summer temperatures were higher than at present (Pigott 1989). The species has survived in the area basically because the trees can live a long time. Evidently the pollen tubes have not evolved the ability to function satisfactorily at 2°C cooler.

Will species be able to migrate fast enough?

If wild species cannot adapt to temperature change, they must either migrate or become extinct. Will they be able to migrate fast enough? Firstly we can ask whether plants migrating after the glacial period were keeping up with the polewards movement of the isotherms. Using present temperatures in fairly level regions of large continents, the average distance north–south to give a 1°C change in mean annual temperature is about 110 km in North America and about 170 km in Russian Asia and in southern South America. The isotherms tend to be further apart in summer, which may be the most important period for living things, but as a first approximation these figures indicate that while the temperature was rising 0.1°C per century in the late-glacial the isotherms would move towards the poles at an average of 0.1–0.2 km yr^{-1}. The species listed in Table 6.4 were thus apparently well able to keep up with this rate, and some migrated faster. If during the 21st century global warming is 3°C per century the isotherms will move at 3–5 km yr^{-1}. There are no known examples of species that migrated that fast after the glacial. One might reply that maybe they had the ability, but the slower warming gave no need for it. However, the discussion (in Chapter 6) of possible mechanisms for rapid transport of seeds suggests that some at least of the rates in Table 6.4 are near the possible maximum. Some animals, for example large birds, should have no difficulty in migrating at 5 km yr^{-1}. Some appear to have altered their ranges in response to the warming of the 20th century. For example between 1900 and 1950 seven bird species, previously of more southerly distribution, including starling and short-eared owl, started breeding regularly in Iceland; in contrast, a high-Arctic species the little auk almost disappeared as a breeding species (Gudmundsson 1951). Chapter 4 describes how the dominant fish in the English Channel has changed twice during the 20th century; some authorities attribute this to temperature changes affecting species that were near the edges of their ranges, but this cannot be conclusively proved.

Figure 2.6 provides a more detailed prediction of migrations that may be required if species are to keep up with climate change. This is based on predictions by a model of regional climates if the atmospheric CO_2 concentration doubles. Such predictions for individual regions are still unreliable, but the map indicates several important general points.

1 If the CO_2 concentration doubles in about 90 yrs (see Fig. 2.2),

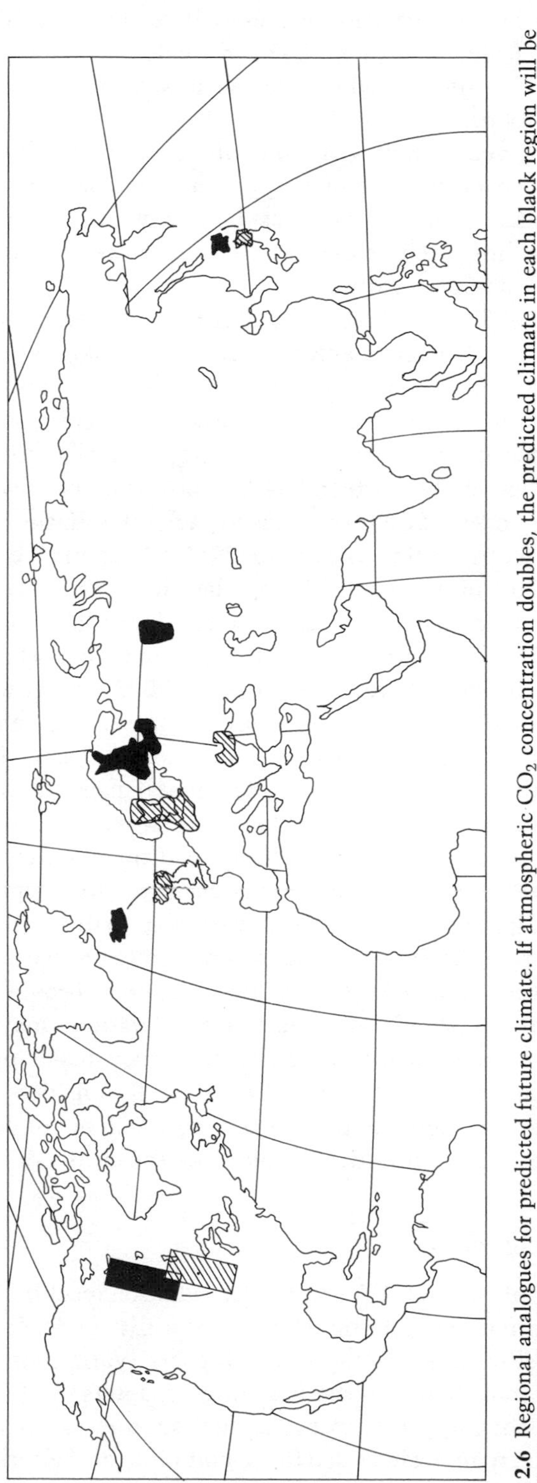

Fig. 2.6 Regional analogues for predicted future climate. If atmospheric CO_2 concentration doubles, the predicted climate in each black region will be similar to the 1990 climate in the hatched region to which it is joined by a line. Future Finland has its 1990 analogue in southern Sweden and northern Germany. From Parry & Carter (1990).

some of the migration routes will require rates of migration several times faster than the 3–5 km yr^{-1} calculated above.

2 Some species will have barriers in their way, such as seas or mountains.

3 It gives examples of a present region having two future analogues, but also of two regions at present far apart having adjacent future analogues. This helps to explain why in the past species have not always migrated at the same rate or in the same direction, and may not do so in the future.

In the past the species composition of communities changed

The different migration abilities of species within communities will raise problems. Because, after the Ice Age, species did not migrate at the same rate or even in the same direction, the species composition of communities changed. For example, hemlock (*Tsuga canadensis*) and beech (*Fagus grandifolia*), which today are common species found together in the northern hardwood forests of New England and New York, occupied different areas to each other in the late-glacial, hemlock near the Atlantic coast around North Carolina, beech near the coast of the Gulf of Mexico. In Europe beech (*Fagus sylvatica*) and deciduous oaks (*Quercus* spp.) often grow together today; but in the late-glacial oak was in southern Spain, Italy and Greece, whereas beech was virtually absent from Europe until 9000 yrs BP, when it first became abundant in Serbia and Bulgaria. These and many other examples are shown in the distribution maps of Davis (1981) and Huntley & Birks (1983). This raises fundamental questions about how integrated ecosystems are. Are there 'keystone' species, whose disappearance will have a major knock-on effect, resulting in the extinction of many other species? This question is discussed in Chapter 8.

Should conservationists encourage or discourage migration?

A key question for conservationists will be: should we attempt to maintain the existing species complement of nature reserves and other wildlife areas, even though the climate is becoming unfavourable for them? Or should we encourage migration, maybe even artificially transporting species to make sure they reach new areas that have become suitable for them? Chapter 8 considers in some detail the creation of new 'natural' areas, the problems of deciding what to aim for, as well as questions about how the aims can be achieved.

Conclusions

- Only about 0.1% of the solar radiation reaching the earth is captured as primary production. However, locally, under favourable conditions, the efficiency can be much higher. (Variations in productivity are important topics in the next three chapters, on food and timber.)
- Growing plants to use as fuel sources can make a limited contribution to replacing fossil fuels; non-biological alternative energy sources are likely to contribute much more.
- The increase of atmospheric CO_2 year by year comes mainly from

burning of fossil fuels, though clearing of forests makes some contribution.

• Planting new forests is not a practical long-term method of slowing down this CO_2 increase.

• If nothing is done to limit production of CO_2 and other greenhouse gases, the mean world temperature will be about 3°C higher by AD 2100.

• If this predicted CO_2 and climate change occurs, on a world scale it should be possible to maintain crop production, though the crops grown in particular areas will need to be changed.

• Wild species will need to migrate in order to survive, and it is not known for most of them whether they can migrate fast enough.

Further reading

Photosynthesis and productivity:
Lange et al. (1983)
Taiz & Zeiger (1991)

Energy sources (fossil fuels, biomass, others):
Lawson *et al.* (1984)
Gates (1985)
Hall *et al.* (1986)
Chapman (1989)

Global climate change and its effects:
Schneider (1989)
Houghton *et al.* (1990)
Jackson *et al.* (1990)

Species migration:
Delcourt & Delcourt (1991)

Chapter 3: Farming: The Ecology of Food Production

Questions

- How many people per hectare can low-input farming systems support? Could such systems support present and future world populations?
- Are there serious disadvantages in long-term use of inorganic fertilizers?
- How long will the world's supply of phosphate for fertilizers last?
- Can we maintain adequate food production without inorganic fertilizers? How?
- Does conversion of natural vegetation to farmland increase soil erosion?
- How can farming practices be altered to maintain soil structure and reduce erosion?
- How can grazing systems be managed to give high animal production without damaging the plant productivity?
- Can overgrazing lead to irreversible changes to grazing lands?

Background science

- Farming systems used in the past. Their efficiency at energy capture.
- Ways of using biological nitrogen fixation to provide nitrogen for crops.
- Promotion of crop growth by mycorrhizas.
- Phosphorus balance of farms under various systems.
- Soil organic matter and its influence on soil structure.
- How grazing animals respond to characteristics of the sward.
- How different plant species respond to grazing.
- The influence of different grazing regimes on the structure, species composition and productivity of the vegetation.
- Unstable states and break points in grazing systems.

World demand for food will increase

This chapter is about the production of food, as its title uncompromisingly declares. It is not about how to conserve wild animals and plants in farmland: that topic is deferred until Chapter 8 (Conservation). One fact that must underlie any consideration of food production is the increasing human population of the world (Fig. 1.1). Although some countries in the developed world (notably U.S.A. and the European Community) are producing too much of some basic foods, in other countries shortages and sometimes severe famines are occurring. On a world scale we must be concerned about food shortages in the future. The underlying theme of this chapter is how to maintain and if possible increase food production, but to do it in a sustainable manner, so that production can be continued long-term. This chapter is about food production today and in the future, in developed and less-developed countries, in temperate and tropical climates. So it is about modern agricultural systems; but it also considers systems used in the past, to see what their limitations were, and also what we can learn from them for farming in the future.

Limitations imposed by climate

The way that food can be produced in an area has always been limited by the climate. Climate is part of the reason why food surpluses and famines occur simultaneously: food surpluses occur only in regions with predictable climate and adequate water supply. (I am not saying that is the only requirement for growing enough food, but it is one of them, or has been up to now.) Box 3.1 gives a simplified classification of the major traditional food production systems in relation to rainfall. The farming system that could be practised in an area had a crucial effect on the type of civilization that developed. Of the five systems listed, three require the people to move around, only two (irrigation farming and infield–outfield) allow permanent villages and towns. If you have to move all your belongings, and you do not have any means of moving them except to carry them yourself, then you do not accumulate many objects. For example, nomadic pastoralists and shifting cultivators have rarely gone in for writing.

In the first two systems listed in Box 3.1 much of the human food came from grazing animals. Today there is still more land area devoted to grazing than to crop production (Table 1.1), and management of grazing is one of the main topics of this chapter. In the other three systems crops were grown, and for this to be sustainable a system that maintains the nutrient status of the soil is essential. Irrigation farming was successful if the river water was nutrient-rich: each year the fields were supplied with nutrients leached from other regions, often far away. Without irrigation, some other method of maintaining soil fertility was needed; shifting cultivation and the infield–outfield system were designed around this need (although their inventors of course lacked modern knowledge of plant mineral nutrition). Later in the chapter I discuss nutrient supply, especially phosphorus, in these and in more modern farming systems.

Box 3.1 Major farming systems of the world in the past, in relation to climate.

1 Rainfall total per year low	
(a) Rainfall unpredictable	Nomadic pastoralists
(b) Annual variation more predictable	Transhumance
(c) River water available	Irrigated crops
2 Rainfall total higher	(a) Shifting cultivation (swidden farming)
	(b) Infield–outfield farming

Description of the farming systems

Nomadic pastoralists obtain most of their food from grazing animals, often cattle, sheep or goats. The people move with their animals in search of pasture. This system was in use more than 3000 yrs ago on the steppes of eastern Europe and Siberia; it is still practised on the southern fringes of the Sahara.

Transhumance. The people migrated annually with their flocks from winter to summer pastures. In Spain they migrated over hundreds of kilometres along special 'drove roads' from winter pastures near the coast to summer pastures inland (Ruiz & Ruiz 1986). Shorter migrations took place in mountainous regions, between lowland winter pastures and higher summer pastures.

Irrigated crops. Some early irrigated farming used areas which were naturally flooded each year by rising river level. Examples: the Nile and Indus Valleys, probably also some Amazon floodplain. Other regions required irrigation canals, pumping or other engineering. Examples: lowland China, and the Tigris–Euphrates 'fertile crescent'.

Shifting cultivation. A patch of forest is cut down and burnt. Crops are grown for a few years. Then the patch is left for natural forest regeneration to occur, and the farmers clear another patch of forest. After some years they return to again cut the regenerated forest patch. So the whole area used by a group will contain patches of forest in various stages of regeneration. There may be a single, static village, or the people may move from time to time. Shifting cultivation can also be practised in savanna and grassland areas (Nye & Greenland 1960).

Shifting cultivation was practised about 8000 yrs ago in northern China. It was being used in eastern U.S.A. at the time of arrival of the first Europeans, and the system was adopted by European settlers (Williams 1989). It is still used today in the tropics, especially in forest areas (Eden 1990).

The infield–outfield system. The area used for crop growth (the 'infield') remains the same for many years, though the crops grown within it may be rotated. Another area, usually larger (the 'outfield'), supplies organic matter and mineral nutrients to the crop fields and so helps to maintain their fertility. The outfield can be forest, open woodland, heathland or grassland. There are alternative ways of arranging the transfer of nutrients:

1 Animals are grazed in the outfield but are moved to parts of the infield (e.g. every night) so that some of their dung and urine falls on it.

2 Plant material (e.g. hay from meadows, leaves from trees) is harvested from the outfield and used to feed animals or as bedding for them. Their dung, or bedding containing the dung and urine, is then spread on the infield.

3 Plant material or turf from the outfield is added directly to the infield, without animals being involved.

The infield–outfield system was widely used in northern Europe for several thousand years. In England it was in use until the 19th century (Cobbett 1830). It is still used in various parts of the world, for example in Zimbabwe (Swift *et al.* 1989) and Nepal (Ives & Messerli 1989).

Further reading: Russell (1967), Simmons (1989).

The structure of this chapter

To summarize the structure of this chapter: it starts by comparing some farming systems, past and present, especially how much energy they could capture per hectare and hence how many people they could support; after that it concentrates on three topics, mineral nutrients (especially phosphorus), soil erosion and grazing management. Crop pests and diseases are considered in Chapter 6.

Energy capture by farming systems

Converting solar energy to food energy

Food provides us with energy, with protein, with various specific organic chemicals (e.g. vitamins) and inorganic elements (for example, calcium, iron and iodine). This chapter is not about human diet, and it mainly considers food as a source of energy. Thus we are considering the efficiency of plants and animals as convertors of energy of solar radiation into chemical energy in forms that human beings can assimilate. The previous chapter provided some background information on energy. Table 2.1 showed that the total net primary productivity of the world's vegetation is about 0.1% of the total incoming short-wave radiation from the sun, and that production of food for people is in turn only about 0.5% of the total net primary productivity. This suggests that there is plenty of scope for increasing food production, but we need to consider in more detail the productivity of various farming systems and what limits them.

Table 3.1 shows the food energy per hectare provided by various contrasting farming systems. The lowest energy production was by the pastoralists, a group of Turkana people in northwest Kenya who moved with their grazing animals over an area of 7500 km^2. More than half their food energy came from milk, as is common among migrant pastoralists. Maize in U.S.A. produces about 4000 times as much food energy per hectare, but even so it is only about 0.2% efficient at converting solar radiation to food energy. One message from the table is the low efficiency of animal farming as a way of producing food energy. This is shown most clearly by comparing cattle and wheat in England (U.K.).

Under 'harvested food energy' the figures in column (a) take into account only the actual area under crops that year. These three figures are closely similar, but that is partly coincidental; for example, estimates of wheat yields in England suggest that average yields in the early 19th century (when the example for Table 3.1 was taken) were three- to fourfold higher than in the 13th–14th century (Stanhill 1976). Calculated in this way (column (a)), these examples of shifting cultivation and infield–outfield farming gave yields about one-seventh to one-fifth of the high-input wheat and maize examples. However, this is not the important comparison: the infield could not maintain its yields without the outfield, nor could production continue in Papua New Guinea without shifting the cultivated fields, so the whole area

Table 3.1 Food production by various farming systems, expressed in energy terms

<table>
<tr><th></th><th colspan="2">Harvested food energy (GJ ha^{-1} yr^{-1})</th><th colspan="2">People supported (per ha)</th><th></th></tr>
<tr><th></th><th>(a)</th><th>(b)</th><th>(c)</th><th>(d)</th><th>Notes</th></tr>
<tr><td>Low-input systems</td><td></td><td></td><td></td><td></td><td></td></tr>
<tr><td>Migratory pastoralists, Kenya, 1981–82</td><td></td><td>0.025</td><td>0.01</td><td>0.004</td><td>1</td></tr>
<tr><td>Shifting cultivation, Papua New Guinea, 1962–63</td><td>19</td><td>1.5</td><td>0.5</td><td>0.3</td><td>2</td></tr>
<tr><td>Infield–outfield, southern England, 1826</td><td>14</td><td>7.4</td><td>2</td><td>1</td><td>3</td></tr>
<tr><td>Infield–outfield, southern India, 1955</td><td>18</td><td>8.0</td><td>2</td><td>1</td><td>4</td></tr>
<tr><td>High-input systems</td><td></td><td></td><td></td><td></td><td></td></tr>
<tr><td>Pasture cattle, lowland England, 1980s</td><td colspan="2">3.5</td><td>0.7</td><td>0.6</td><td>5</td></tr>
<tr><td>Wheat, U.K., 1986–90</td><td colspan="2">97</td><td>20</td><td>17</td><td>6</td></tr>
<tr><td>Maize, U.S.A., 1986–90</td><td colspan="2">104</td><td>18</td><td>18</td><td>6</td></tr>
</table>

Columns

(a) Calculated by (energy in food from arable crops)/(land area under arable crops that year).

(b) Calculated by (total food energy produced)/(total land area used by village). 'Total land area' includes forest fallow or outfield.

(c) At actual food energy use per person in the area at that time.

(d) If food energy use per person the same as U.S.A. mean for 1987–89 (15.4 MJ person^{-1} day^{-1}).

Notes

1 Most of the food came from herded cattle, sheep, goats, donkeys and camels, plus a little from growing sorghum and from wild plants and animals. From Coughenour *et al.* (1985).

2 About one-tenth of the area usable by the village was cultivated at any one time, the remainder was regenerating forest fallow. Meat was obtained by feeding some of the crop produce to pigs, plus a small amount of hunting in the forest. From Bayliss-Smith (1982).

3 Wheat and barley grown on the infield; additional food from cattle, sheep and pigs which grazed the outfield. From Bayliss-Smith (1982), based on contemporary description of a village in Wiltshire by Cobbett (1830).

4 Infield irrigated rice + dry-land millet. No fertilizers or other inputs apart from irrigation. Cattle grazing the outfield provided milk. From Bayliss-Smith (1982).

5 Values typical of well-managed beef production on heavily fertilized pasture. Based on Lazenby (1981).

6 Mean yields for whole country, from *FAO Production Yearbook 1990*, converted to energy values using information from Pimentel & Pimentel (1979).

used by the village must be included in the calculations. This was done for column (b), and it shows that the two infield–outfield examples gave yields an order of magnitude lower than the modern wheat and maize, and the shifting cultivation lower still. In both of the infield–outfield examples there was some nutrient input by irrigation—in the Indian village from a canal, in the English village by natural flooding of water meadows. The Indian rice fields would also have N input from N-fixing cyanobacteria. However, it is not certain whether either of these farming systems was sustainable long-term. Both were in a period of change, which had resulted in the cultivated infield area being increased until it exceeded the outfield. In the Indian village the building of the irrigation canal 15 yrs earlier had allowed new crops to be grown. In the English village the communal open fields and grazing lands had been enclosed to become private property and the proportion of infield had been extended. The next section gives more information on the nutrient balance of infield–outfield systems and whether they are sustainable long-term.

How many people per hectare?

Table 3.1 shows how many people could be supported per hectare by each of the systems, taking into account all the land used, not just the amount under cultivation that year. It is calculated for column (c) on the basis of actual food energy use per person in that area at that time. That varied from 5.3 $MJ\,day^{-1}$ for the Kenyan pastoralists to 15.4 $MJ\,day^{-1}$ in U.S.A. If everyone's food energy supply had been brought up to that of the average U.S. inhabitant in 1987–89, the number of people supported per hectare would be lower (column (d)).

The number of people supported by these different systems can be compared with the number of people that the world needs to feed now and in the future. It is virtually certain that the world's population will soon reach 6 billion (Fig. 1.1). The world's combined area of cropland and permanent grazing land is at present 4.8 billion ha (Table 1.1). Suppose that we need to support 6 billion people on 5 billion ha, that is an average of 1.2 people per ha. In practice large areas of present rough grazing will support less than that, so higher yields will be required from other areas. Comparing with figures in the 'people supported' columns in Table 3.1 gives some indication of the farming systems we need for the future. A substantial part of our future farmland will need productivities nearer to the bottom two lines of the table than to the other five lines. Animal production can certainly make some contribution in the future, and low-input systems should not be ruled out; but these on their own will not support the population of the future. Pastoralism, shifting cultivation and the infield–outfield system have all broken down in the past in various areas because the human population increased. The density of grazing animals became too high, the fallow period became too short, the outfield became too small, and the system was no longer sustainable. Later in this chapter there is more discussion of the conditions under which these systems are sustainable. Although we cannot return to these systems as our main means of food production, we may be able to learn from them when planning for the future.

Fuel energy inputs

Modern farming methods require use of fuel energy. Table 2.5 gave figures for the energy used in production of ethanol from sugarcane. Pimentel & Pimentel (1979) gave energy data for production of a variety of foods in developed countries during the 1970s. The results can be divided into two groups. Foods where the energy return on investment is above 1 (i.e. there is more energy in the food than the fuel energy used in its production) include: cereals, food legumes (e.g. beans), potatoes and sugar. Foods whose energy return on investment is below 1 include: vegetables, fruit, beef and fish (whether from fish farms or ocean fisheries). Thus the foods that are our main sources of energy generally require less fossil fuel energy to grow them than we get out of them, whereas the reverse is true for many foods we eat for other purposes (e.g. for protein, roughage or because they taste nice). It is not necessarily bad to use 2 MJ of coal to produce 1 MJ of food: we cannot eat coal. Nevertheless, because there will be increasing pressures in the future to reduce use of fossil fuels (see Chapter 2), we shall need to give more attention to energy inputs. The enormous increases in food production per hectare that have occurred during the last hundred years have entailed great increases in energy input; we need to ask whether further increases in productivity in the future can be brought about in ways that require less energy.

Mineral nutrients

Nutrient inputs and outputs

All plants and animals require a suite of elements including N, P, K, Ca and others. Fast growth is bound up with adequate supplies to the plant or animal of these essential nutrients. In natural ecosystems cycling of nutrient elements within the system plays an important part in providing the supplies for uptake by the species there; but there are bound to be losses from the ecosystem, and these have to be balanced by inputs if the system is to be sustainable long-term. In farming there is additional, unavoidable loss of nutrients from the farm in the food product harvested. Table 3.2 shows, as examples, N losses from three types of farm, two receiving N fertilizer, the other receiving no artificial inputs. On those farms inputs of N dissolved in rain and as gases were 14–17 kg ha^{-1} yr^{-1}, so clearly some additional N input is required if the N status of the soil is to be maintained. Nutrient inputs to farmland are not a new idea: as explained earlier, a key aim of the shifting cultivation and infield–outfield systems (see Box 3.1) was maintaining the soil fertility of the cropland.

Farming using inorganic fertilizers

Nowadays the normal method in developed countries for supplying nutrients to crop plants is inorganic fertilizers. The very high yields of modern farming, compared with traditional systems (e.g. Table 3.1) are

Table 3.2 Losses of nitrogen (kg ha^{-1} yr^{-1}) from three types of farm in Europe

	Wheat, Britain, receiving fertilizer (1970s)	Dairy farming, Netherlands	
		No fertilizer or supplementary cattle feed (1937)	Using fertilizer and supplementary cattle feed (1970s)
Leaching	10	11	44
Gaseous losses	18	56	230
Harvest: removed in plant and animal produce	77	19	115

Data from Frissel (1978).

partly, though by no means entirely, because of use of fertilizers. This can work satisfactorily for more than a century, as shown by experiments at Rothamsted, England (Jenkinson 1991). There wheat, barley and hay have each been grown continuously since the mid-19th century with inorganic fertilizers as the only human input of nutrients. Yields have been maintained, and in the case of wheat have substantially increased because of use of improved varieties and of modern herbicides for weed control. This is the longest-running test of growing crops with inorganic fertilizer as the only non-natural nutrient input, but similar experiments have been run successfully for decades in other parts of the world. In this section I consider two questions: (1) Are there any serious disadvantages in use of inorganic fertilizers? and (2) Are there any feasible ways of maintaining high yields without using fertilizers? I apply these questions first, fairly briefly, to nitrogen, and then in more detail to phosphorus.

Nitrogen

Suggested disadvantages of substantial use of inorganic N fertilizers are listed in Box 3.2. Nitrates in the diet pose some health risks, and many countries have set maximum nitrate concentrations for drinking water. Amounts of nitrate in seeds are usually negligible, but the concentrations can be high in leaf material, e.g. cabbage and lettuce, and in edible roots such as carrot and beetroot. In some experiments inorganic N fertilizer caused a marked increase in nitrate concentration in foliage but farmyard manure did not (Vogtmann *et al.* 1984). The chapter on pollution (Chapter 7) gives information on nitrate leaching from farmland, its effects and how these can be minimized.

Energy cost of N fertilizer

Almost all N fertilizer is derived from ammonia, which is made by combining N_2 and H_2 gases at high temperature and pressure. The supply of raw materials is therefore unlimited. The reaction does, however, use a lot of energy. In six farming systems in U.S.A., U.K. and Japan, involving high fertilizer inputs, N fertilizer production used 10–50% of the total energy input for food production (Pimentel & Pimentel 1979). On the two sugarcane farms in Table 2.5 the figures were 22% and 23% (of total on-farm energy use). In a time of concern

Box 3.2 Potential disadvantages of using inorganic nitrogen fertilizers.

1 High nitrate concentration in food can be a health risk.
2 Leaching losses can give high nitrate concentration in water:
(a) Health risk from drinking water,
(b) Eutrophication of lakes giving algal blooms and harm to fish.
3 High fuel energy use to make ammonia from N_2 gas.
4 High cost of fertilizer 'at the farm gate' in countries with poor transport systems.

about use of fossil fuels this is one argument for reducing N fertilizer production.

It is sometimes stated that inorganic fertilizers damage soil. This statement needs to put in a more precise form before it can be discussed; I have therefore omitted it from Box 3.2, but it is considered in the section on erosion.

In many poorer countries the final disadvantage in Box 3.2, cost of fertilizer at the farm gate, will be the most compelling. Although use of N fertilizers will certainly continue for some time, their disadvantages are sufficient to make it worth considering what are the alternatives. I assume throughout this chapter that world food production has to be

Box 3.3 Ways of using nitrogen fixation to provide nitrogen input for crops.

Methods already in use.

Most involve legume–*Rhizobium* symbiosis.

1 Grow legumes as food crops (e.g. peas, beans, soybeans). Disadvantage: seed yields much lower than cereals.

2 Use legume to supply N to other crops. Disadvantage: less space for desired crop.

Mixed cropping: growing a leguminous crop with a non-leguminous crop, usually as alternate rows. Or the legume can be a tree species ('alley cropping').

Rotation: grow a leguminous species every few years and leave the residue.

Green manure: harvest plant material from a sown legume or from grassland containing legumes, and apply it to fields where the crop will be grown.

Graze domestic animals on pasture containing legumes, then pen them on the field where the crops will be grown, to provide manure.

Feed domestic animals indoors with hay containing legumes; collect their dung and urine to apply to the fields.

Cyanobacteria. Some N-fixing species live on or near the soil surface. Can be important contributors to N input in rice paddy.

Possibilities for the future.

1 *Involving non-symbiotic N-fixing bacteria.*

Some N-fixing species are found in the rhizosphere, where they can obtain organic substrate from root exudates and dying cells. Amounts of N fixed are low, however (Giller & Day 1985).

Inoculating with a suitable mixture of microbial species can result in substantial N fixation using cellulose, e.g. straw, as energy source (Lynch & Harper 1985).

A N-fixing bacterium has been found in large numbers in stems of sugarcane and in sweet potato. It may use the concentrated sucrose solution of the phloem as energy source (Boddy *et al.* 1991).

2 *Other possibilities.*

It has been suggested that genetic engineering techniques could allow a *Rhizobium*–cereal symbiosis to be produced, or even the insertion of a functional N-fixing gene into the cereal plant itself. These developments are unlikely to happen soon (if ever).

Further information: Dixon & Wheeler (1986).

maintained at least at present levels, so some nitrogen input to farms is essential. Organic manure can maintain yields as high as inorganic fertilizer. For example, in the long-term wheat plots at Rothamsted (Jenkinson 1991), from 1852 to 1986 the grain yield from a plot receiving farmyard manure at 35 tons $ha^{-1} yr^{-1}$ remained similar to that from a plot receiving inorganic NPKMg fertilizer supplying 144 kg N $ha^{-1} yr^{-1}$. The problem is that the nitrogen in the dung and urine must come from the animals' food, and ultimately from the air. We need to consider how much land is required for that fixing of N, and what other gains there are from that N-fixing land. Box 3.3 summarizes methods for providing N for crops, all of them based on biological N fixation. Some of these methods are not new. For example, including a legume in a rotation has been recommended for several thousand years (e.g. Virgil 29 BC). Infield–outfield systems often involved a rotation on the infield, which could include a legume. The methods can almost certainly be improved. Selection and genetic manipulation of legumes and their associated rhizobia, with the aim of increasing yields, is one promising way forward. I do not discuss this subject any further. The main message is that there are plenty of options.

Phosphorus

Availability of soil P to plants

Phosphorus as a plant nutrient is in several key ways a contrast to nitrogen. Much attention needs to be paid to the *availability* of soil P to the plant. In soils there is P in the mineral material, in organic matter, adsorbed on to surfaces, and dissolved in soil water. This P is in various physical and chemical states. Although it is common to talk about P being in 'labile' and 'non-labile' forms, or 'available' and 'non-available' to plants, there is in fact a whole range of states, more freely and less freely available, interconvertible at various rates. Even phosphate in soil solution is not all readily available to plant roots: it may be too far away. Nutrient ions must move to root surfaces before they can be taken up by plant cells. Phosphate moves in soil mainly by diffusion. The rate of diffusion of an ion depends on its diffusion coefficient. The diffusion coefficient of phosphate in soil is low compared with most other nutrient ions, for example it is about 2–4 orders of magnitude lower than nitrate. The result of this is that a phosphate ion only 1 mm from a root can take days to diffuse to the root surface. Root hairs and mycorrhizal hyphae can substantially increase P acquisition by plants, by increasing the volume of soil that is close enough to an uptake surface for P to diffuse there rapidly. (Further reading on P states in soil and its availability to plants: Nye & Tinker 1977, Wild 1988, Chapter 21).

Mycorrhizas and P uptake

This low availability of much P in soil has led to interest in how to make plants able to acquire more of the P that is there. One possibility is mycorrhizas. Most crop species can form mycorrhizas of the vesicular–arbuscular type (VAM), though some, notably members of the cabbage family, cannot form any sort of mycorrhiza. Mycorrhizas can substantially increase P uptake by some plants. Figure 3.1 shows an example, for two annual species growing on an Australian soil very low in available P, to which various amounts of phosphate had been added. In the absence of mycorrhiza and with low P addition the clover was able to take up little P and grew poorly. The grass, even when non-mycorrhizal, had much greater ability to take up P from these low-P soils. This difference could be at least partly explained by differences in root morphology: ryegrass had finer roots and so about twice as much length per gram of root; it also had more abundant and longer root hairs. These differences would allow ryegrass to exploit a larger volume of soil than clover for each gram of root. Inoculation with VA mycorrhizal fungus produced a very large increase in clover growth and P uptake in low-P soils, but the proportional response decreased in the soils that had received more phosphate. Ryegrass became infected with the mycorrhizal fungus but did not respond to it in any soil. These results, and many others, show that some species can benefit markedly from mycorrhizal infection on soils low in available P. In contrast, other species, including many grass species, become infected with the VA fungus but show little or no increase in P uptake and growth. These differences between species are at least partly explained by differences in root morphology: some species are efficient at P acquisition even without the help of mycorrhizal fungi but others are not, as Fig. 3.1 illustrates. The primary reason why inoculation of crop

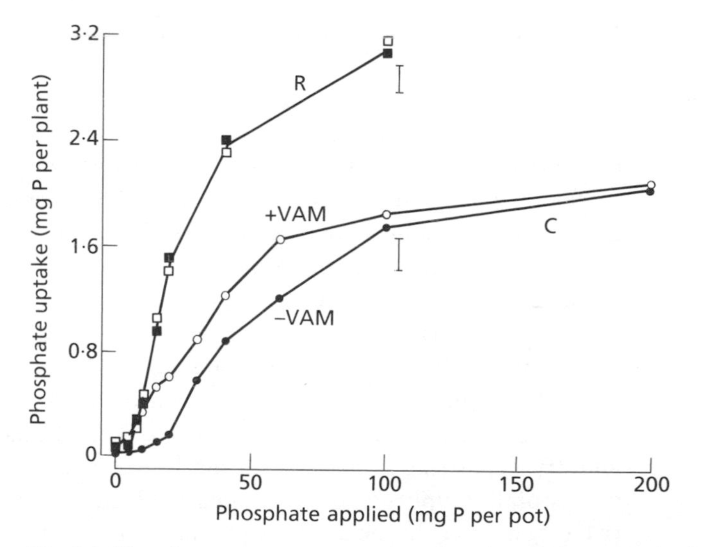

Fig. 3.1 Phosphorus content, at age 5 weeks, of annual ryegrass (*Lolium rigidum*) (R, squares) and subterranean clover (*Trifolium subterraneum*) (C, circles) grown in soil with various amounts of phosphate added, and with (open symbols) or without (black symbols) VA-mycorrhizal infection. Vertical bars: least significant difference ($P < 0.05$). From Bolan *et al.* (1987).

plants with VAM is not widely practised is that VAM inoculum is already present in almost all farm soils. Plants rarely fail to become infected through lack of inoculum. Exceptions occur when soil has been sterilized to kill pathogens, e.g. in citrus orchards. It is possible that other treatments to soil may make the soil less favourable for indigenous fungal species of VAM, so that inoculation with a different fungal species tolerant of the new conditions would be beneficial. An example is where the pH of acid soils has been raised by liming (Hayman & Mosse 1979). There are many species of mycorrhizal fungus, and it may be that some are more beneficial than others to the host plant, for example more efficient at uptake of P and its transfer to the plant. If that is so, it could be beneficial to replace indigenous fungi by more efficient strains. Powell (1982) tested VAM fungi from 20 New Zealand soils for their ability to promote growth of ryegrass and white clover, and was able to select three that were particularly effective. Such research has been hampered because VAM fungi are difficult to work with: they are difficult to identify; no-one has yet found out how to grow them separate from the host plant; they have no known sexual reproduction, making conventional breeding impossible; and they have so far proved difficult material for molecular-genetic techniques. If these technical difficulties can be overcome, there could be opportunity to produce improved strains of VAM fungi analogous to the improved plant and animal varieties that have contributed so much to increases in food production.

Long-term P balance of farms

On the time-scale of days, a growing season or even a few decades the key question about P is not how much is in the soil but how much of it is available to the plants. But on a longer time-scale it is still essential to consider P inputs to farm systems and whether they balance outputs. Table 3.3 gives examples of P inputs and outputs of farmland. Three of the farms also featured in Table 3.2, and comparison of the two tables shows marked differences between P and N. Because of the low mobility of phosphate in soil, P losses by leaching are usually low. If there is erosion the particles can carry additional P with them. But usually the main output of P from the farm is in the material harvested. If the productivity of the farm increases the P removed in harvest inevitably increases too, as Table 3.3 shows. Phosphorus in rain and in fine particles from wind erosion provides the only natural inputs of P to most farms, and the amount is small. The highest input value shown in Table 3.3, 1 $kg\,ha^{-1}\,yr^{-1}$ in rainfall and dry deposition combined, is rarely exceeded. Likens *et al.* (1977) cite rainfall P inputs from various parts of the world, ranging from 0.04 to 0.6 $kg\,ha^{-1}\,yr^{-1}$. The lowest value is their own, from New Hampshire (see Table 5.8). Proctor (1987) cites values up to 2.3 $kg\,ha^{-1}\,yr^{-1}$ from rainforest areas. He also cites 17–27 $kg\,ha^{-1}\,yr^{-1}$ from Venezuela, but the author of those figures later withdrew them (Jordan 1989). This illustrates the point that measuring P in rain-water may sound

Table 3.3 Phosphorus balance and yields of produce of four contrasting farms in Europe

			Dairy farming, Netherlands	
	Cereals, Sweden (1750)	Wheat, Britain, receiving fertilizer (1970s)	No fertilizer or supplementary cattle feed (1937)	Using N-fertilizer and supplementary cattle feed (1970s)
Phosphorus inputs ($kg\,ha^{-1}\,yr^{-1}$)				
Wet and dry deposition*	?	0.3	1	1
Organic manure	11–14			
Fertilizer		25		
Cattle feed				86
Phosphorus outputs ($kg\,ha^{-1}\,yr^{-1}$)				
Leaching	?	0.5	<0.1	<0.1
Harvest	4	17	4	22
Yields of produce ($ha^{-1}\,yr^{-1}$)				
Grain (tons)	0.8	4.4		
Meat (tons)			0.15	0.77
Milk (thousand litres)			2.9	18
Notes	1	2	3	3

* P dissolved in rain, plus P in airborne particulate matter.

Notes

1 Data for village in Scania, southern Sweden, operating infield-outfield system. Calculated mainly from written records of land use, crop production and number of animals. Figures apply to cropped infield only. From Olsson (1988).

2 Data compiled by Frissel (1978) from various sources, to give figures representative of British wheat farms in 1970s.

3 Two farms in the same area of the Netherlands. From Frissel (1978).

straightforward but is not: the concentrations are extremely low, the measurement methods need to be very sensitive and small amounts of contaminants can greatly alter the results. The origin of the dissolved P compounds in rain-water is not clear, and the reasons for these variations between sites are therefore not known. We can conclude that the P input is often within the range $0.1–1\,kg\,ha^{-1}\,yr^{-1}$. This is far too low to balance outputs of P in harvest, from modern high-yield farms or even from more traditional systems (Table 3.3).

Of the four farms in Table 3.3, three had total P inputs exceeding total P outputs; but the fourth, the 1937 Netherlands dairy farm, because it had no P input except wet and dry deposition, was losing more in harvest than its inputs. Presumably this farm relied on P stored in the soil, which could be slowly converted to forms available to the plant. So this system would not be sustainable indefinitely, though many soils contain enough P to go on supplying $3\,kg\,ha^{-1}\,yr^{-1}$

for several centuries. In all studies of P balance one of the unknowns is weathering of rock. Many rocks contain P, but the rate at which it is released by weathering is not known well enough to indicate whether this could contribute significantly to long-term P balances of farms.

Was infield–outfield farming sustainable for P?

The Swedish farm in 1750 relied on manure from cattle to maintain the P status of the cropped fields. The cattle did not create this P, they obtained it in their food, which grew in the outfield. We can ask whether infield–outfield farming has ever been sustainable long-term. Phosphorus is moved, year by year, from the outfield to the infield, but is the P status of the outfield being maintained by natural inputs or is P being 'mined'? Are most of the old forests, heathlands and permanent grasslands of Europe systems whose nutrient status has been depleted by centuries of use as nutrient suppliers for farms? The 18th-century village in southern Sweden (Table 3.3) provides a rare opportunity to answer this question. In this village the meadows and rough grazing used as the outfield were about 10 times the area of land sown to crops in any year. Therefore if the P input in rain and dry deposition was $0.4\,kg\,ha^{-1}\,yr^{-1}$ or higher (which is quite possible), the outfield could provide enough P to balance the $4\,kg\,ha^{-1}\,yr^{-1}$ removed in harvest. The grazing animals were essentially concentrating the P supplied in rain and dust, and the whole system should be sustainable. In practice average removal of P from the outfield was $1.1–1.4\,kg\,ha^{-1}\,yr^{-1}$. This is probably more than rainfall was providing. According to the figures in Table 3.3 this amount of P supplied in manure was more than crop removal required, so if long-term P balance had been the only criterion the farmers should have reduced their number of cattle grazing the outfield. However, this amount of manure might have been necessary to balance N losses. This example serves to illustrate the importance of the size of the outfield area, relative to the infield.

P in shifting cultivation

We can also ask whether shifting cultivation is sustainable for P. After a field is abandoned, forest or grassland is allowed to grow (the 'fallow'). The plants accumulate P and other nutrients. Then they are cut down, left to dry, and burnt. Much of the P in the plant material is retained in the ash and some of it is available to the crops that are then planted. But where does the P in the growing fallow plants come from? Is the P status of the soil being depleted by each successive fallow? Or are there inputs sufficient to maintain its P status? Nye & Greenland (1960) drew together results from investigations of shifting cultivation in West Africa. Their figures show that the rate of increase of P in the vegetation during forest fallow was often higher than could be supplied by rain. For example, at Kade in Ghana the mean rate of P increase in vegetation of a forest was $2.3\,kg\,ha^{-1}\,yr^{-1}$, while the input in rainfall was 0.4. However, their figures also show that the P in the ash from the burnt forest at the end of the fallow period would be far more than the P removed in 3 yrs' crops (a common length of time to grow crops in one patch before allowing forest regrowth). A key fact about P supply to crops in shifting cultivation is that the ash raises the pH, which on many tropical soils increases the availability of P; this seems often to be more important than the total amount of P in the soil (Nye & Greenland 1960, Jordan 1989). One reason for abandoning crop growth on a patch after about 3 yrs is that the pH has fallen again; other reasons include weed invasion, increasing risk of soil erosion, and the need to let sprouts from tree stumps regrow.

If shifting cultivation and infield–outfield are viewed as ways of concentrating P from a large catchment area on to a small field, the message from these examples is that the fields cropped in any year need to be no more than a small fraction of the total area used, if the system is to be sustainable long-term. This is the most serious limitation to these systems, setting the upper limit to the number of people they could feed.

Returning to Table 3.3, both the 1970s farms used artificial P inputs. On the dairy farm no P fertilizer was used, but the cattle were given food that had been bought in, and the P from this passed via their dung to the soil in the farm's pastures. The cattle food was made from plants that would have been grown with the use of fertilizer, so indirectly there was P input from fertilizer. The British farms used P fertilizer on their wheat fields. Both these types of farm had P inputs exceeding outputs, so the P stored in the soils would be increasing. Are there any disadvantages to a farming system dependent on P fertilizer? Unlike nitrate (Box 3.2) high concentrations of phosphate and other P compounds in plants or drinking water are not a known health risk. Phosphate is a major cause of eutrophication; however, most of the P reaching lakes and enclosed seas is not by leaching from farmland, which is slight (Table 3.3). More important sources of excess P in waters are dung of farm animals that are kept indoors, and domestic sewage. (See Chapter 7 for more information about P and eutrophication.)

How long will P supplies last?

The basic ingredient for phosphate fertilizer has to be mined, so unlike nitrogen fertilizer the supply has a limit. Almost all P fertilizer is made from rock phosphate, which is very insoluble calcium phosphates. One estimate of the world reserves that could be extracted with current technology is 34 Gtons (Bockman *et al.* 1990). In 1989 the amount of rock phosphate mined throughout the world was 0.16 Gtons (*UN Industrial Statistics Yearbook 1992*), and if that rate of use continues the reserves would last about 200 yrs. However, world phosphate use is increasing; during the 1980s it increased at about 2% per year. If that rate of increase continues the 34 Gtons reserve will be all used up by AD 2073. Some estimates put the world reserves larger (Williams, R.J.P. 1978), and there are other potential sources of phosphorus, if we can develop methods for extracting them at reasonable cost. Phosphate nodules on ocean floors are one possibility. Nevertheless, it must be worrying if the world's food production is based on a commodity whose supply might be exhausted within a century, and this is a reason for looking for ways of reducing the dependence of agriculture on phosphate fertilizer. A more immediate worry to many countries is that the reserves of rock phosphate are located in only a few regions of the world. In 1988 76% of the world's production of rock phosphate came from four countries, U.S.A., Morocco, U.S.S.R. and China (*UN Industrial Statistics Yearbook 1988*). Since then one of those, U.S.S.R., has split up into separate countries which are in a volatile state politically and economically. This serves to highlight the potential problem for the many countries which do not have any phosphate reserves on their own territories.

Where the main P sources are

Since opportunities for P input to farms, apart from mined phosphate, are limited, an alternative we should consider is improving the cycling of P within the system. Up to now we have assumed that all the nutrients in the harvested product must be lost from the system. An alternative is to consider people as part of the ecosystem. Some of the harvested material will be lost in transport or processing, some will be eaten by pest animals during storage, but much, hopefully, will be eaten by people. Of the mineral nutrients in the food we eat, most

Box 3.4 Information about sewage sludge.

Based mostly on Berglund *et al.* (1984) and Gray, N.F. (1989)

Common concentrations of elements (mg g^{-1})

N	10–80
P	5–40
K	up to 7
Cd	0.002–1.5
Cu	0.2–8
Pb	0.05–4
Ni	0.02–5
Zn	0.6–20

Advantages in use as manure include:

1 Much of the P (often about half) is in water-soluble forms readily available to plants. However, only about 10–20% of the N is in inorganic, plant-available forms.

2 The organic matter content is high, which has favourable long-term effects on soil structure (see Erosion section).

Disadvantages in use as manure

1 Contains parasitic animals, bacteria and viruses that are pathogenic to humans (though far fewer than in raw sewage).

2 Contains heavy metals, in varying amounts (see above). These leach only slowly from soil, and can reach concentrations harmful to plants and to animals and people that eat them (see Chapter 7).

3 Much human faeces is produced in towns, far from the farms to be fertilized. Because the concentration of N and P in sludge is much lower than in inorganic fertilizer (e.g. $Ca(H_2PO_4)_2$ ('triple superphosphate') 270 mg P g^{-1}, NH_4NO_3 350 mg N g^{-1}), the costs of transport per unit of P and N are much greater.

Sewage as a P source

leaves our bodies in faeces and urine, so why not use sewage as manure? Application of raw human excrement to farmland is still widely practised in many tropical countries, but it promotes spread of some serious human diseases, including cholera. Preventing human faeces from contaminating drinking water has been one of the most important contributions to improved health in developed countries. In these countries sewage is treated; the solid product is called sewage sludge. It contains N and P, some of it in forms available to plants, so it has potential as an organic manure. In some experiments heavy applications of sewage sludge have given crop yields as high as inorganic N-P fertilizer (Berglund *et al.* 1984). Box 3.4 summarizes advantages and disadvantages involved in using sewage sludge as manure. All the three problems listed are at present serious deterrents; but they may well be soluble in the long-term. The heavy metals come mainly from industrial sewage. If sewage sludge is in future to be widely used as manure it may be necessary to separate domestic and industrial sewage much more than happens at present. (Rock phosphate also contains Cd, but the ratio of Cd:P is about two orders of magnitude lower than in sludge.) To reduce transport costs may require more concentrated fertilizer to be made from the sludge.

Soil erosion

Soil erosion can sometimes be dramatically obvious. On 11 May 1934 'New York was obscured in a half-light' (according to the *New York Times*) by a great cloud of dust blown from the farmlands of the mid-West more than 1000 km away. But erosion is not new. Much of the sedimentary rock in the earth's crust was formed from the products of previous erosion. The volume of sediment now on the floors of the oceans is so large that it must be the product of millions of years of erosion (Howell & Murray 1986). Soil is not only lost but is also formed, by weathering of rocks, mostly at the soil–rock interface. I am not aware of good estimates of the rate of this formation, so I cannot tell whether on a world scale soil formation balances soil loss.

Rates of erosion

Milliman & Meade (1983) used measurements of amounts of sediment carried in major rivers to estimate the total amount of soil lost each year from the continents to the oceans by water erosion. Averaged over the world's land surface it is 1.5 tons $ha^{-1} yr^{-1}$. Since the bulk density of soils averages not far from 1, this is about equal to 1.5 $m^3 ha^{-1} yr^{-1}$, or a thickness of 0.15 $mm\ yr^{-1}$. This may not sound much. However, we must take into account that this is only the amount that reaches the oceans; other soil will be eroded from one place and deposited in another part of the same continent. As mentioned earlier, the fertility of seasonally flooded farmland beside the Nile depended on soil washed down from areas higher up the Nile Valley and its tributaries. This average erosion rate of 0.15 $mm\ yr^{-1}$

conceals great variation between sites. Morgan *et al.* (1984) gave a table of many published rates of erosion from various parts of the world, with various types of ground cover, including forest, grassland, crops and bare soil. The rates range from 0.005 to 439 tons $ha^{-1} yr^{-1}$, about equal to 0.5 µm–4 cm yr^{-1}.

How do changes in land use affect erosion?

The key questions for this section are: how much has soil erosion been influenced by people and particularly by farming practices? and what should we do about it? Over the last few thousand years land areas have been converted from forest, grassland and other natural vegetation cover to farmland. Has this increased erosion? There is a widespread belief that the answer is *yes*. For example, lands around the Mediterranean were probably much more forested in classical Greek and Roman times than they are now (Thirgood 1981). Many historic sites there are today surrounded by rocky hillsides, and a visitor may wonder whether they formerly bore deep soil which has been eroded away. We need more critical evidence. Three ways of measuring rates of erosion are:

1 Change in top-soil thickness. This requires measurements some years apart.

2 Troughs set into the soil to collect surface water and sediment running off down-slope.

3 Measuring water flow and sediment concentration in streams and rivers to determine erosion from a whole catchment (watershed).

Methods 2 and 3 measure mainly water erosion, not wind erosion.

Effect of felling a forest: an experiment in U.S.A.

An experiment to investigate the effect on erosion of felling temperate deciduous forest was conducted in Hubbard Brook Forest, New Hampshire, U.S.A. (Bormann & Likens 1979). This mixed deciduous forest covered several steep-sided valleys. Late in 1965 one complete catchment (watershed) was clear-felled; the cut trees were left on the site, and for the next 3 yrs regrowth of vegetation was prevented by application of herbicides. Water flow and suspended sediment were measured in the outflow stream from this catchment and a neighbouring uncut control catchment, so the mean soil loss per

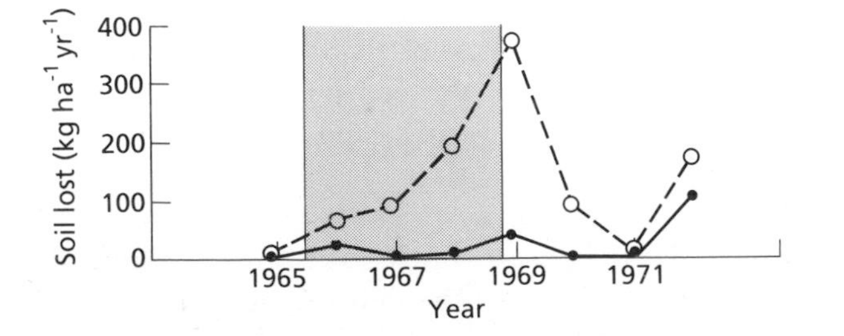

Fig. 3.2 Soil erosion from two valleys in Hubbard Brook Forest, New Hampshire, U.S.A., measured by particulate matter in stream water. ●, undisturbed control forest. ○, forest clear-felled in late 1965 (first vertical line), regrowth prevented for 3 yrs (shaded area), then allowed to occur. From Bormann & Likens (1979).

hectare could be calculated. The mean rate of loss from the control forest during these years was 25 kg ha^{-1} yr^{-1}, about 2.5 μm yr^{-1} in thickness terms. So in spite of the steep slopes erosion was extremely slow from the undisturbed forest. Figure 3.2 shows the annual loss of particulate matter from the two catchments. During the 3 yrs that regrowth was prevented the soil loss from the cut forest increased each year, reaching more than 10 times the rate from the control. Evidently there was some progressive change going on during those 3 yrs that led to erosion increasing year by year rather than stabilizing at a new, higher rate. After 3 yrs herbicide application was stopped, regrowth of trees and other plants commenced, and within 2 yrs erosion had returned to a rate similar to the control. The authors do not explain the marked increase in erosion loss in 1972; but it occurred in both catchments and was not a result of the previous felling.

Replacing forest by crops

These results show that felling forest can lead to a great increase in rate of erosion. But they also show that regrowth of vegetation on the site can quickly reduce erosion again. So we now need to consider what happens if forest is clear-felled and replaced by a farm crop. In an experiment in Mexico (Maass *et al.* 1988) patches of tropical deciduous forest were cut and maize planted, a common practice in the area. Soil erosion was measured by collecting all the soil washed downslope to the bottom of each plot. In control patches of forest the erosion was less than 0.2 tons ha^{-1} yr^{-1}, whereas loss from the maize plots averaged in the first year 100 tons ha^{-1} yr^{-1}. This very high rate, more than 100 times faster than the cleared watershed at Hubbard Brook (Fig. 3.2), can be attributed to the region in Mexico having very steep slopes (23° in the experimental plots), poorly structured soil and outbursts of heavy rain.

Davis (1976) provided evidence on how erosion in a small area of Michigan changed when the first European settlers cleared forest and replaced it by farmland. Before Europeans settled in the area the

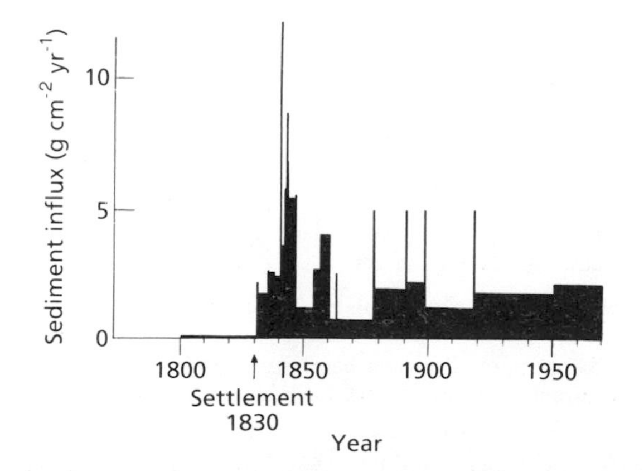

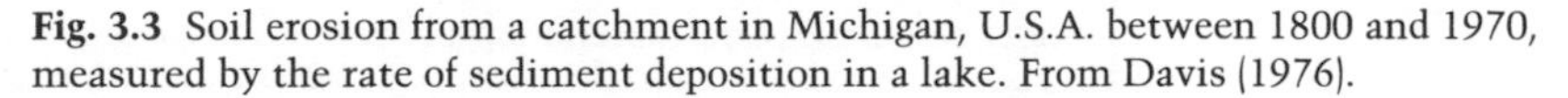
Fig. 3.3 Soil erosion from a catchment in Michigan, U.S.A. between 1800 and 1970, measured by the rate of sediment deposition in a lake. From Davis (1976).

vegetation was deciduous forest with some pine, as shown by pollen records. After the settlers arrived, about AD 1830, the forest was quickly cleared for farming. In 1976 it was mostly meadows and cornfields. The area drains into a small lake which has no outflow, so the sediment in the bottom of the lake can be used as a measure of erosion. The layers in the sediment could be dated by pollen and ^{14}C (see Box 2.6). Figure 3.3 shows the rate at which sediment reached the lake, from 1800 to 1970. Before 1830 erosion averaged $90\,kg\,ha^{-1}\,yr^{-1}$. About the time of clearance there was a very large peak of erosion, and then it settled to a steadier rate with some fluctuations and short bursts. There is no clear indication that changes in farming practice since 1880 have altered the erosion rate in a consistent way. The average rate from 1900 onwards was about $900\,kg\,ha^{-1}\,yr^{-1}$, about 10 times the pre-clearance rate, though still only about $0.1\,mm\,yr^{-1}$. Measurements in other parts of the world also show that erosion can be faster, by a factor of 10 or more, from arable land than from forest.

Ways of reducing soil erosion

To provide a basis for considering how soil erosion can be reduced, Box 3.5 summarizes the main factors that influence the rate of erosion. Something can be done to influence each of these four categories of factors. Wind can be reduced locally by shelter-belts, hedges and fences; there is, however, no practical way of reducing rain impact on a field scale. Slopes can be levelled by terracing, and this is widely done in parts of Asia. It used to be widespread in Europe, too, and the terraces, some more than 1000 yrs old, can still be seen in various parts. Terracing on steep slopes is not well compatible with modern mechanized farming, but on less steep slopes ploughing along the contours or planting bands of grass along contours can reduce erosion. In the remainder of this section I concentrate mainly on erosion control related to the other two categories in Box 3.5, soil structure and vegetation cover.

Box 3.5 Factors affecting the rate of soil erosion.

Wind and rain. A few short periods of exceptionally high wind or heavy rainfall can cause much of the total erosion during a year, so total annual rainfall or mean wind speed are often poor predictors of erosion rate.

Landform. Steepness is particularly important; the length of a slope and its shape can also have an effect.

Soil structure. In many soils the size and strength of aggregates (crumbs) is particularly important, since small particles are more easily washed or blown away. Large pores between aggregates allow rapid infiltration of rain water; when rainfall exceeds infiltration surface run-off occurs, which is when most water erosion takes place.

Vegetation. Above-ground canopy, roots, microorganisms associated with roots, and dead plant material on the soil surface can all reduce erosion.

How soil structure affects erosion

Soil structure is a complex subject, involving fundamentals of soil physics and chemistry beyond the scope of this book. One relevant fact of basic physics is that the rate of flow of a liquid through a tube is proportional to the fourth power of the tube radius. So if 1 ml of water flows through a tube 0.1 mm in diameter, the amount that flows through a tube 1 mm in diameter in the same time, under the same pressure gradient will be 10 000 ml. If we compare one tube of radius 1 mm with 100 parallel tubes of radius 0.1 mm, even though the cross-sectional area of the two systems is the same, 100 times as much water will move through the single large tube as through the 100 small ones (under a given pressure gradient). This is why a non-structured clay soil, with pores mostly less than 10 μm diameter, has a permeability to water orders of magnitude lower than a clay soil with a well-developed crumb structure and hence pores of diameter 1 mm or more. This is turn will mean that much of the rain that falls on the unstructured soil will flow off along the surface, causing erosion, whereas most of the rain falling on the well-structured soil will percolate downwards, except during particular heavy downpours. Soils with large water-stable aggregates are therefore much more resistant to

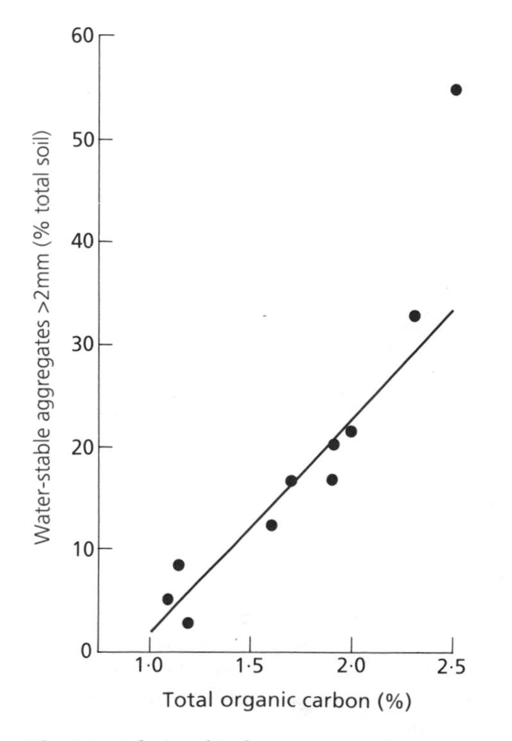

Fig. 3.4 Relationship between organic matter content and abundance of water-stable aggregates more than 2 mm in diameter, in soils from adjacent sites in South Australia that had carried different crops, rotations or pasture for 50 yrs. From Tisdall & Oades (1980).

erosion by water. ('Water-stable aggregates' means crumbs that do not break down when wet.) They are also less prone to wind erosion, because the larger the particle the stronger the wind needed to move it. A well-developed crumb structure also promotes root growth because the soil has a wide range of pore diameters and so tends to hold water in the finer pores but to have air in the larger ones. There has therefore been much interest in how to maintain a good crumb structure in arable soils.

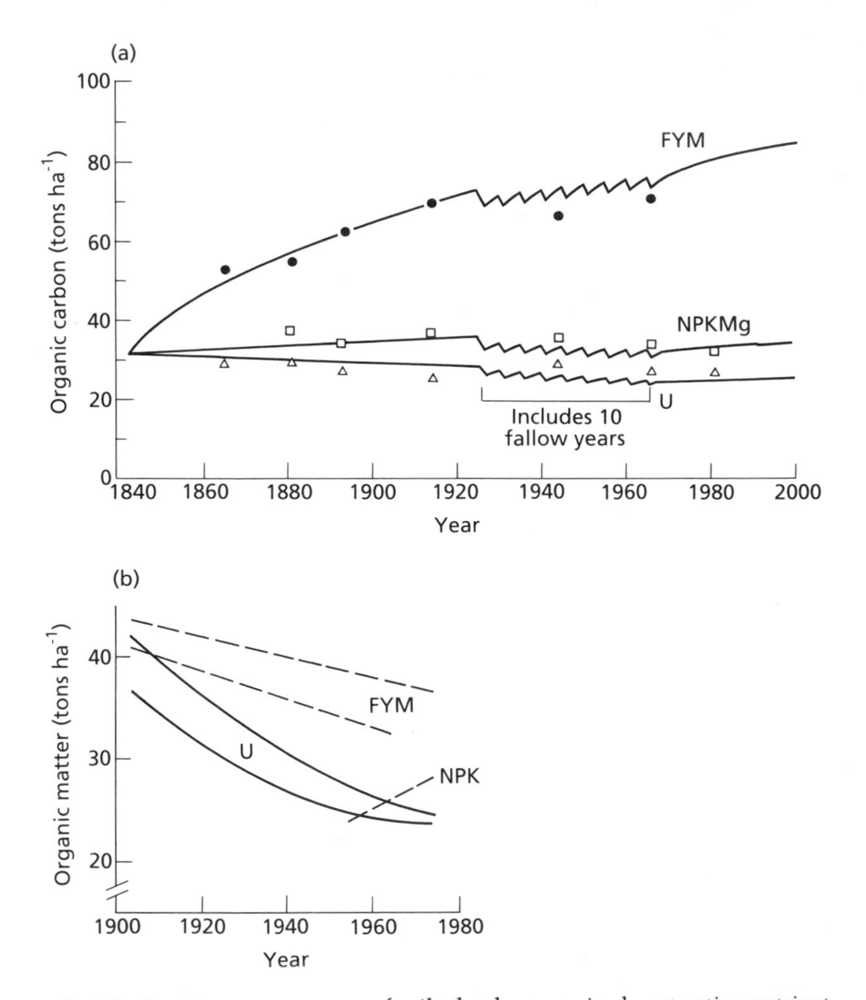

Fig. 3.5 Organic matter content of soils that have received contrasting nutrient additions in long-term experiments in U.K. and U.S.A. (a) Rothamsted, southern England, top 23 cm of soil. Winter wheat grown since 1843. The plots received annually farmyard manure (FYM), inorganic fertilizer (NPKMg) or no addition (U). The symbols show measured values, the lines are fitted by a model. From Jenkinson (1991). (b) Urbana, Illinois, top 17 cm of soil, maize grown since 1876. The treatments were: unfertilized throughout (U); unfertilized until 1904, thereafter farmyard manure, lime and phosphate (FYM); unfertilized until 1955, then inorganic NPK and lime (NPK). The lines are statistical best fits. From Odell *et al.* (1982).

How farming practices alter amounts of soil organic matter

It has been found in various parts of the world that higher organic matter content in soil tends to be associated with greater abundance of water-stable aggregates. Figure 3.4 shows an example of this. This has led to interest in how farming practices, including inorganic fertilizers, affect soil organic matter content. In plots at Rothamsted Experimental Station in England on which winter wheat has been grown since 1843 (Jenkinson 1991), with inorganic fertilizer application or with no fertilizer, the soil organic matter content has remained virtually unchanged, whereas application of farmyard manure has produced a marked increase (Fig. 3.5(a)). The abundance of water-stable aggregates over 0.5 mm in diameter has increased substantially in the plot given farmyard manure (Wild 1988, p. 437). Very different results have been obtained from plots on the campus of the University of Illinois (Fig. 3.5(b)). In the unfertilized plots soil organic matter has decreased substantially since 1904; addition of farmyard manure has slowed this decline but not reversed it. In contrast, after inorganic fertilizer additions were started to some previously unfertilized plots, their organic matter has increased.

Whether soil organic matter increases or decreases, and how fast, depends on the balance between inputs and losses of organic matter. Inputs come from:

1 Residues of the crop—roots, stem, leaves—left after harvest, and from any cover crops or plants grown in the rotation as a green manure.

2 Organic manures added, such as farmyard manure or sewage sludge.

Organic matter is lost by its decomposition by soil microorganisms. Gains and losses were measured in a long-term experiment in Oregon (Rasmussen & Collins 1991). The fields alternated winter wheat and bare fallow, and various treatments were applied that gave contrasting

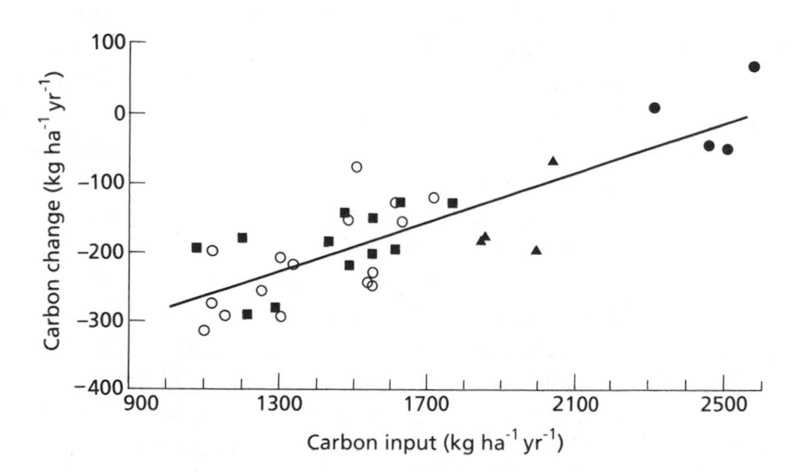

Fig. 3.6 Rate of change of organic carbon during 21 yrs in wheat fields in Oregon. Treatments: ●, straw and cattle dung added; ▲, straw and pea shoot material added; ■, straw added; ○, straw burnt on field. From Rasmussen & Collins (1991).

amounts of organic matter input (Fig. 3.6). The points fall approximately on a straight line fitting the equation

$$\Delta C = 0.18C_i - 460$$

where ΔC is the rate of change of organic C in the soil, and C_i is the rate of organic C input. The units are $kg\,ha^{-1}\,yr^{-1}$. This can be interpreted as showing that each year 18% of the added organic C was incorporated into soil organic matter, the remainder being converted by microbial respiration to CO_2 within the year; and each year 460 kg ha^{-1} of C from the soil organic pool was also respired to CO_2. So for the soil organic matter content to remain constant at this site about 2500 kg of organic C (about 6 tons of organic matter) needs to be added each year to balance the loss by microbial breakdown. In this experiment all treatments apart from addition of farmyard manure led to decline in soil organic matter. The fallow years would contribute to this, since during bare fallow microbial breakdown of soil organic matter continues while there is no input from crop residues. We can now see how it is possible for plots receiving only inorganic fertilizer to increase in soil organic matter while others receiving farmyard manure fail to do so (Fig. 3.5(b)). The inorganic fertilizer, by increasing crop growth, increases the crop residues left at the end of each season. If plots receiving farmyard manure have lower crop yield the total organic matter input can be less.

How organic matter holds soil crumbs together

Up to now we have considered all the organic matter in the soil as a potential contributor to aggregate stability. But soil organic matter is made up of chemicals differing greatly from each other in properties. Some are broken down by microorganisms within days or weeks, while others are so resistant to microbial attack that they remain for many years. In some soils some of the humic material is more than 1000 yrs old, as shown by radiocarbon dating (Jenkinson & Rayner 1977). Humus includes complex polymers whose chemistry is still inadequately understood. Some organic components are likely to contribute much more than others to binding crumbs together. Tisdall & Oades (1982) have suggested that in aggregates up to about 0.25 mm diameter the component soil particles are held together mainly by natural glues. Some of these are polysaccharide-based mucilages produced by bacteria and by roots, but there are others whose chemical nature is unknown. In aggregates larger than 0.25 mm Tisdall & Oades (1982) suggest that these glues become increasingly inadequate to hold the aggregate together, and they suggest that fungal hyphae and roots play an increasingly important role in maintaining the structure. Some of the fungi are heterotrophs, living off soil organic matter and root exudates, others are mycorrhizal. There is still no satisfactory method for determining what proportion of the hyphae in soil are mycorrhizal. Since VAM fungi will not grow on agar media, normal dilution-plating methods of assessing fungal abundance do not record them at all.

Table 3.4 Root and fungal development and abundance of water-stable aggregates in an experiment in which ryegrass (*Lolium perenne*) was grown in pots, some of which received treatments to reduce mycorrhizal infection

	VAM reduced	Control	Statistical significance*
Root length ($m\,g^{-1}$)	0.77	0.76	N.S.
VAM infection in roots (% of root length)	8.9	21.1	Sig.
Hyphal length in soil ($m\,g^{-1}$)	6.3	12.5	Sig.
Water-stable aggregates >2 mm diameter (% of total soil)	10.3	14.7	Sig.

* Sig. indicates that the values differ significantly between the treatments ($P < 0.05$). N.S., not significant.
From Tisdall & Oades (1979).

Tisdall & Oades (1979) provided some evidence that fungal hyphae can contribute to binding larger aggregates. In a pot experiment they applied treatments to reduce the amount of mycorrhizal infection in perennial ryegrass. This reduced the amount of fungal mycelium in the soil, though not the length of root (Table 3.4); it also reduced the abundance of water-stable aggregates.

The soil under old grassland has often been found to have a better-developed crumb structure than under arable crops. At one site in Illinois the soil under an old remnant of prairie was found to have 70% of its weight as water-stable aggregates 1 mm or more in diameter, in contrast to only 15% in a nearby field that had been under arable crops for many years. When prairie species (grasses and forbs) were planted on some of the arable land to re-create prairie, within 5 yrs the abundance of water-stable aggregates was similar to that in the old prairie (Jastrow 1987). One reason for this effect of grassland species on soils is that they have denser root systems than annual crop species, and the roots are present all through the year. It is also possible that they support more mycorrhizal fungus and that more mucigel and other gums are produced by the roots or their associated microbes.

Influence of crops and cultivation methods on erosion

In regions prone to erosion much attention has been paid to how cultivation methods can be altered to reduce it. Methods such as windbreaks and terracing have already been mentioned. Here I briefly consider possible changes to the way the crop itself is grown.

Growing a cover crop to maintain a complete plant cover can sometimes dramatically reduce erosion. For example, in experiments conducted in Tanzania (Temple 1972) in an area prone to heavy rain storms, the mean erosion from areas planted to coffee bushes, with the ground kept clean-weeded, was $38\,m^3\,ha^{-1}\,yr^{-1}$, equivalent to about $4\,mm\,yr^{-1}$. Growing a cover of leguminous plants under the coffee reduced the mean rate to $0.2\,m^3\,ha^{-1}\,yr^{-1}$.

Table 3.5 Abundance of aggregates, and soil loss by erosion during a simulated rain storm, from plots on which grain sorghum was grown with conventional tillage, or grain sorghum was grown in summer and clover in winter with no tillage

	Sorghum, conventional tillage	Sorghum–clover, no-till	Statistical significance
Water-stable aggregates (% of soil dry weight)	50	87	Sig.
Soil loss in 1 h (tons ha^{-1})			
If residues of sorghum left on soil surface	2.1	0.4	Sig.
If residues removed	4.5	1.3	Sig.

Sig. indicates a statistically significant difference ($P < 0.05$) between the two columns.
Data from West *et al.* (1991).

Low-till and no-till farming

In traditional farming methods the land was not only ploughed and harrowed to prepare it for sowing of seeds, but was later cultivated to keep down weeds. Modern herbicides can control weeds and have made possible 'low-till' and 'no-till' farming systems, in which seeds are sown into the remains of the previous crop and disturbance of the soil is greatly reduced. Table 3.5 summarizes results from an experiment in Georgia, U.S.A., in which grain sorghum grown in summer with conventional cultivation was compared with sorghum grown in summer but with a clover cover crop to cover the ground for the rest of the year, and no tillage. The crops were grown in this way for 4 yrs; then erosion was measured when there was no cover of living sorghum or clover, by a device that simulated an extremely heavy rain downpour for 1 h. The experiment involved three changes in crop management: the tillage, the cover crop and the presence or absence of sorghum residues during the 'rain storm'. Table 3.5 shows that these combined could reduce erosion by a factor of 10 (from 4.5 to 0.4 tons ha^{-1}). Leaving the crop residues on the soil surface made a substantial contribution to reducing erosion. However, erosion was slower in the no-till plots even when the residue treatment was the same, and at least part of the cause of this would be the greater abundance of water-stable aggregates. Physical disturbance of soil can reduce the amount of live VA fungus, as assessed by infection of plants subsequently grown in the soil (Jasper *et al.* 1987, Evans & Miller 1988); this may be due to breakage of the hyphae. This suggests that one means by which cultivation of soil alters its structure is by damaging the mycorrhizal mycelial network.

Management of grazing animals

Efficiency of grazing animals as energy convertors

People have for thousands of years kept herbivorous animals as sources of meat and milk. Inserting a second trophic level in the farm ecosystem inevitably lowers productivity per hectare. The efficiency with which animals convert their food into growth can be expressed as (production/consumption), where 'production' is the energy content of the new animal tissue produced by growth and reproduction, plus other products such as milk and eggs; and 'consumption' is the energy content of the food eaten by the animals. For terrestrial herbivorous mammals this efficiency is commonly about 5% or less, though it can be as high as 17% for intensive dairy farming (Brafield & Llewellyn 1982, Coughenour *et al.* 1985). It follows that animal farming, viewed as a method of converting solar energy to food energy for people, must have an efficiency lower than most systems for production of plant food. Table 3.1 gives some comparisons that illustrate this.

Why use animals as food?

So why do we use animals as food? It is possible for people to live, grow and remain healthy without eating any meat; many people do. Fewer people live without any animal product (milk, milk products, eggs), but it can be done. However, there is no sign that animal farming is decreasing, on a world scale. During the 25 yrs from 1965 to 1990 the number of cattle in the world increased by 22%, sheep by 16% and goats by 49% (*FAO Production Yearbooks*). This is less than the increase in the human population (61%), but nevertheless a clear indication that animal farming has continued actively. Table 1.1 shows that a quarter of the world's land surface is grazing land, more than twice as much as is devoted to crop production. So growing animals for food involves a considerable commitment of land. In most infield–outfield farming systems grazing animals played a key role in maintaining the fertility of the arable land (Box 3.1); so they were not competing against the arable crops for space, but on the contrary were essential for crop production. It was not essential to eat or milk the animals, but it was common sense to use them for food since they were there. Today in some countries cattle and sheep are grazed on sown pasture on fertile land where arable crops could be grown, so they are competing with crops for land. Cattle and sheep are sometimes fed on grain, and pigs and poultry often are, an even more direct competition with plant food production for people. However, much of the land on which cattle and sheep graze is not suitable for arable cropland, for example because it is too steep, the soil is unsuitable for cultivation (e.g. too rocky), or the climate is unsuitable (e.g. low and erratic rainfall). Much of this area has low net primary productivity, and one of the functions of grazing animals is to act as concentrators of energy: the meat in one cow may incorporate energy and nutrients from plants spread over several hectares. A further major advantage of cattle and sheep is that they can, with the help of their rumen bacteria, digest cellulose. Because humans cannot digest cellulose we grow crops for materials such as starch, sugar, fat and protein, and this has led to an emphasis on seed-producing crops, most of which are annuals. Because grazing animals can eat leaves and stems, most pastureland can be composed mainly of perennial plants, which tend to be more efficient than annuals at capturing solar radiation on a yearly basis, and to make the soil less prone to erosion.

Managing two interacting trophic levels

Because animal production involves two trophic levels, its management is inevitably more complex than that of arable land. The animal is the final product, but the manager must also be concerned about the productivity and nutritional value of the plants. Intensive animal production can involve ploughing, sowing one or more desired plant species, and applying fertilizer when necessary; so the farmer has considerable control over the soil conditions and the species composition. However, in 'rough grazing' the unfavourable environment makes such practices uneconomic. Often the only management techniques that can be used involve the animals: altering the stocking density, perhaps moving them from one area to another during the year. The density of grazing animals can, as we shall see, have major and sometimes irreversible effects on the pasture and hence on the productivity of the whole system. It is on this aspect of grazing management that I concentrate in this section.

One animal species, one plant species

Photosynthesis in relation to amount of foliage

Much of what the grazing mammal eats is leaves. Therefore the animal is destroying part of the photosynthetic apparatus of the plants, and management of grazing has to cope with this conflict inherent in the system. Figure 3.7 shows an example of the relationship between the photosynthesis per unit area of ground by a grass sward and the

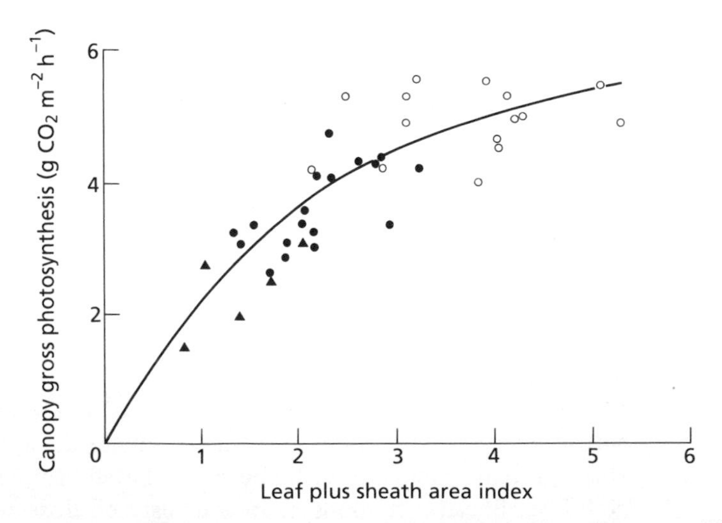

Fig. 3.7 Relationship between rate of photosynthesis and amount of foliage in a field of perennial ryegrass in southern England. Each point refers to photosynthesis over a short period; the incoming solar radiation was approximately the same for all. 'Leaf plus sheath area index' = (Area of leaf + sheath)/(Area of ground). The symbols refer to different grazing regimes. From Parsons *et al.* (1983a).

amount of green leaf + sheath area per area of ground (here abbreviated to leaf area index, LAI). Photosynthesis is closely related to the amount of incoming radiation absorbed. When LAI is low photosynthesis is approximately proportional to LAI, but as LAI increases above 1 there is more and more overlap between leaves so the curve levels off towards a plateau photosynthetic rate. Such information can be useful for planning grazing regimes. Because of the shape of the curve, some removal of herbage by the grazing animals, say reducing LAI from 5 to 3, may have only a small effect on primary productivity, but each further successive unit of LAI removed (from 3 to 2, from 2 to 1) will have a greater effect.

Rates of production and consumption

In Fig. 3.8(a) line P_1 shows the expected influence of the grazers on net primary productivity of pasture. This is part of the curve from Fig. 3.7 reversed left-to-right, on the assumption that the more animals there are per hectare the more of the foliage they will remove. This diagram assumes that as the stocking density increases the amount eaten by each animal remains the same, so the total amount consumed (C) forms a straight line through the origin. This diagram summarizes a basic conflict in the grazing system: the animals depend on photosynthesis to provide their food and yet their own feeding reduces photosynthesis. High secondary productivity is associated with rates of primary productivity below the maximum the site could support. The area between the two continuous lines (P_1 and C) represents primary production not consumed by the grazers, so it is used by

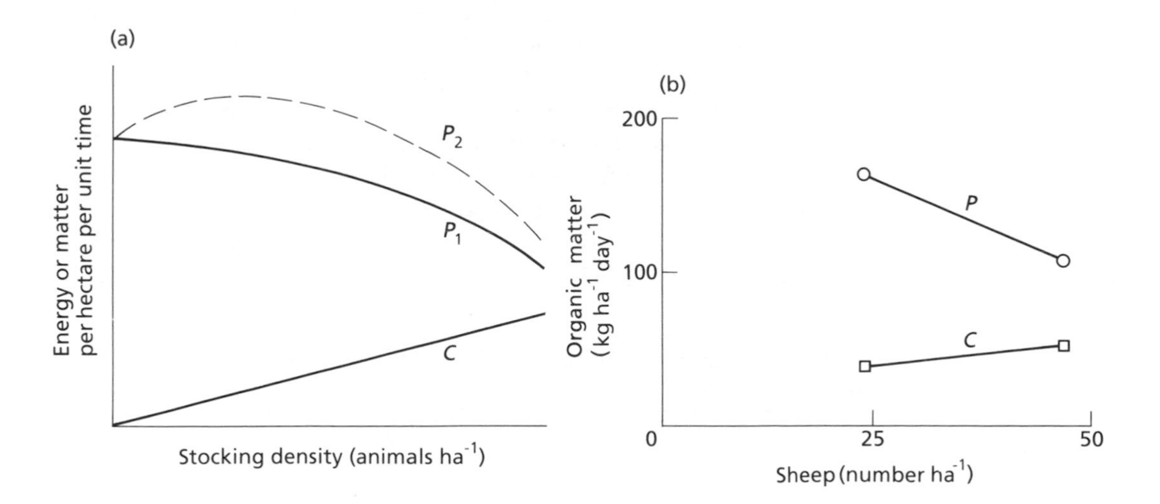

Fig. 3.8 Relationship between stocking density of grazing animals, net primary productivity of sward (P) and amount consumed by animals (C). (a) Theoretical. (b) Experimental results for sheep on perennial ryegrass pasture in England. From Parsons *et al.* (1983b).

decomposer organisms. In terms of mammal production it is wasted, though it may have important effects on the soil.

Measured data to confirm that Fig. 3.8(a) is correct have not often been produced. It is not easy to measure primary productivity when animals are eating some of the production, or to measure consumption by grazing animals. Figure 3.8(b) shows results from the same experiment in southern England that produced the data for Fig. 3.7. Sheep were grazed at two densities on perennial ryegrass sward. Photosynthesis was measured by enclosing small areas of the sward in a transparent chamber and measuring CO_2 uptake. Herbage consumption was calculated from measurements of faeces and of digestibility of the herbage. Measurements were made only during the spring–summer growing season. The leaf area index was usually 2–3 with the lower stocking density and 1–2 at the higher density. Figure 3.8(b) provides confirmation of the key fact that as stocking rate increases the total consumption per hectare goes up while the net primary productivity per hectare declines. For the moment we avoid the important question of what happens to the right of Fig. 3.8(a) as the lines converge.

Can animals increase plant growth?

So far we have assumed that the animals affect the plants only by reducing the photosynthetic area. Are there any ways in which the animals could increase plant growth, so as to partly offset the damage by grazing? If so, this could lead to primary productivity responding approximately as the upper line, P_2, in Fig. 3.8(a). Results of this sort were reported from an experiment in New South Wales, in which sheep were grazed at three densities on perennial grass–clover pasture. Photosynthesis was measured, by gas exchange, every few weeks over a year. Table 3.6 shows that the plots which received the middle intensity of grazing gave the highest total photosynthesis over the year, even though they did not on average have a higher amount of shoot and leaf material than the more lightly grazed plots. The relationship between stocking density and photosynthesis conforms approximately to line P_2 in Fig. 3.8(a).

There are several possible ways in which grazing could increase photosynthesis. Caldwell *et al.* (1981) and Nowak & Caldwell (1984)

Table 3.6 Measurements on sward in New South Wales, in paddocks with different numbers of sheep

	Stocking density (sheep ha^{-1})			Standard error
	10	20	30	
Mean dry weight of above-ground plant material ($g\,m^{-2}$)	183	160	71	16
Photosynthesis during 48 weeks ($kg\,CO_2\,m^{-2}$)	3.86	4.87	3.77	0.42

Data from Vickery (1972).

investigated how clipping affects two *Agropyron* spp., bunchgrasses that are important in semi-arid grazing lands of the American Intermountain West. Following clipping, photosynthesis by the remaining leaves, per unit leaf area, was increased up to twofold. This was associated with increased soluble protein in the leaves. It has been found for leaves of other species that higher protein concentration and higher rate of photosynthesis tend to go together. This is at least partly because some of the protein is enzymes involved in photosynthesis. Increased protein concentration also raises the nutritive value to the grazing animal. Returning to the *Agropyron* experiments, we can ask where the extra protein or its N would have been if the plants had not been clipped? Would it have been in other parts of the plant? In other words, was the increased photosynthetic capacity at the expense of other parts of the plant? For example, clipping shoots often results in reduced root growth, which may not have any marked effect immediately, but in the longer-term would probably affect the plant's nutrient and water status. Similar limitations apply to other short-term experiments that have shown clipping increasing photosynthesis or shoot growth: the increase may have been at the expense of growth or functioning of other parts of the plants. (However, this does not apply to the experiment of Table 3.6, since the grazing regimes had been in operation for 6 yrs before the measurements started.)

This increase in photosynthesis or growth following grazing does not always occur, in fact the majority of experiments have found reduced growth following clipping or grazing (Ellison 1960, Gordon & Lindsay 1990). Even when there is an increase, its effects cannot entirely overcome the removal of foliage: in Fig. 3.8(a) the dashed line bulges up, but it must curve down to the right as intensity of grazing increases further.

How grazing animals respond to sward structure

In Fig. 3.8 there is a disagreement between parts (a) and (b). Part (a) assumed that the consumption *per sheep* would be the same at any stocking density, so *C* is a straight line through the origin. But part (b) shows this was not so in practice: line *C* does not extrapolate to the origin, and in fact consumption per sheep was 29% lower at the higher stocking density than the lower. This leads us to consider the behaviour of grazing animals, and how they respond to swards of different heights and structures. Black & Kenney (1984) investigated this by making artificial grass swards; they fixed freshly cut grass tillers into holes in boards, at various spacings. In this way they produced 'swards' of various heights and densities, but made up of tillers that were very similar. These swards were offered to sheep for 30 seconds each. The amount consumed by the sheep was mainly determined by their intake per bite; they consumed more if the turf was tall and if it was dense, though the relationship was not entirely a simple one. Figure 3.9 shows that the sheep had a maximum intake rate of about 6 g min^{-1}. If the grass tillers were very close together the sheep could achieve

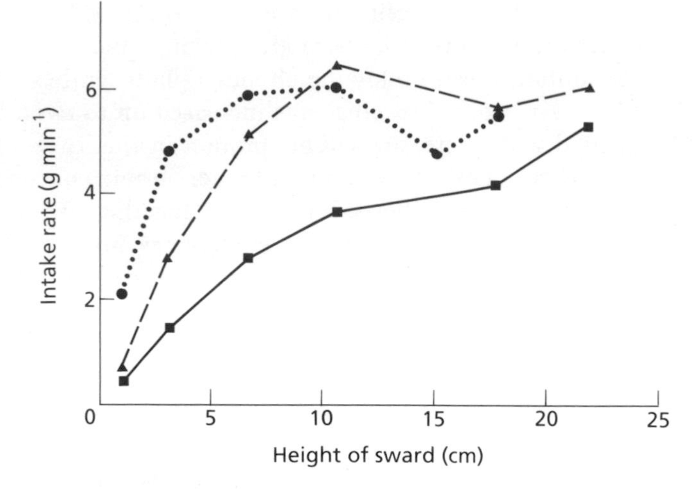

Fig. 3.9 Rate of herbage consumption by sheep from artificial grass swards of different heights and densities (■, 1623; ▲, 6495; or ●, 25 980 tillers per m^2). From Black & Kenney (1984).

their maximum rate even from short turf, but as tiller density decreased they required taller turf to achieve maximum intake.

Jaramillo & Detling (1992) showed that cattle grazing on semi-arid prairie in Wyoming responded to urine patches. Simulated urine (a solution of urea plus some mineral salts) was applied in late May in small patches each simulating one 'urination event'. In these patches the N concentration in above-ground tissue of a dominant grass, *Agropyron smithii*, greatly increased, though the grass showed little response in growth. The cattle grazed the urine patches preferentially: a greater percentage of tillers were grazed than elsewhere, and they were grazed to nearer the ground. The cattle were presumably responding to the chemical status of the foliage, since there was little difference in tiller density between the urine patches and elsewhere.

These examples show that if, in order to maintain high primary productivity, the manager wishes to maintain a sward of a particular structure—a particular height, tiller density or leaf area index, for example—the grazing behaviour of the animals will need to be taken into account.

One animal species, many plant species

When pastureland contains more than one plant species, one important aim of management will be to maintain a balance between the species that is favourable for the grazing animals. Sown pastures can have more than one species: clover as well as grass or more than one grass species. But here I concentrate on grazing land where species have never been planted.

Grazers select particular plant species

Cattle and sheep are willing to eat many plant species; nevertheless when a choice is available they can be selective. Table 3.7 shows an example. The diet of sheep grazing outdoors in Scotland was measured by inserting a tube into the oesophagus, removing samples of what they had just eaten, and identifying the plants by features of the epidermis visible under a microscope. The results show strong selection by the sheep of grasses and sedges in preference to heather. These results are from the summer, when selection was most extreme; in spring and late autumn the sheep showed much less preference among different plants, perhaps because there was less food available. Measurements made on samples from the oesophagus showed that the plants preferred by the sheep were more digestible than the heather.

How grazers alter the species composition

This selective grazing by the animals could alter the species composition of the pasture. If the species favoured by the animals are more nutritious, e.g. because they have more protein, less fibrous material or less tannin, continual selection could, by reducing the abundance of these species, reduce the overall nutritive value of the pasture. In tropical savanna regions cattle eat grasses but rarely eat leaves of shrubs and trees. At high stocking densities this can lead to vegetation more strongly dominated by woody plants, with much less grass and so less valuable for cattle grazing (Walker *et al.* 1981). An example often cited from Britain concerns change in abundance of grass species in upland pasture on acid soils. *Festuca ovina* and *Agrostis* spp. form pasture of high nutritive value to sheep. *Nardus stricta* has more fibrous leaves and so is less palatable and less nutritious. Under persistent heavy grazing the preferential selection of *Festuca* and *Agrostis* by the sheep can lead to increase of *Nardus* and hence reduction in the value of the pasture. Figure 3.10 shows results from long-term experiments in two upland areas grazed by sheep for many years, in which plots were then fenced off to exclude sheep. After 24 yrs *Nardus* had almost disappeared from the ungrazed plot, showing that its abundance in the heavily grazed pasture outside the exclosure was dependent on grazing. In contrast, the very palatable *Agrostis vinealis*

Table 3.7 Contribution of heather, grasses and sedges to diet of sheep in Scotland

	Calluna vulgaris (heather)	Grasses + sedges
Percentage in sward*	50	<1
Percentage in diet*	1	24
Digestibility	0.46	0.71

* In July.
Data from Grant *et al.* (1987) and Hodgson *et al.* (1991).

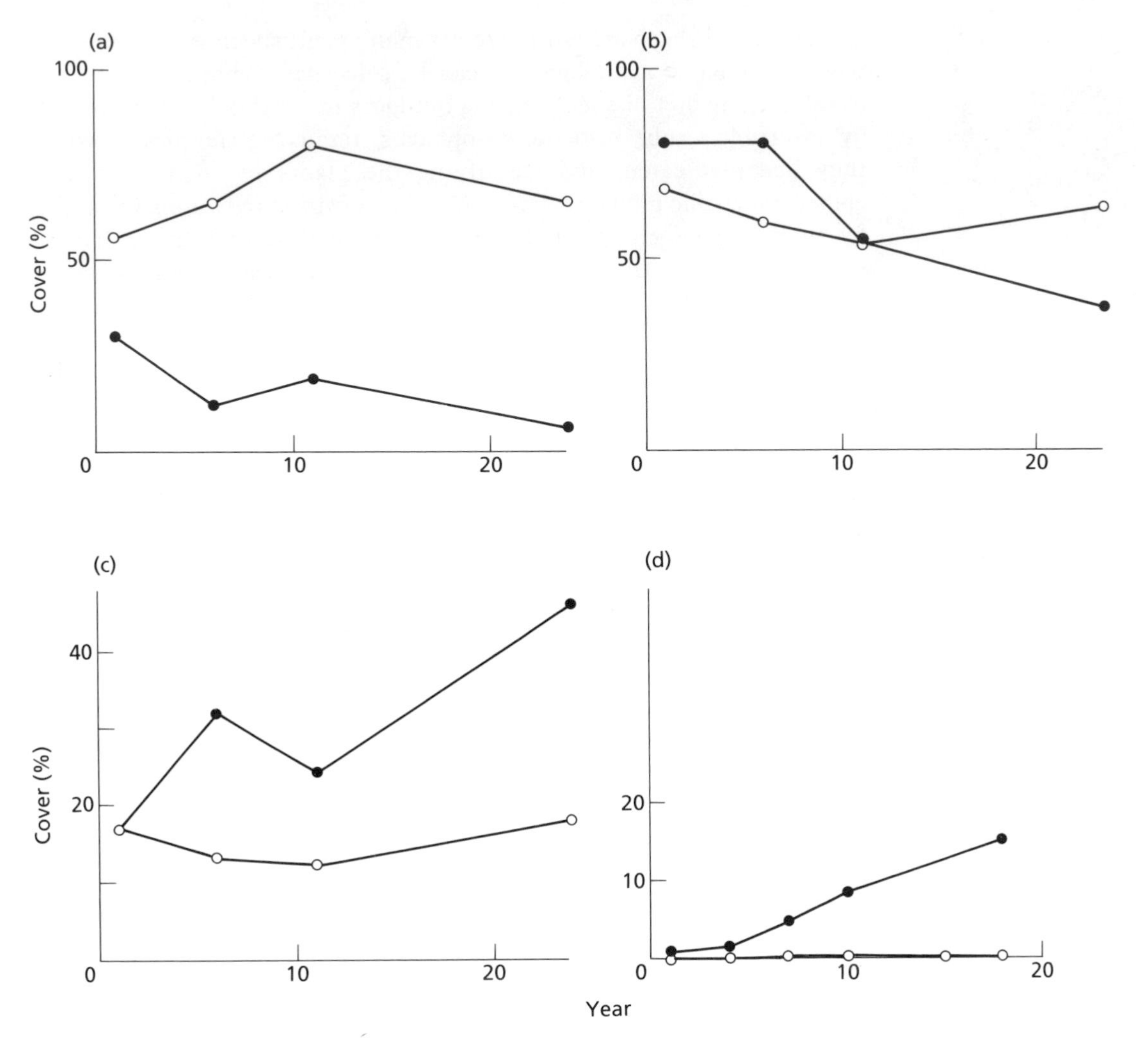

Fig. 3.10 Abundance of species at two sites in upland Britain. ○, grazed by sheep; ●, sheep excluded from year 0 onwards. (a)–(c) Three grass species, *Nardus stricta* (a), *Festuca ovina* (b) and *Agrostis vinealis* (c), in Snowdonia, North Wales. From Hill *et al.* 1992. (d) *Calluna vulgaris* (heather) in northern England. From Marrs *et al.* 1988. Note difference in scale between upper and lower graphs.

was sparse in the grazed area and increased when grazing was excluded. However, *Festuca ovina*, which is also favoured by sheep, was abundant in the grazed area and declined somewhat in the exclosure, i.e. it behaved more like *Nardus* than like *Agrostis*. This shows that palatability to the grazers is not the only factor that determines how a plant species will respond to grazing. Two other factors likely to be involved are:

Why species differ in their response to grazing

1 The stature of the plant, since species that can grow tall in the ungrazed pasture will be successful competitors for light.

2 Where the meristems are, since plants with apical meristems will be more damaged by grazers than those with basal meristems.
Both of these are likely to be important in the way grazing affects the balance between *Calluna vulgaris* (heather) and herbaceous species. Figure 3.10(d) gives results from another exclosure experiment, in northern England, where *Calluna* was extremely sparse on the grazed hillside, but increased steadily once grazers were excluded. In the absence of grazing *Calluna* can outcompete most grasses; but when sheep are present, although they strongly select grasses in preference to *Calluna* (see Table 3.7), even light grazing destroys the apical meristems of *Calluna* and hence reduces its growth and abundance.

Morphology of the plants was also important in the response of American prairie grasses to cattle grazing. Effects of grazing on the plants were much studied in earlier decades in various parts of the prairies. The techniques were crude by modern standards, but the principal results were so clear that few people would wish to dispute them. Figure 3.11 shows results comparing different grazing regimes which, as in most of these older studies, were not instigated by the researchers; there is no information on how long they had been happening. The example given is from unreplicated adjacent areas, but the paper also gives similar results from five other sites in western Kansas. Each of the four most abundant grasses responded differently

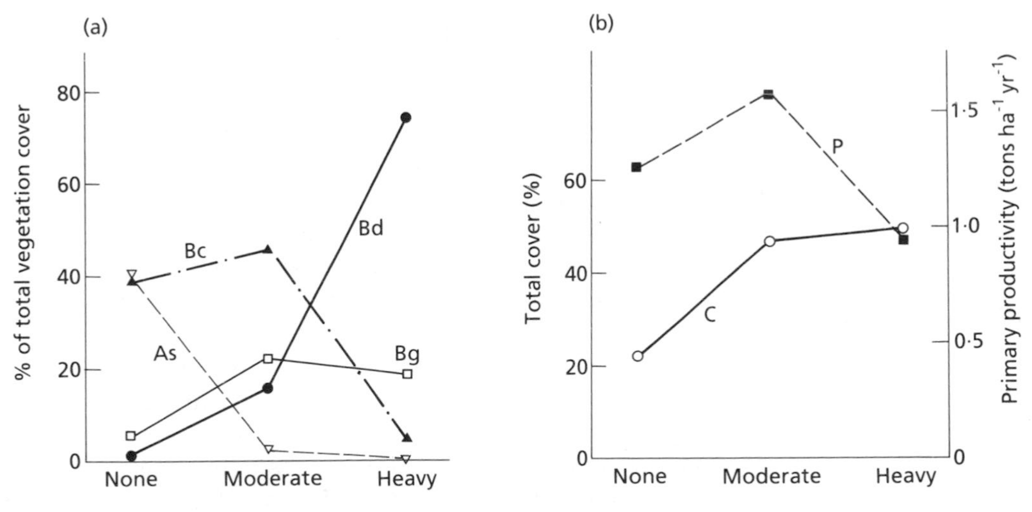

Fig. 3.11 Data on vegetation of three adjacent areas of prairie in western Kansas, U.S.A., subject to no grazing, moderate or heavy grazing by cattle. Data of Tomanek & Albertson (1957). (a) Percentage of total vegetation provided by the four most abundant grass species. As, *Andropogon scoparius* (little bluestem); Bc, *Bouteloua curtipendula* (side-oats grama); Bd, *Buchloe dactyloides* (buffalo grass); Bg, *Bouteloua gracilis* (blue grama). (b) Above-ground dry weight production (P) and total ground cover (C) by all species combined.

to grazing (Fig. 3.11(a)). The two dominant species in the ungrazed prairie, *Andropogon scoparius* and *Bouteloua curtipendula*, were fairly tall and there was much ground between them that was bare or covered by dead plant material. These two were almost completely replaced in the heavily grazed area by lower-growing, more spreading grasses, and the percentage of the ground covered by vegetation increased substantially under grazing (Fig. 3.11(b)). Perhaps as a result of this increased cover, shoot production was higher on the moderately grazed area than the ungrazed; however, that was not true of most of the other five sites.

Before the introduction of cattle the prairies had been only lightly grazed by native large herbivores such as bison. Many of the large herbivores present at the end of the full-glacial died out quite suddenly about 11 000 yrs BP, probably because of hunting by people (see Conservation chapter). The subsequent light grazing allowed dominance in many areas by bunchgrasses, tall species whose apical meristems are high above the ground and so easily damaged. In contrast, buffalo grass (Bd in Fig. 3.11), not only has apical meristems near the ground, but is capable of rapid lateral spread by stolons. More recent research has mostly confirmed this importance of morphology. Caldwell *et al.* (1981) compared various morphological and physiological characteristics of two bunchgrasses of the Inter-mountain West; the native *Agropyron spicatum* is more sensitive to grazing than *A. desertorum*, which was introduced from Eurasia. Several characters of the species could contribute to this difference in grazing tolerance. Clipping increased the photosynthesis rate of remaining leaves about twofold in *A. desertorum* but less in *A. spicatum*. However, by far the largest difference was that *A. desertorum* responded to clipping by producing many new side-branches (tillers) and so could quickly grow much new leaf area; whereas *A. spicatum* showed this scarcely at all.

Importance of the timing of grazing

The *timing* of grazing can also influence the balance between species. An example is the balance between shrubs and perennial grasses (including the *Agropyron* spp. mentioned above) in the sagebrush vegetation that covers large areas of Wyoming, Idaho and neighbouring states. The winters are cold, but plant growth is also limited by water shortage in summer. The grasses provide important grazing in spring, but the shrubs are also important, particularly to provide continued food supply into the autumn when the grass foliage has died back; so management aims for a balanced abundance of grasses and shrubs. Heavy grazing tends to result in increase of shrubs and decrease of grasses. In a 13-yr experiment in Idaho, some plots were grazed by sheep in spring only, some in autumn only, and some never. Spring grazing continued the trend found for all-year grazing, an increase in the dominant shrub and decrease of the dominant grass (Fig. 3.12). In contrast autumn grazing produced similar results to complete exclusion of grazers, a modest recovery of the grass and fall in the shrub. These

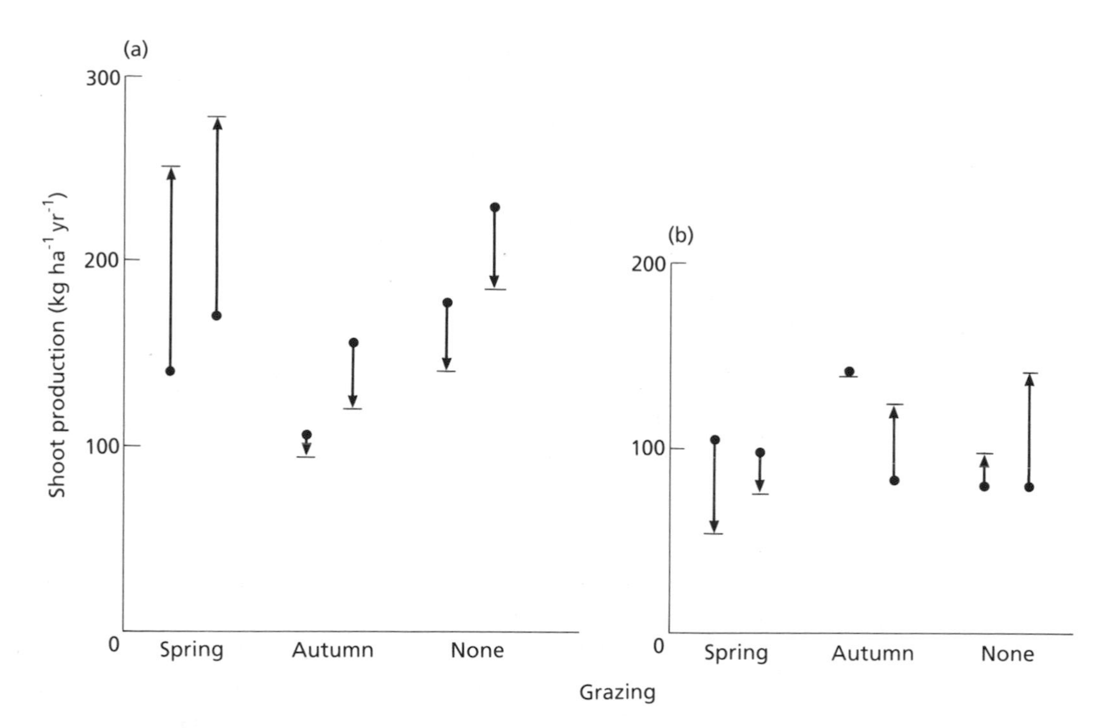

Fig. 3.12 Changes in abundance of (a) *Artemisia tripartita* (sagebrush) and (b) *Agropyron spicatum* (a bunchgrass) between 1950 (●) and 1964 (–) in a controlled-grazing experiment in Idaho, U.S.A. The two plots receiving each treatment from 1950–64 had previously been under different grazing regimes. Data from Laycock (1967).

responses relate to the grasses having most green foliage and making most of their growth in the spring, and therefore being most damaged by spring grazing; whereas the *Artemisia* retains foliage into the autumn, and so is less dependent on photosynthesis in spring. This research showed that, in order to maintain a balance between the shrub and the grasses, particular attention needs to be paid to the amount of grazing allowed during spring.

Using several animal species

In East African areas where several large herbivores coexist in savanna, they differ in the timing of their seasonal migration and in the species or parts of the plants that they eat (Crawley 1983, Pratt & Gwynne 1977). Giraffe eat almost solely trees, wildebeest grass almost solely, but some others, e.g. Thomson's gazelle, will eat both. This results in fuller use of the primary production than by a single herbivore species, and tends to maintain a balance of plant species. One might wonder whether a mixture of animal species would also be a more efficient way for people to exploit grazing land. The migratory pastoralists of northwestern Kenya who featured in Table 3.1 use cattle, sheep, goats and camels as food sources. The cattle eat almost entirely grasses, the camels almost entirely material from woody plants; the diet of sheep and goats is more mixed, and varies more through the year (Coppock *et al.* 1986). These differences in diet would be expected to increase the total amount of food the pastoralists can obtain from their animals, in a region with low and variable rainfall and hence periods of very little plant growth. Whether mixtures of animals would be beneficial in other areas where woody and herbaceous plants grow together, for example North American sagebrush, is an interesting question. As far as I know it has not been tried. The inhabitants of developed countries tend to be unadventurous in what sorts of meat they will eat. If camel steaks appeared in your local supermarket, would you buy them?

Overgrazing

In Fig. 3.8 the stocking density is kept within a range where consumption by animals is less than plant productivity. What if the stocking density is increased further: will the C and P lines meet and then cross? Clearly consumption cannot exceed production long-term; but on a day-to-day basis plant growth rate is bound to vary in relation to such factors as incoming solar radiation, temperature and soil water status, and there are likely to be periods of the year when consumption exceeds production. Consumption at a particular time is related to the plant biomass present rather than productivity at that time (e.g. Fig. 3.9), and Fig. 3.13 makes use of that. It is based on some graphs by Noy-Meir (1975), who presented many more alternatives and discussed them in much more detail.

Stable and unstable states in grazing systems

In Fig. 3.13 the animal stocking density is assumed to remain constant. The horizontal axis could represent total plant biomass per hectare, or above-ground biomass only or green material only. The dashed lines, C, represent a relationship between consumption and standing plant biomass similar to that in Fig. 3.9, and also found in field measurements. In the horizontal plateau region the animal is consuming as much in a day as its digestive system can process. The continuous line, G, shows growth, the rate of increase in plant biomass, i.e.

Growth = net primary productivity − death

where 'death' is due to any cause except grazing. Basically curve G declines to the left because of inadequate interception of radiation by the small foliage area, and to the right because of substantial death of tissue within the large standing biomass.

To take an example, consider line C_2 in part (a) of Fig. 3.13, and suppose the biomass is to the left of point r. Then growth exceeds consumption, so the biomass will increase. It will continue to increase

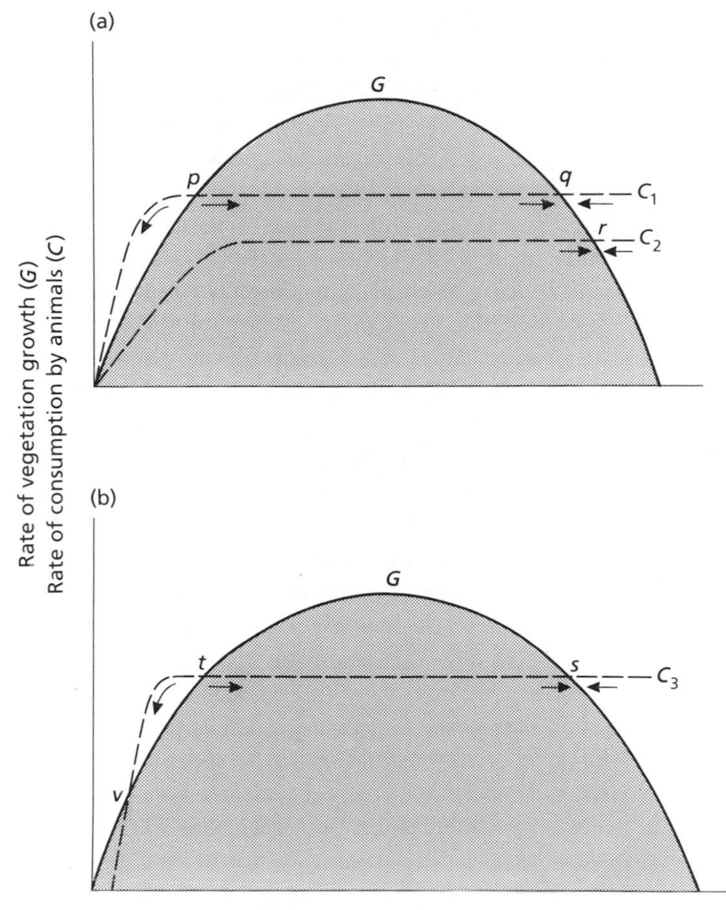

Fig. 3.13 Possible relationships between growth rate of vegetation (*G*), amount of plant material consumed by grazers (*C*) and standing biomass of vegetation; all are per unit ground area. Modified from Noy-Meir (1975).

as long as line *G* is above line *C*, i.e. until they meet at point *r*. There growth and consumption are equal so the biomass should not change further. Point *r* is a stable equilibrium: if the biomass is either increased or decreased it will in due course return to *r*; e.g. if the biomass is above *r* consumption will exceed growth so the biomass will decline back to *r*. Now consider line C_1. This, too, has a stable point, at *q*. However, point *p* is an unstable point. Provided the biomass is above *p* it will increase up to *q*; but if it is below *p*, even slightly, it will decrease further. This is because consumption exceeds growth, so biomass is reduced, and in this range reduced biomass causes further reduction in growth. Lines C_1 and C_3 show two alternatives for the region to the left of this unstable point. With line C_1, once the biomass has got slightly below *p* the vegetation is doomed: the animals will continue to eat it faster than it grows, until it has all gone. C_3 assumes that the plants have some productive parts that are not grazed, e.g. very close to the ground. Then a new stable point, v, can occur. So this system has two stable points, v and s, supporting different amounts of plant biomass, plant growth and animal consumption. If the system is in state v it cannot recover to state s unless one of the curves G or C_3 is changed: e.g. the stocking density could be reduced until line C_3 falls entirely below line G in the $v-t$ region.

You might think that there is no problem for the stock manager: provided the plant biomass is above p or t the system is stable. But supposing the weather changes, e.g. there is a drought, and the plant growth falls. If the biomass was in the $t-s$ region but the G curve fell to below the C_3 line, biomass would decrease, and by the time the rainfall recovered it might be to the left of the unstable point at t.

Break points

This simple graphical model predicts that grazing systems can have a 'break point' beyond which a major change takes place to a system with lower productivity and capable of supporting fewer grazing animals or perhaps none at all; and that recovery to the previous, more productive state may then be difficult to achieve. The shapes of the curves in Fig. 3.13 conform individually to observed and measured results, but how they intersect is crucial. Is there any evidence that such break points do occur in the real world? In real examples major change in the vegetation involves changes in species composition as well as biomass. One type of effectively irreversible change is when some species disappear altogether. In Fig. 3.11 one of the grass species, *Andropogon scoparius*, that dominated the ungrazed prairie, had disappeared from the heavily grazed area. If that happens over a large area removal of grazing animals would probably not lead to the species' return, so an irreversible change would have occurred. Such losses of species do occur. Large areas of the American west that formerly carried perennial grass–shrub vegetation are now mostly covered by annual species introduced from Eurasia (Mack 1986). It is widely assumed that heavy grazing allowed this invasion of alien species, though critical evidence is lacking.

Irreversible changes that have occurred

An area in southern New Mexico that seems from 1858 survey descriptions to have been mostly 'good grass' was by 1963 largely dominated by low shrubs, particularly mesquite (*Prosopis juliflora*) with mobile sand (Buffington & Herbell 1965). Cattle have been grazed in the area since the 19th century. Buffington & Herbell discussed factors influencing the vegetation there, and were not able to suggest any likely cause other than cattle for the change in vegetation. Yet after a 260-ha fenced area was set up in 1931, and all cattle excluded from it, they report that 'the mesquite sand dunes have advanced completely across the enclosure'. Unfortunately they give no further details. Possibly by 1931 mesquite was sufficiently established that the effect of cattle was no longer necessary for its spread, in other words the break point had already been passed. Walker *et al.* (1981) concluded, on the basis of a model, that there was no stable grass–mesquite mixture: once invasion of mesquite had begun its progress to dominance was inevitable. One contributing factor was that reduced grass cover led to formation of a soil pavement and increased soil erosion. Walker *et al.* argued that in semi-arid savannas in Africa woody plants and grasses can form a stable mixture at some intensities of cattle grazing, even though the cattle do not eat the woody plants; but if cattle density increased beyond a critical level a break point would be passed and the community would change to dominance by woody plants. Unfortunately these predictions from Walker *et al.*'s model have not yet been supported by much experimental evidence. They proposed that coexistence of woody plants and grasses was maintained in part by the woody plants drawing much of their water from soil layers too deep to be reached by grass roots. However, subsequent research in one area of South Africa (Knoop & Walker 1985) showed this was not so. A major difficulty in proving that overgrazing has caused observed changes in vegetation is that often weather is also involved. As Fig. 3.13 illustrated, plant growth and animal consumption could be in a stable relationship for years, but then a period of unfavourable weather could push the system past a break point so that even a return to the previous more favourable weather conditions would not allow it to recover.

Desertification

There has been much written about desertification (see Le Houerou & Gillet in Soulé 1986, Verstraete & Schwartz 1991). This imprecise term means essentially that in an arid or semi-arid area the vegetation becomes much sparser and its productivity lower. This may be followed by soil erosion leading to formation of desert pavement, mobile sand or other changes that make recovery of the vegetation slow or impossible. Desertification around the Sahara has received particular attention. Overgrazing is one likely cause of desertification, but other human activities such as excessive collection of firewood, and shifting cultivation with too short a recovery period, have probably contributed in some areas. All three are the result of increasing human population density. In spite of its importance for millions of people, the scientific evidence on desertification is frustratingly inadequate. In one review of the subject Middleton (1991) concluded that although Sudan is one of the best-studied parts of the world for desertification, the evidence is conflicting even on whether desertification has occurred there at all.

Conclusions

- Farming systems differ greatly in their productivity and in the number of people they can support per hectare. Low-input systems of the past would not, on their own, be able to feed the present world population.

- Inorganic N fertilizers have several disadvantages, though crops have been grown with them successfully for more than a century. Biological N fixation offers several alternative ways of providing N input to farms, though not necessarily achieving the same yields as with fertilizer.
- Most of the P in soil is not available to plants. Mycorrhizas can increase the P-capture ability of plants; but so far inoculation of farmland with mycorrhizas has been little practised.
- The world's readily usable sources of phosphate for P fertilizer might be exhausted within one or two centuries.
- Sewage sludge could be a major P source for the future, but several difficulties would first have to be overcome.
- Converting forest to farmland can increase the rate of soil erosion 10-fold or more.
- The amount and chemical nature of soil organic matter affects crumb structure and hence resistance to erosion. Inputs from crop residues, as well as organic manures, contribute to maintaining soil organic matter.
- Rate of erosion can be reduced 10-fold by altering farming methods.
- Interaction between grazing animals and their food plants involves: (1) response of feeding rate to the structure of the vegetation; (2) selective use of different plant species by the grazers; (3) effects of canopy structure on photosynthesis rate; and (4) differences between plant species in their response to grazing. Information on these should help in planning the intensity and timing of grazing so as to achieve high, sustainable production.
- Overgrazing can lead to irreversible changes in vegetation.

Further reading

History of food production:
Russell (1967)
Loomis (1984)
Simmons (1989)

Energy in food production:
Pimentel & Pimentel (1979)
Bayliss-Smith (1982)

Soil:
Wild (1988)
Brady (1990)

Nutrients in soil and uptake by plants:
Nye & Tinker (1977)

Soil organic matter, nutrient cycling:
Stevenson (1986)

Mycorrhiza:
Harley & Smith (1983)

Erosion:
Morgan (1986)

Grazing:
Crawley (1983)
Huntly (1991)

Chapter 4: Fish from the Oceans

Questions

- Why is fish productivity in most of the oceans so low?
- Is it possible to fish the oceans in a sustainable way?
- How can we decide how much fish can be caught this year without reducing the future catch?
- The great year-to-year variation in the number of new recruits to the catchable fish stock makes it much more difficult to predict how many can safely be caught. How can we cope with that problem?

Background science

- Primary productivity and food chains in the oceans.
- Growth and death of fish in relation to their age.
- Mathematical models: using them to predict how the amount of fish caught this year will affect future fish populations.
- How well do the models fit real data?
- Large year-to-year variations in mortality of eggs and larvae, and their possible causes.

During most of its existence the human race has obtained its food by hunting and gathering, mainly on land. Hunting on land still goes on, for meat, for other animal products such as furs, and for recreation. But in this book I concentrate on hunting in the oceans for fish. The oceans cover nearly three-quarters of the surface of the earth (Table 1.1), and on a world scale fish from the oceans are important providers of human food. The amount of fish caught from the oceans has increased year-by-year since the end of the Second World War: *FAO Fishery Statistics Yearbooks* show that in 1989 more than five times as much fish was caught, worldwide, as in 1948. The rate of increase was, however, slower in the 1970s and 1980s than before: the catch increased 40% between 1979 and 1989. The need to control fishing so as to provide sustainable yield has been recognized for decades and has attracted much attention from fisheries scientists. This has been an active area of applied ecology research.

The great complexity of the ocean environment

In this chapter I ignore coastal ecosystems such as estuaries and coral reefs, but even so we are dealing with systems of great complexity. Intensity of solar radiation is bound to decrease with depth in the water, and temperature and nutrient concentrations often change markedly with depth, too. Horizontal variation occurs in relation to distance from land, but there are also patches in the oceans, at scales from centimetres to hundreds of kilometres, differing in physical factors and abundance of organisms. The water and the living things in it are moved horizontally by currents, vertically by upwellings and downwellings, and by turbulence. In this chapter I can give only some very abbreviated background information on the ecology of the oceans, before concentrating more specifically on fisheries management.

Productivity and food chains

The total fish catch worldwide in 1989 was close to 100 million tons fresh weight (*FAO Fishery Statistics Yearbook*). Taking the food energy content as 1.6 GJ per ton fresh weight (Pimentel & Pimentel 1979), the food energy in this catch was 160×10^6 GJ. Averaged over the whole area of the oceans (Table 1.1), this comes to 0.004 GJ $ha^{-1} yr^{-1}$. This may be compared with examples of animal production on land given in Table 3.1. It is three orders of magnitude lower than modern animal production on pasture; and even migratory pastoralists in a semi-arid region can obtain substantially higher production. Is fish production in the oceans really so low, or are we not exploiting it fully?

Primary production in the oceans

Almost all the photosynthesis in the oceans is carried out by phytoplankton, which are unicellular cyanobacteria and algae living in the 'photic' zone, the upper 100 m or so where solar radiation is sufficient to support photosynthesis. Primary productivity of plankton has most often been measured by placing a small volume of sea-water, with its suspended organisms, in a transparent flask, hanging it at a particular depth in the ocean, and measuring either oxygen production or the incorporation of ^{14}C from dissolved $^{14}CO_2$ into organic matter. Many measurements have been made at individual points on individual days. Scaling these up to primary productivity over the whole year over large areas of ocean is difficult, but at present most of our knowledge of productivity is still based on these methods. Remote sensing from satellites can now provide maps of phytoplankton abundance and may in future be able to give estimates of primary production at scales from kilometres to whole oceans (see Box 2.2).

Why much of the fish stock is near land

The ^{14}C and oxygen-production measurements have shown wide variations in primary productivity. For example, in the Indian Ocean to the southeast of South Africa production was found to range from 14 to 1080 $mg\,C\,m^{-2}\,day^{-1}$ (Dunbar 1979). The highest rates were near to the coast and the lowest near to the Sub-tropical Convergence, where there would be a downwelling of water. In tropical oceans the water tends to be permanently stratified, with warmer water near the surface. This happens in temperate oceans in summer, but breaks down in winter when the surface waters cool. If there is only limited mixing between the upper and lower waters the photic zone becomes depleted of mineral nutrients, which then severely limit plankton productivity. For this reason primary productivity is much higher where there is upwelling, because the currents bring more nutrient-rich waters up from below. Many regions of upwelling are near to land. Large parts of the open oceans have much lower productivity, though there are also bands of upwelling across the major oceans near the equator. This strong influence of upwelling on productivity has major practical implications: it results in much of the oceans' fish stock being close to land, where it is relatively easily accessible by boat trips from the land; and because it is concentrated more fish can be caught per unit effort. If the world's fish stock were evenly spread throughout the oceans, so much more effort would be required to catch each fish that ocean fishing would be a lot less attractive as a way of obtaining food. The concentration of fish near shores has also aided the management of fish stocks. Since the late 1970s the ocean extending 200 miles (320 km) off the shore of each country has been the 'Exclusive Economic Zone' of that country, meaning that it legally controls the fishing within that zone. This covers most of the world's fishing, since most fishing is within 200 miles of shore, leaving only a small proportion in international waters where agreements on control of fishing require negotiations between governments.

Comparing productivities of oceans and land

Many of the measured primary productivities in the oceans are within the range 50–1000 $mg\,C\,m^{-2}\,day^{-1}$ (Longhurst & Pauly 1987, Barnes & Hughes 1988), which is approximately equal to 0.5–10 tons dry weight $ha^{-1}\,yr^{-1}$. Table 2.2 compared this with some primary productivities on land. In the most productive regions of the oceans the primary productivity is similar to that obtained from average modern farms and forestry plantations. However, this upper end of the productivity range is found only in a small proportion of the world's ocean, whereas vast areas of open ocean are towards the lower end of the range. Furthermore, we cannot make direct use of the primary production for human food: the phytoplankton is far too diffuse. We have to rely on the natural food chains to concentrate the energy.

The size of plankton affects where they can live

Box 4.1 gives a simple and widely used classification of ocean species by where they live, whether or not they are photosynthetic and how big they are. Size determines what can eat what, but it has other important effects (Fogg 1991). Larger cells sink more rapidly through still water. They also have reduced ability to acquire nutrients, because the unstirred layer immediately around each cell forms a barrier across which the nutrient ions have to diffuse. Plankton can partly overcome these problems by being actively motile, or (in some cyanobacteria) by controlling their buoyancy by gas vesicles. But the most favourable places for microphytoplankton (which are the larger phytoplankton) are upwelling regions, where the upward water movement can counteract their sinking and also provide nutrients. In the nutrient-poor regions elsewhere much of the primary productivity is provided by nano- and picophytoplankton.

Food chains supplying food to fish

Figure 4.1 shows a generalized food web for the oceans. The smaller the primary producers the smaller the animals that can eat them, and hence the more links there are in the food chain from them to fish. Phytoplankton, even when healthy, release dissolved organic matter which is probably the main energy source for the very abundant planktonic bacteria. This adds a further step into the food chain. The

Box 4.1 A simple ecological classification of organisms of the oceans.

Benthic. Living in, or on the surface of, the bottom deposits.
Pelagic. Living in water above the bottom.
Plankton. Small suspended organisms. Although some can swim actively, all are small enough to be much affected by currents.
Nekton. Larger, actively swimming animals. Includes fish.

Classification of plankton by size

Size range (μm)		Main groups	
		Photosynthesizers	Heterotrophs
<0.2	Femtoplankton		Viruses
0.2–2	Picoplankton	Cyanobacteria	Bacteria
2–20	Nanoplankton	Flagellates	Flagellates
20–200	Microplankton	Diatoms, desmids	Ciliates
200–2000	Mesoplankton		Crustaceans
>2000	Macroplankton		Larvae of fish

Further information: Barnes & Mann (1991).

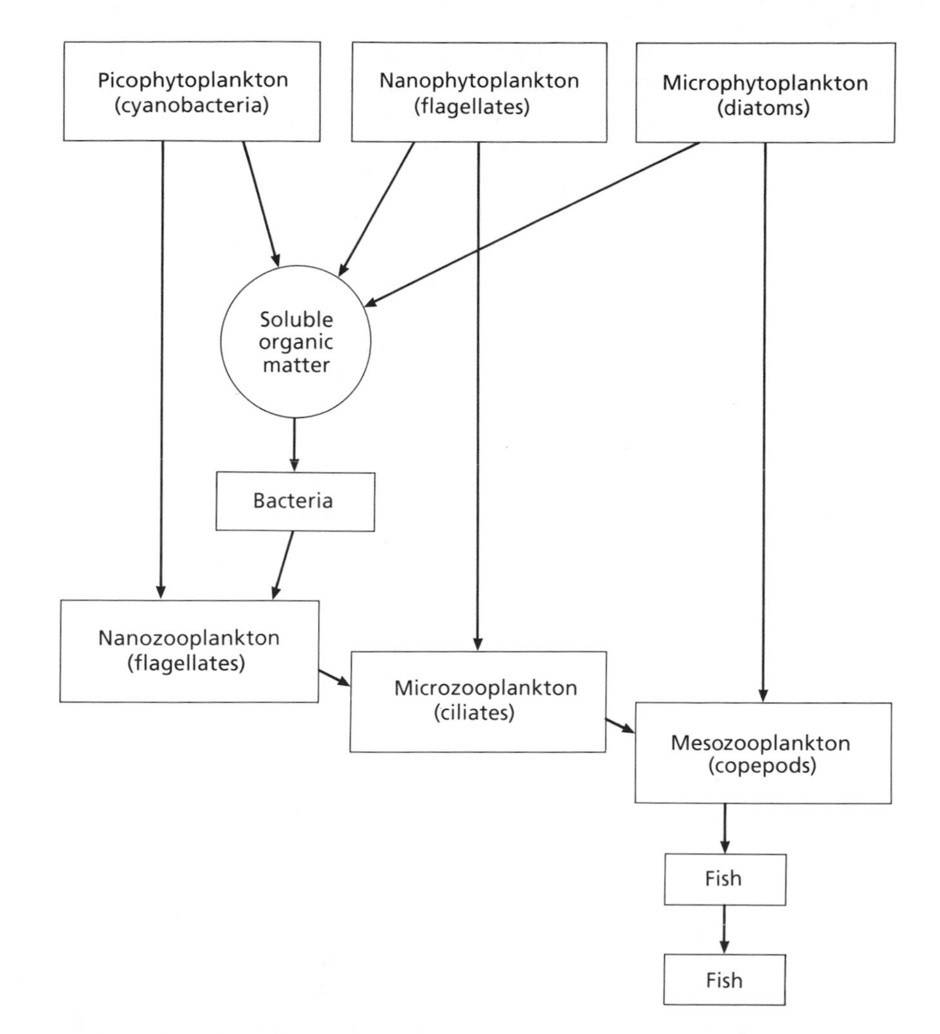

Fig. 4.1 Generalized food web in the oceans. Only the links leading to fish are shown. In most parts of the oceans all the steps shown here occur to some extent, but their relative importance varies. For definition of types of plankton see Box 4.1. A major component of each plankton type is given in brackets, but these are not the only groups involved. Based on Barnes & Mann (1991).

classic approximation is that each step in a food chain is 10% efficient, i.e. for each 100 J of energy in the food a fish consumes it makes growth or reproduction containing 10 J. There is still great uncertainty about the efficiency values in many marine food chains; we know that the efficiencies vary substantially, being sometimes well above 10%, sometimes well below. However, we can safely generalize that at a given site the more steps there are in the food chain, the lower will be the productivity of the end species. So there are two causes, which reinforce each other, for the great variations in fish production between different parts of the oceans: in regions of nutrient-rich upwelling the primary productivity is higher and it is also provided by larger organisms, so there are fewer steps in the food chain. Although a few fish can use large phytoplankton as food, most are at the third, fourth, fifth or even higher trophic level in the food chain.

Figure 4.2 shows foods chains leading to three fish species. Two of these (Peruvian anchoveta and North Sea herring) are less important commercially than they once were, because their abundance has declined greatly, for reasons we shall consider later. These three examples show that many fish species cannot be assigned to a particular trophic level. Anchoveta obtains much of its food from phytoplankton but some from zooplankton, so its trophic level might be stated as 2–3. The adult herring is 3–4, and the Pacific tuna 3–5. So although we cannot assign an exact whole number to each species, there are differences between them in the number of steps in their food chain. This will affect the yield of the desired fish, because of the loss of energy at each step.

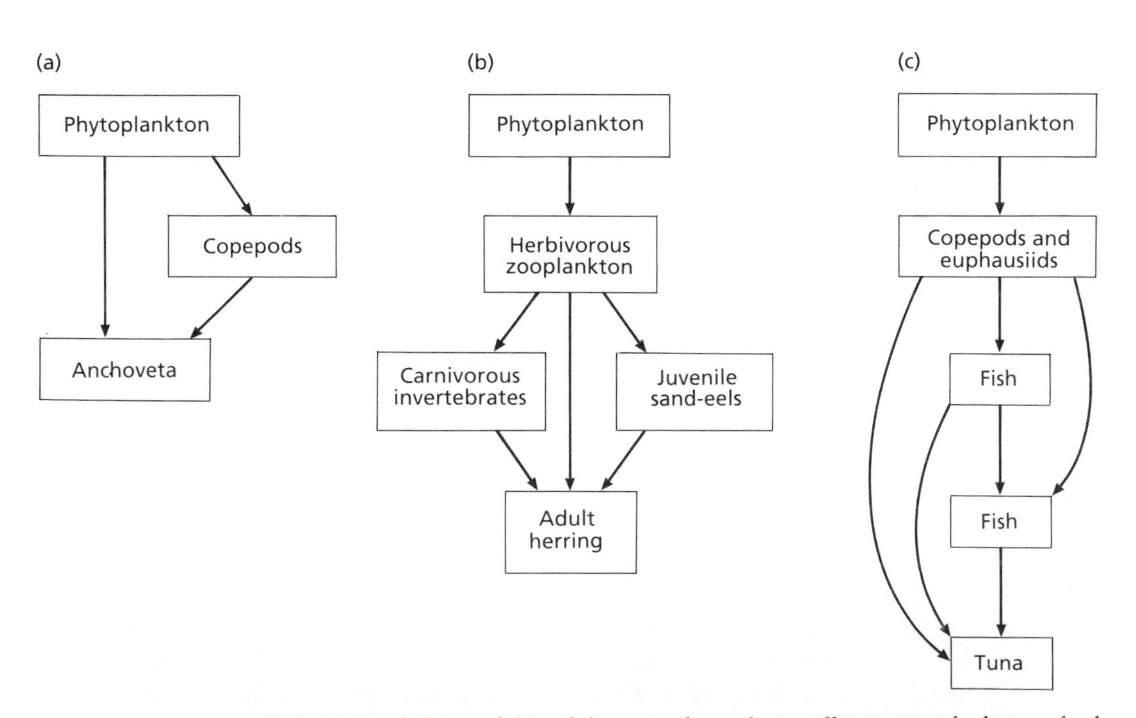

Fig. 4.2 Food chains of three fish species formerly or still important for human food. All simplified from the original authors; only the links which contribute significantly to the food supply of the harvested fish are shown. Copepods and euphausiids are crustacean zooplankton. (a) Anchoveta off Peru, before 1972. From Walsh (1981). (b) North Sea herring. From Hardy (1959). (c) Pacific tuna. From Longhurst & Pauly (1987).

Fish life history

This chapter concentrates on teleosts (bony fish). Whaling also raises important questions for applied ecology, but there is no space to deal with them here. Most teleosts lay enormous numbers of eggs, often hundreds of thousands or several millions per female fish per year. These hatch into minute larvae, which often live in a different part of the ocean from the adults and eat different food; many of them eat phytoplankton, and are effectively part of the zooplankton. Unlike most mammals, adult fish do not reach a stable size, but continue growth throughout their lives. Figure 4.3(a) shows an example. Such graphs rely on ability to determine the age of a fish. In many species this can be done from bands in the scales, bones or otoliths (small bony structures in the ears).

For planning fisheries management it is important to know the mortality rate of fish at different ages. Box 4.2 shows the three basic ways in which the mortality of an animal species may be related to the animals' age. The human species has a survivorship curve of Type I, i.e. mortality risk increases with age. If you, Dear Reader, are aged about 20 and you have a grandmother who is still alive, you know that you have a better chance than she does of still being alive in 10 yrs' time. Your granny knows it too. Many other mammals have Type I survivorship curves; but most fish do not. The egg and larval stages have much higher mortalities than the adults, so in that sense fish have a Type III curve. But within each of the three stages, egg, larva, adult, the relationship is about Type II; so there is no 'normal' life-span for adults. Figure 4.3(b) shows survivorship of eggs and larvae of plaice in the southern North Sea during their first 4 months, comparing 5 yrs. As in Fig. 4.4, the vertical axis is on a log scale, so a straight line would indicate constant mortality risk. Notice the enormous mortality: in three of the years less than one larva per thousand original eggs was still alive after 120 days. But notice also the big difference in mortality between years. Figure 4.3(c) shows that mortality risk in adults, at any rate after age five, was unrelated to age. Of adults alive at the start of any year, about half died within the year, i.e. the half-life was about 1 year. In contrast, in the larval stage the half-life was about 10 days in 1962, 1968 and 1971. (The mortality of adults, but not of larvae, was partly due to fishing.)

Management of fish stocks

It has been accepted for decades, and perhaps centuries, that fishing should aim for sustainable yield. We want to take the maximum weight of fish, consistent with being able to continue to do so each year in the future. Achieving a sustainable high yield has sometimes proved difficult, and not only during the 20th century. The Hanseatic

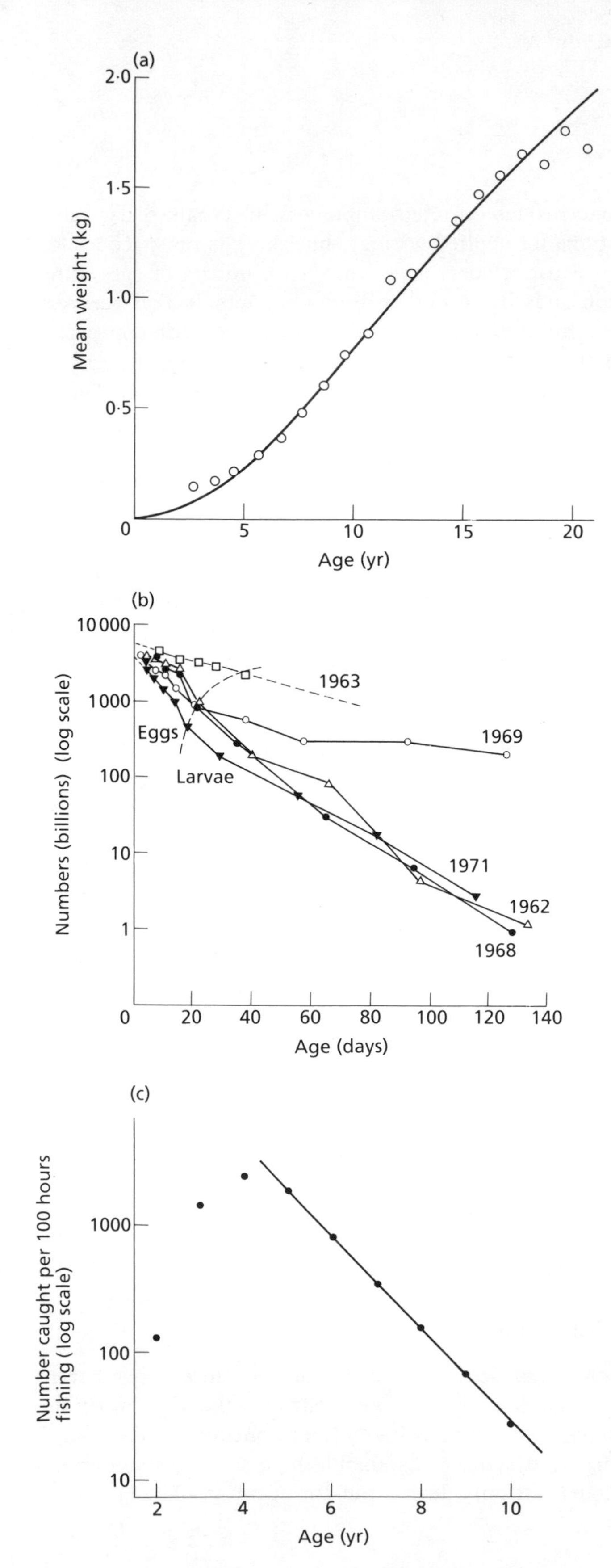

Fig. 4.3 Growth and survivorship of North Sea plaice, in relation to age. (a) Fresh weight per fish. (b) Numbers of eggs and larvae surviving. (c) Numbers of adult fish. The first three points were lower because some fish were still in the nursery area. (a) and (c) are means for 1929–38, from Beverton & Holt (1956). (b) From Cushing (1981).

Box 4.2 Mortality in relation to age.

Mortality risk is the chance that an individual will die within a specified period, e.g. within the next year. Species have been classified into three groups (often called Deevey types I, II and III), depending on how the mortality risk changes with age.

I Mortality risk increases with age.
II Mortality risk unrelated to age.
III Mortality risk decreases with age.

The three types can be represented graphically. In Fig. 4.4 (below) each population starts with 1000 individuals, and the line shows how many are still alive at a particular age. Because the vertical axis is on a log scale, the slope of the line represents the mortality risk, so if mortality risk increases with age (Type I) the curve gets steeper.

Further information: Begon *et al.* (1990), Chapter 4.

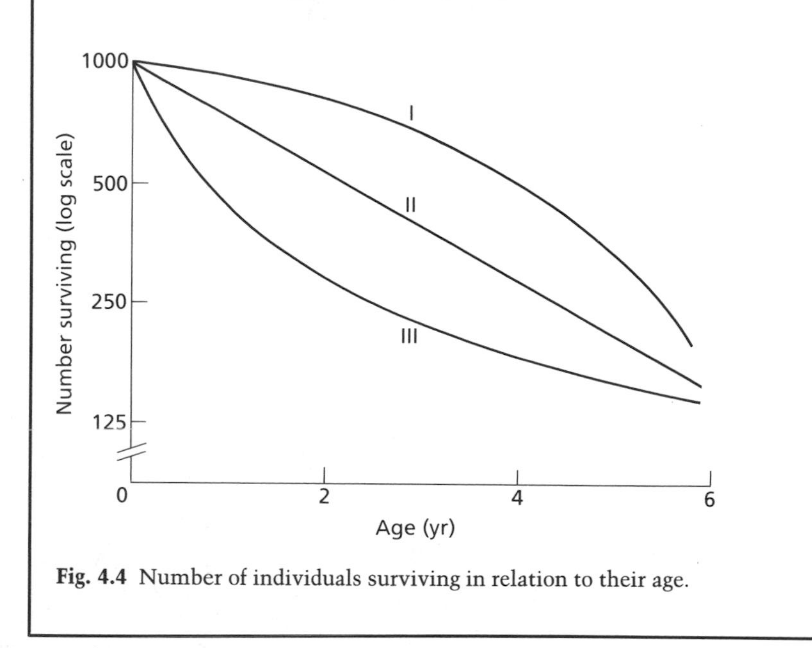

Fig. 4.4 Number of individuals surviving in relation to their age.

League of towns in north Germany organized herring fishing in the Baltic Sea from the 12th century onwards. But between 1416 and 1425, for reasons unknown, the herring stock of the Baltic virtually disappeared, never to return (Hardy 1959).

Figure 4.5 shows the annual catches of three commercially important fish over about 30 years, chosen to show three markedly different time-courses. More cod was caught in the North Sea in 1985 than 30 years earlier, but there were some marked fluctuations in between. Herring catches declined almost to nothing over a 13-yr period. In

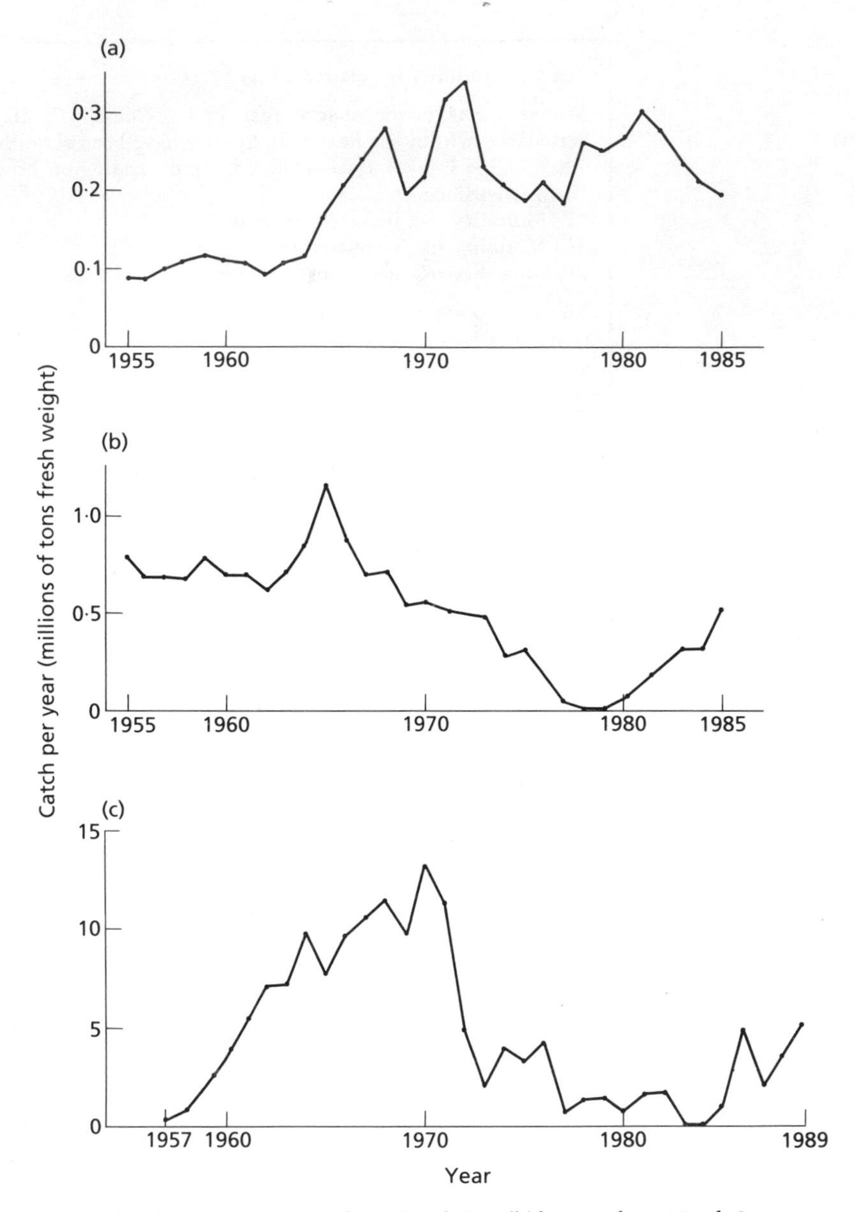

Fig. 4.5 Yearly catches. (a) Cod from North Sea; (b) herring from North Sea; (c) anchoveta off coast of Peru. Source of data: (a) and (b) Garrod (1988), (c) *FAO Fishery Statistics Yearbooks.*

contrast, anchoveta catches rose over about 13 years, then crashed suddenly. Catches of herring and anchoveta recovered partially during the 1980s. The amount of fish caught is dependent not only on the amount of fish present (the 'stock'), but also on the 'effort' expended, which depends on the number of fishing boats active in the area, the number of days per year each is at sea and what fishing gear they use.

The fishing effort for cod in the North Sea increased between 1955 and 1985, but the fluctuations in catch in the late 1960s and early 1970s were due to natural variations in the stock. The decline in catch of North Sea herring from 1965 to 1978 was not due to reduced effort. On the contrary, the effort increased fairly steadily from the late 1940s until 1976, accompanied by a decline in the stock. After 1976 fishing was greatly reduced, and the stock began to recover. Fishing for anchoveta off Peru built up rapidly from the mid-1950s until 1970, when the stock suddenly declined dramatically. Causes of changes in fish populations will be discussed later; but the examples in Fig. 4.5 serve to show that the aim of a sustained high yield of fish has not always been achieved.

The economics of ocean fishing

The fishing boats that put out to sea are owned by individuals or firms, but the oceans and fish stocks are not. This makes the economics of ocean fisheries fundamentally different from, for example, farming where the farmer owns not only the equipment but the animals and plants, and often the land as well. If one individual (*A*) decides to buy improved netting so that one boat increases its catch, that may increase *A*'s profit that year. However, it may also reduce the catch by other boats, and perhaps *A*'s own catch next year, though the effect on each boat will be slight. Economists have predicted how they would expect ocean fisheries to be operated in a free-market economy, without any other controls (Common 1988, Pitcher & Hart 1982). We need to know whether allowing free-market economics alone to regulate ocean fishery would result in long-term survival of the fish stock and sustainable yield. Or would the most favourable course financially for a fisherman be to fish the stock to extinction, sell the boat and invest in something else? The answer to the last question is sometimes *yes*, but for many likely situations the prediction is that falling profits would cause some fishermen to leave before the fish became extinct, and a balance would be reached. The economists predict that under any likely conditions the free market will result in a lower amount of fish caught than could be obtained by managing solely for maximum sustained catch. Put crudely, this is because it is often worthwhile for someone, for example individual *A* mentioned above, to invest in new netting or a new boat to increase his or her catch even if the total future catch from that fishing ground is thereby reduced; *A* will be better off even if everyone else is worse off. The total catch will also be lower, the economists predict, than it would be if each fisherman owned a walled-off piece of ocean and its fish stock. On the basis of such arguments the governments of most major fishing nations have for decades accepted that management of ocean fishing should not be left solely to a free-market economy, and that control by individual governments and agreement between governments is necessary. Governments have looked to ecologists to provide them with the information and predictions upon which their control of fisheries can be based. The rest of this chapter is concerned with that ecological aspect. It does not dwell on how far the world's governments and their international organizations have or have not been able to put the ecologists' recommendations into practice.

Deciding how much fish should be caught

Using models in ecology

Since the 1950s fisheries management has been strongly influenced by mathematical models. I have already referred to the use of models, in previous chapters. Chapter 2 mentioned, but did not explain, general circulation models used for predicting future climate. Figure 3.13 presented, in graphical form, a model of the interrelations of grazing animals and vegetation. That model did not aim to make precise quantitative predictions (indeed there are no numbers on the graphs), but rather to provide understanding and to make predictions about the sorts of things that might happen, e.g. if grazing were increased. It could be called a *strategic model*. In contrast, fisheries models are expected to make quantitative predictions, e.g. of how much fish we can catch this year without gravely reducing the fish remaining for future years. Models that aim for quantitative predictions are often called *simulation models*. If you are not very keen on mathematics you may find the prospect of mathematical models alarming. However, in this age of computers the mathematics is often simple; there may be many steps in the model, but each one arithmetically simple. Often the biggest challenge is not in the mathematics but in the basic design of the model, which is a job for biologists. The model must incorporate the key features of the real system without becoming too complex. If it is too simple it ends up telling you nothing useful; if it is too complex you find that essential information on which to base the model is not available. Another challenge to biologists is in *validation* of the model, meaning checking its predictions against real observations. This is bound to be difficult: a model is most useful if it predicts something that is not easily measured. Chapter 2 presented predictions from models of how increases in greenhouse gases will alter the world's climate. To check these fully rigorously would require several replicate worlds, in some of which greenhouse gases were increased while others were kept as controls.

It has been generally assumed that the only means by which we can influence ocean fish is by our fishing activities: how many people fish in which fishing grounds on how many days, at what times of year, using what equipment and mesh size. In theory it might be possible to manage the fish population in other ways, e.g. adding nutrients to increase primary productivity, or selectively removing unwanted fish species that were competitors with the ones we want to harvest, or that feed on them. Such active management is possible in practice in lakes, and may be in future in screened-off portions of the ocean. In this chapter I stick to conventional ocean fisheries management. The aim of the models, therefore, has been to provide a basis for control of fishing activities. A very influential family of models are *surplus yield models*, first proposed in the 1950s and modified in various ways since. They are described by Pitcher & Hart (1982) and in other fisheries texts. Figure 4.6 is used to explain the basis of these models. It applies to one species of fish in one fishing area. 'Stock' means the total number of fish large enough to be catchable, and 'recruitment' means the number of fish that join the stock per year by growing large enough to be catchable; both stock and recruitment can alternatively be expressed as biomass. If each adult female fish lays the same number of eggs, and the percentage mortality from egg stage to recruitment is always the same, then recruitment should relate to stock by a straight line through the origin, e.g. line *E*. Surplus yield models assume that, on the contrary, recruitment is *density-dependent*, so that as the stock increases the number of recruits *per adult stock* decreases, e.g. because of increased larval mortality. Line *R* gives an example of one possible relationship of recruitment to stock. The shape of this curve has been much debated, with many alternative equations proposed. Curves such as *R*, which slope down to the right and may ultimately reach zero, are known as Ricker curves, after Ricker (1954); with Beverton & Holt curves (Beverton & Holt 1956), to the right of point *M* the curve becomes a level plateau, as in Fig. 4.9(a).

Surplus yield models

Maximum sustainable yield

In Fig. 4.6 maximum recruitment is at point *M*, predicting that *maximum sustainable yield* will be obtained by keeping the stock at that level. The graph does not yet tell us how many fish can be taken

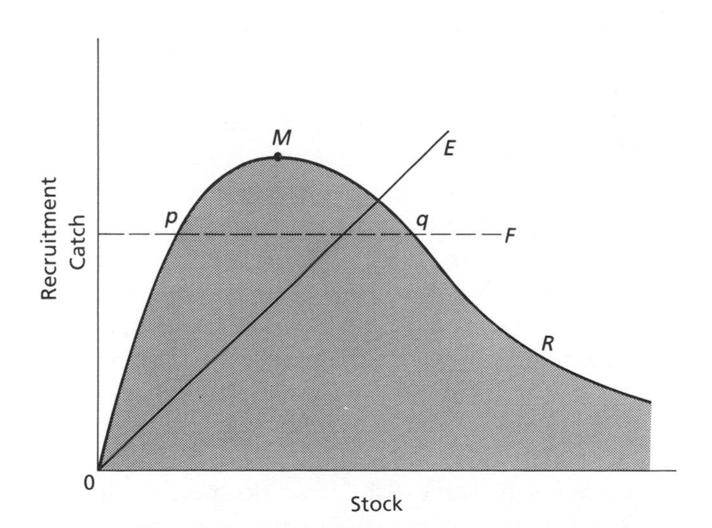

Fig. 4.6 Relationships between stock (total number or weight of fish), recruitment (number or weight of new fish) and catch, used as a basis for surplus yield models. The scales on both axes are linear arithmetic (not log). The meaning of the letters is explained in the text.

each year, because we have no information on natural mortality. If we pretend for the moment that there is no mortality except through fishing, the graph can illustrate two alternative strategies for control of fish catch. One is to decide on a *fixed quota*, the maximum amount of fish to be taken this year, for example line F. Lines R and F bear a strong resemblance to G and C_1 in Fig. 3.13, and share the important features that q is a stable point but p is an unstable point (a 'break point'). So if the stock level is anywhere to the right of p it should move towards q. But if, for any reason, the stock level falls below p, continued fishing at rate F will drive it to extinction. The Peruvian anchoveta (Fig. 4.5(c)) provides an example of a fishery that was being exploited at near the maximum sustainable yield, and which was pushed past an unstable point by a year of unfavourable water conditions. More details on this example are given later.

Stable and unstable points

An alternative method of controlling fishing is by *effort*, e.g. by the total number of boats allowed to operate or the length of the season. Fresh-water fishing and hunting for some land mammals are controlled by issuing licences to individual people, which is a way of limiting effort. Line E shows the simplest prediction of the catch with a constant effort. This has an advantage over a fixed quota that as stock decreases catch decreases too, and for most shapes of R curve there is no unstable point, predicting less danger of a population crash.

Surplus yield models assume that at point M of Fig. 4.6 there is surplus yield, in other words the recruitment is more than is needed to balance natural mortality, and this surplus yield can be removed by the fishermen without endangering the stock. We need to predict, for each fish species in each area, what the surplus yield is. A difficulty in practical application of a model of the type shown in Fig. 4.6 is that it requires knowledge of the size of the stock, i.e. the total number or weight of fish of the species (of catchable size) in a large area of ocean. This can be estimated, but not easily. Numbers of fish can be estimated by mark-and-recapture methods: individuals are caught, marked and released, some of the same species are caught soon afterwards, and the proportion of them marked is used to estimate the total population (Pitcher & Hart 1982, Krebs 1985). This can work well for animals in small habitats with clear boundaries, e.g. fish in a lake or small mammals in a wood, but it is difficult to apply to ocean fish, with the much larger areas and ill-defined boundaries. Another possibility is to estimate egg numbers by sampling in the spawning area, and to calculate the number of adult fish from this, from known mean number of eggs per female. Sonar (sound-echo) has been used successfully to determine numbers of some fish; it works best if they form dense localized shoals. All of these methods are time-consuming and often difficult to apply. The classic surplus yield model of Schaefer (1954) for predicting the maximum sustainable yield requires only two sorts of information, both routinely available in many countries: the total catch and the total effort. 'Total effort' could be expressed as total number of days the fishing boats were at sea, if they were all the same size and had the same equipment. In practice the boats differ, so measurement of effort is more complex, though not different in principle. Figure 4.7 shows the type of relationship predicted. The precise shape of the curve can vary and need not be symmetrical. The predictions apply to an equilibrium situation, so the effort does not just affect this year's stock but has affected previous years' stocks and hence recruitment. If effort is great enough it wipes out the stock and the catch becomes zero. If effort and catch data are available from enough past years to produce such a curve, it enables the maximum sustainable yield, M, to be determined.

One serious limitation of surplus yield models is that catch is just treated as total weight of fish, irrespective of size. One of the things fishermen can control is the minimum size of fish caught, by the size of mesh used. Increasing the mesh size sometimes increases the catch, on a long-term basis: if the small fish not caught this year are still in the sea next year they will be larger, so the total weight of the catch will be increased. In practice some of them will have died, so the optimum mesh size depends on two things, the rate of growth of the fish and their rate of natural mortality. This shows that a model incorporating age and size could be useful, and *dynamic pool models* allow this. They are based on the equation

Dynamic pool models

$$S_2 = S_1 + (R + G) - (M + F)$$

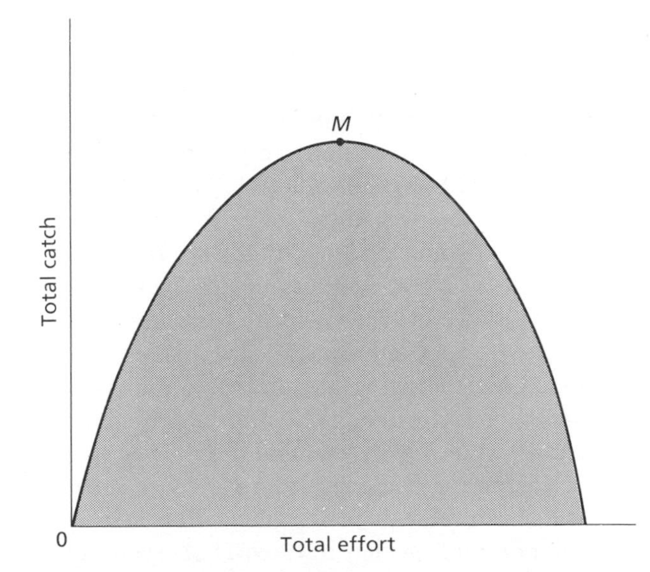

Fig. 4.7 Relationship between catch and effort used as a basis for predicting maximum sustainable yield (M).

where S_1 and S_2 are the biomasses of the stock in two successive years, R = recruitment, G = tissue growth, M = natural mortality, F = mortality due to fishing. In its original form (Beverton & Holt 1956) S was the biomass of all ages taken together, but it is more useful to treat each age-class separately. In the model F expresses the fishing effort (or its result), and if there is adequate information on R, G and M the model should allow predictions of how various fishing strategies (resulting in various values of F) will affect $S_2 - S_1$, i.e. how much the fish stock next year will increase or decrease. Much attention is given to predicting recruitment (R), because it can vary so much (e.g. Fig. 4.3(b)), and therefore be such a major cause of variation in the stock. However, data on growth (G) have also been collected for major species, and it can vary substantially between areas and between years (see Rothschild 1986, Fig. 6.1 and Garrod 1988, Fig. 8.2). Figures for natural mortality (M) are sparser, because mortality due to fishing (F) is often about as high or higher, and the two are difficult to separate.

Comparing the models' predictions with real data

Now we come to the crunch: validating the models and using them. Do curves such as R in Fig. 4.6 fit the real data? Can we determine the shapes of the curves, or the values for the equation above, accurately enough to use the models to achieve sustainable yield? Many examples have been published of observed recruitment plotted against stock size, to see whether they fall on a curve like R of Fig. 4.6. Collections of such graphs are given by Cushing (1981), Pitcher & Hart (1982), Rothschild (1986) and Laevastu & Favorite (1988). Figure 4.8 shows one example, with a Ricker curve fitted. You can see that the real data are widely scattered; at the same stock density the recruitment could vary more than 10-fold. Yet the points do not appear quite random: there is some indication of a downward trend to the right, as though high stock densities were setting some limit to recruitment. I think this is a fair example of stock/recruitment relationships. Some published data show even poorer fit, some a bit better, but I have seen no example that gives a really close fit to any smooth curve. So we have a problem. Sustainable fishing depends on maintaining the stock. In order to maintain the stock the number removed by fishing, and lost by natural mortality, must be balanced by new recruits. If we cannot predict recruitment with even modest accuracy, do we have a scientific basis for fisheries management?

Vast numbers of eggs, then very high mortality

The feature of fish biology that makes recruitment so variable is that each female fish lays enormous numbers of eggs (of the order of a million for the cod of Fig. 4.8), and then there is enormous mortality of eggs and larvae. As an example, Fig. 4.3(b) shows that the number of plaice larvae that hatched from eggs was virtually the same in 1968 and 1969, but after that mortality differed: it was about 7% per day in 1968 but only 1% per day in 1969. That sevenfold difference in mortality produced a difference of more than 100-fold in numbers of larvae by the time they were 120 days old. This high fecundity plus high mortality

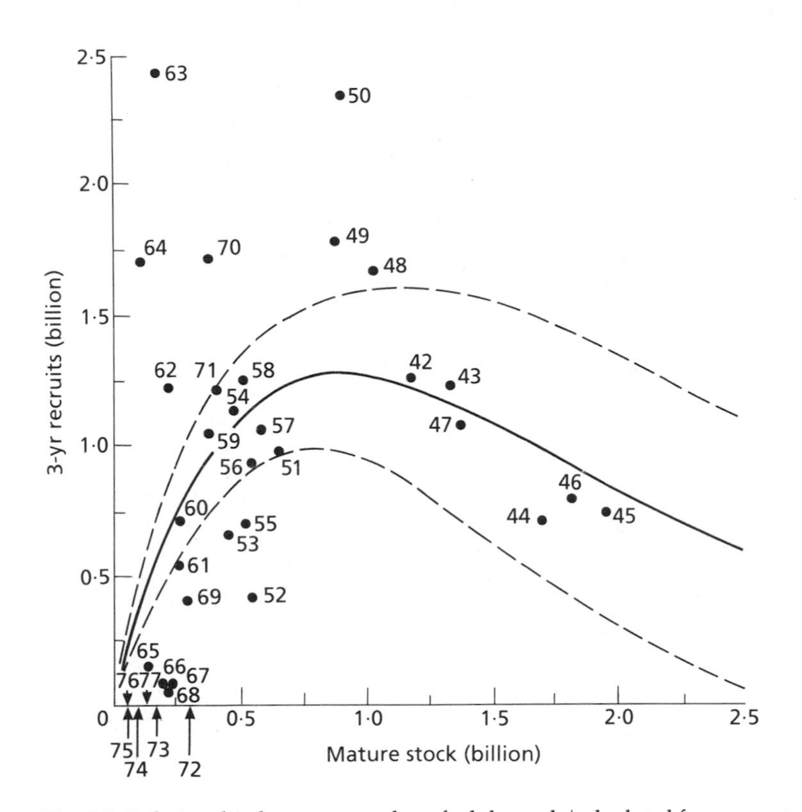

Fig. 4.8 Relationship between number of adult stock (calculated from counts of eggs) and number of new recruits, for cod in the Arcto-Norwegian region of the Atlantic Ocean. Points are for individual years, indicated by the numbers. Smooth curves: — best fit to a Ricker model, --- 95% confidence limits. From Cushing (1981).

is fundamentally different from mammals and birds, where there are few young per parent per year but low mortality. In Chapter 8 we shall be considering population fluctuations in mammals and birds (e.g. Fig. 8.6), and whether these pose risks to the survival of the populations. Modelling that situation for mammals and birds is basically more straightforward than for fish.

How to make predictions in spite of great variability in recruitment

In the face of this enormous year-to-year variation in fish recruitment, there have been three sorts of approach to providing useful predictions of what the catch size should be.

1 Try to discover the causes of the variations in recruitment.

2 Predict the amount of recruitment next year from the size of the pre-recruitment age group this year.

3 Insert a random element into the model.

Stochastic models

I will discuss these in reverse order. Alternative 3 says, in effect, 'we do not know why recruitment varies and we are not going to try to find out why, we will just consider it as random variation'. To a scientist this may look like admitting defeat, but as a practical strategy for management it may have advantages. Figure 4.6 represents a *deterministic model*: if we know the amount of stock we can read off curve *R* the amount of recruitment that will occur, as a precise number without any range of uncertainty. In contrast, if we look at the real data in Fig. 4.8, if the stock level is about 0.5 the graph offers a range of possible numbers of recruits, from 0.4 to 1.2. A *stochastic model* takes into account such uncertainties. A deterministic model provides an answer to questions such as: if we take this much fish this year what will the stock size be next year? A stochastic model provides answers to: if we take this much fish, what is the probability that the stock will be lower next year than this year? What is the probability that it will be at least 50% lower? that it will be 99% lower? So the model can assess risks. Such figures for risks may be more valuable to decision-makers than a predicted population size without any indication of its uncertainty. The calculations do not necessarily involve complex mathematics. *Monte Carlo* models require the computer to perform the calculations hundreds or thousands of times, each time with different input values. For example, lots of different combinations of recruitment, growth and mortality values could be tried, to see what percentage of them led to prediction of reduced stock size. For the calculations to be useful, the values fed in need to be realistic, e.g. we need to have information about the probability of recruitment values lying within different ranges. So stochastic models can be more demanding of basic biological information than deterministic models.

How much unexplained variation can we cope with?

The limitation to the usefulness of stochastic models of fish recruitment has been shown by Koslow (1992). He started with a Beverton–Holt deterministic model, i.e. a stock-recruitment curve similar to one of the smooth curves in Fig. 4.9(a). The egg production per adult was set at one million. He then tested the effect if different amounts of year-to-year variation in mortality occurred. Mortality was assumed to occur in the egg, larva and young adult stages, until the time the adult was large enough to join the stock. The amount of year-to-year variation was indicated by its coefficient of variation (CV = standard deviation/mean). If the CV was 0.05 a plot of stock against recruitment showed a wide scatter of points (Fig. 4.9(a)), but it was still possible to fit a statistically significant curve of best fit. The scatter has some similarity to the real data of Fig. 4.8. By the time the CV was 0.12 (Fig. 4.9(b)) the scatter was so great that no significant best-fit curve was possible. Koslow quotes data for two fish species, one of them North Sea plaice (see Fig. 4.3) that are known to have year-to-year variations in mortality of eggs or larvae that have a CV greater than 0.12. The message is that stochastic models can be useful provided the random variation is not too large. If the random element in the model is very large it so dominates the results that you might as well use a random number generator on its own to predict next year's recruitment.

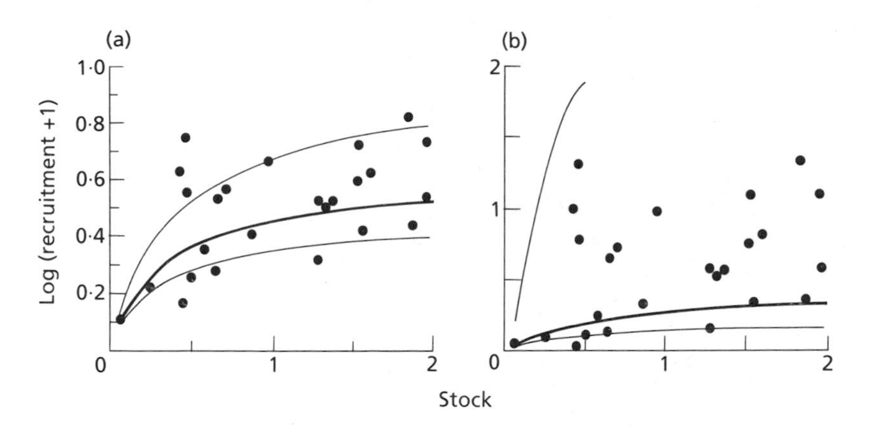

Fig. 4.9 Predictions of recruitment in relation to stock size, from a model in which the number of eggs was fixed at 10^6 per adult fish per year, but mortality of eggs, larvae and young adults was allowed to vary randomly. Each point is for an individual year. On each graph the central curve is the statistical best fit, the outer curves are its 95% confidence limits. Coefficient of variation of mortality: (a) 0.05, (b) 0.12. From Koslow (1992).

Counting young adults

The second suggestion was to count numbers of fish in the age class before they are large enough to be caught in the nets. Most commercially important fish are present in the sea as young adults for several years before they reach catchable size, i.e. before recruitment. Most of the mortality between egg and recruitment takes place in the egg and larval stages (see Fig. 4.3) and therefore most of the opportunity for year-to-year variation is during that stage. In a species where fish are recruited to the catchable stock at age four (for example), the number of 3-yr-olds is likely to be a much better predictor of next year's recruitment than is the present catchable fish stock. The main practical deterrent to using this method is that the pre-recruit fish are (by definition) not routinely caught, so determination of their numbers requires special sampling.

Possible causes of variations in recruitment

The other method of trying to improve predictions of future recruitment, stock size and catches is to try to find out why they vary. This is the scientifically rigorous approach, and the one that should in the long run provide the soundest basis for fisheries management. Changes in populations from year to year can occur for two sorts of reason: because of changes in the environment, e.g. unfavourable weather, or because of properties of the living things and the way they interact. In a stable environment some populations will settle to a stable, constant abundance, but others vary in abundance indefinitely. An example of persistent cycling is the lynx and snowshoe hare in Canada and Alaska (Fig. 4.10(a)). The exact mechanism producing this cycle is still, after

Persistent population cycles

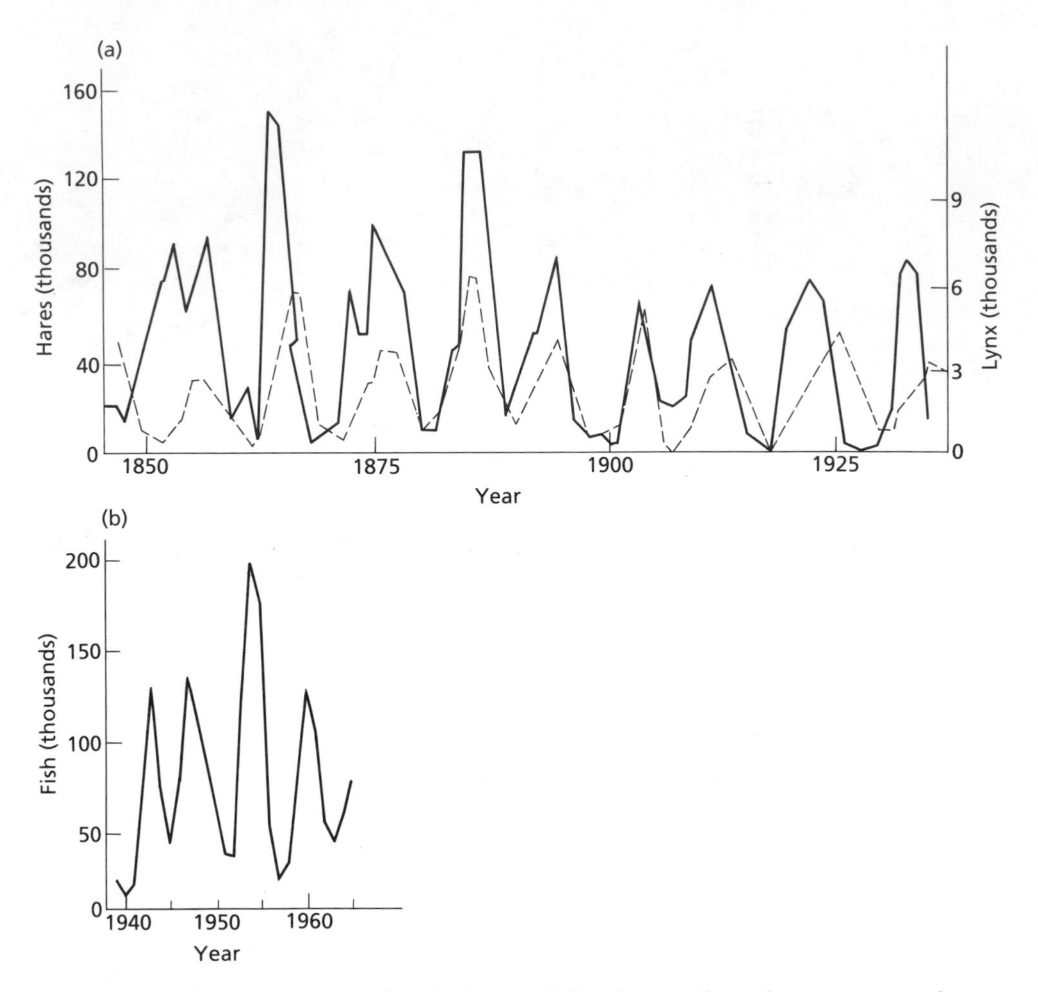

Fig. 4.10 Examples of cycles in animal abundance with nearly constant periods. (a) Numbers of snowshoe hare (—) and lynx (---) pelts collected by the Hudson Bay Company. From Begon *et al.* (1990). (b) Numbers of rock sole in the year-class reaching catchable size in northern Hecate Strait off western Canada. From Laevastu & Favorite (1988).

much research, not fully clear, but it probably involves interactions between three trophic levels, the lynx, its prey the hare, and the hare's forage plants (Keith *et al.* 1984, Sinclair *et al.* 1988). At peak hare populations shortage of plant food weakens hares enough to make them more susceptible to predation by lynx. As hare numbers decline lynx begin to suffer from food shortage; decrease in their numbers and recovery of the plants in due course allows the hare population to start recovering. The full story may well be more complicated than this, but the key point is that the cycles are self-maintaining, without any variation in the environment. Some other land animals that are hunted also show regular cycles, e.g. grouse in some areas of Britain.

Such cycles are detectable with certainty only if they are *stable limit cycles*, i.e. there is a lag somewhere in the system that keeps the periodicity of the cycle nearly constant. If the fluctuations are less regular they could be direct responses to variation in the environment, or oscillations triggered by some temporary variation in the environment. Another possibility is *chaotic dynamics*. Some simple mathematical models (May 1981) predict that if a species has discrete, non-overlapping generations the numbers can change from one year to the next in quite irregular ways. Furthermore, a small change in the number of offspring per adult can alter this pattern greatly. The result in the real world would be population changes from year to year that were extremely difficult to distinguish from random variation; to predict more than a few years ahead would be essentially impossible. Not surprisingly, it is uncertain whether chaotic dynamics does often occur in real populations. If it occurs in fish species attempts to predict recruitment each year would probably have to be abandoned.

Chaotic dynamics

If a fish species shows regular cycles in the numbers of recruits, this would provide a useful basis for prediction of future recruitment. Some of the long-term records of fish catches do show fluctuations, some regular, others less regular. Fig. 4.10(b) shows a 6-yr cycle in the numbers of sole in a particular year-class, the recruits to the fishery. An example of a regular cycle of fish catch is the herrings landed at Bornholm in the Baltic Sea, which showed a 3–4-yr cycle between 1885 and 1925 (Laevastu & Favorite 1988, Fig. 7.2). Whether these cycles are caused by interactions among the species, as in the hare–lynx example, is not known. Some fish cycles can be explained by a very favourable year for survival of larvae leading to a large year-class entering the adult phase and producing a large amount of spawn before many of them are caught. This is particularly true of fish that spawn only once and then die, e.g. Pacific salmon. However, one would expect such an oscillation to damp out with time, for example because of competition between year-classes.

It is safe to assume that some at least of the year-to-year variation in mortality of eggs and larvae is due to variation in the environment. This could affect the eggs and larvae directly, it could alter their supply of food or it could affect the amount of predation they suffer. As an example of a direct effect, in an experiment using controlled conditions Alderdice & Forrester (1971) showed that mortality of Pacific cod eggs increased sharply below 3°C: at 3°C 90% hatched, at 2°C only 10% did. Decreased water temperature has been suggested as the reason for the marked decline in the catches of West Greenland cod in the late 1960s (Garrod 1988). There are some examples where predation is known to be the major cause of death of larvae, others where starvation is more important (Bailey & Houde 1989). In the classical recruitment models, e.g. Fig. 4.6, mortality is higher at higher densities, and this might suggest competition for food. However, such density-

Effects of temperature changes

dependence does not rule out predation being a major cause of death. Higher densities of larvae could increase predation, e.g. by allowing predators to aggregate and feed more rapidly. High densities can also increase cannibalism, which is sometimes a major cause of death in larvae. Effects of environmental conditions, food supply and predation can be interlinked: if larval development is slowed by shortage of food or lower temperature this results in the larvae remaining in that stage longer before they can metamorphose to the less predator-prone adult phase. Bailey & Houde (1989) showed what a large effect this could have, taking into account known mortality rates per day and lengths of the larval stage. For example in favourable conditions cod can complete the larval stage in 37 days, but in unfavourable conditions it could take three times as long. Taking a realistic mortality rate of 8% per day, this threefold extension of the larval stage would reduce the number surviving to become adults by a factor of 400.

Timing of the spring bloom

In most temperate oceans there is a 'spring bloom' of phytoplankton production. During the winter, mixing of the waters has replenished nutrient concentrations near the surface. In spring, warming of the upper waters leads to the establishment of a thermocline, and the more stable, warmer photic layer allows the rapid growth of phytoplankton. Their numbers later decline again, because of nutrient depletion and because of a bloom of zooplankton. This bloom of phytoplankton followed by zooplankton offers the period of maximum food supply for fish larvae that feed on them, and there are important fish species of temperate waters that have a burst of egg laying at about the time of the bloom. However, the timing of the spring bloom varies from year to year, e.g. by as much as 6 weeks in the North Sea (Grahame 1987, Cushing 1990), whereas the timing of egg laying varies less. This can lead to large variation from year to year in the degree of 'match' between the timing of larval and plankton development, which could be a major cause of differences in mortality from year to year (Cushing 1990).

Population changes in the English Channel, summarized by Southward (1980), may have been indirect responses to temperature. Changes in the species composition of the zooplankton were noted in the 1920s, and were followed by a drastic decline in herring in the 1930s, pilchard replacing it as the dominant fish. In the late 1960s and early 1970s the plankton reverted to a state similar to pre-1930; pilchard disappeared but was replaced by mackerel, not herring. Southward (1980) discusses possible causes of these changes, and favours climate as the driving force. The water temperature in the Channel rose to a peak about 1940 and then fell, in parallel with the air temperature shown in Fig. 2.5. Although the change was less than 1°C it may have been enough (perhaps aided by changes in water currents) to alter critical balances between competing species. Herring was near the southern edge of its distribution in the Channel, pilchard has a more southerly range and is adapted to warmer conditions. In such a complex system, with several trophic levels, competition within each trophic level and no fixed boundaries between ocean regions, attempting to relate population changes to a single climatic factor is bound to be difficult.

The cause of the Peruvian anchoveta crash

The sudden decline of the Peruvian anchoveta fishery after 1970 (Fig. 4.5(c)) can be more convincingly related to ocean temperature (see Longhurst & Pauly 1987 for a clear account of this oft-described event). In this upwelling region in the eastern tropical Pacific the fish yields were extremely high, and this encouraged the rapid increase of the fishing fleet during the 1960s. A surplus yield model predicted a maximum sustainable yield of 10 Mtons yr^{-1}, and the actual catch was near this from 1964 to 1971 (Fig. 4.5). This catch might well have been sustainable if the weather and ocean currents had remained constant from year to year, but in 1972 an El Niño Southern Oscillation Event occurred. This is a complex interaction of atmosphere and oceans that occurs in the Pacific every few years (see Longhurst & Pauly 1987, pp. 55–61). The result off the coast of Peru is that the water near the surface is warmer than usual, and lower in nutrients. This led to extremely poor recruitment of anchoveta in 1972. In the absence of fishing the stock might have soon recovered, but with fishing near or above the maximum sustainable yield the result was a collapse in the stock. Subsequently sardine increased substantially. There are other regions of the Pacific in which the two species had coexisted, but previously off Peru anchoveta had dominated, until in 1972 the balance between them was altered. In 1982–83 there was another, more extreme, El Niño event which further reduced the anchoveta stock.

How the anchoveta crash affected other species

Walsh (1981) has estimated the energy flows through the food chain in this region in 1966–69 and 1976–79, before and after the anchoveta crash. Primary productivity changed little. The very high yield of anchoveta was due to the very high primary productivity of this upwelling area (over $1000\,g\,C\,m^{-2}\,yr^{-1}$), and the short food chain, anchoveta obtaining about five-sixths of its energy by eating phytoplankton and only one-sixth from copepod zooplankton. Sardine, in contrast, eats only zooplankton, so the food chain length was effectively increased when sardine replaced anchoveta as the dominant fish; this was part of the reason for the much-reduced total fish catch. One might have expected that zooplankton would increase after their competitor for phytoplankton food, the anchoveta, decreased, but this appears not to have happened. Instead, in 1976–79 about half the phytoplankton production was estimated to end up as dead phytoplankton which sank through the water column. Some of this supported increased benthic production, mainly microbial, but oxygen supply was inadequate to support full use of the increased organic input, the benthic region became anoxic, and much of the organic matter became incorporated into the sediment. In other words, after anchoveta ceased to be the major fish about half of the primary production was 'wasted', not leading to any secondary production. Thus the decline of the anchoveta led to major changes in many other species, too. Fishing for anchoveta has resumed at a lower level (Fig. 4.5), but whether the ecosystem will ever return to its state of the 1960s is unclear. This will depend partly on what intensity of fishing is carried out.

We have seen examples where interaction between commercial fish species is important—anchoveta and sardine off Peru, herring–pilchard–mackerel in the English Channel. This has encouraged the development of multi-species models. Instead of each fish species being considered separately (e.g. Fig. 4.8), the model can take into account that several commercially important fish may interact, by competing for food or by one eating another, and that increased fishing of one may therefore influence the stock of another.

Fish supplies in the future

The attempts to exploit the fish of the world's oceans in a sustainable manner have been partially successful. The clearest evidence of this is that fishing still continues at an undiminished rate, when viewed on a world scale. There have, however, been conspicuous failures in some areas, where the stock of a particular species has declined drastically because of overfishing; Fig. 4.5 showed two examples. The future of fish production may be influenced by difficulties in sustainable management of fishing which this chapter has outlined. But it must also be influenced by the basic ecological inefficiency of the ocean ecosystems as convertors of solar energy into food energy and as producers of protein. As this chapter has explained, the low fish productivity is partly due to the low primary productivity of much of the oceans, but also to the long food chains. Fish are the only carnivores commonly eaten by people: the land animals we eat are almost invariably herbivores, so the food chain is shorter.

As the expanding world population looks for additional sources of food, including protein, we shall be looking for ways to obtain more from the oceans. The way forward is almost certainly not by getting more from fisheries of the conventional sort, but towards growing fish under more controlled conditions. Growing fresh-water fish in fish farms is already well established. Fish can be grown in controlled regions of salt waters too, in large cages or in fenced-off shallow coastal regions (Pitcher & Hart 1982). In this book I do not discuss fish farming, because I want to maintain a balance between different subjects; but there can be little doubt that it will increase in importance in the future. On land hunting has become less important over the last few thousand years as farming has increased. Whether that will happen in the oceans in future remains to be seen.

Conclusions

- Primary productivity varies greatly between different parts of the oceans. Regions of high primary productivity also tend to have shorter food chains. Much of the world's fish catch therefore comes from localized regions, and large parts of the oceans contribute very little.
- Most teleost fishes lay enormous numbers of eggs, and there is enormous mortality of eggs and larvae, which may vary greatly from year to year.
- Mathematical models, especially *surplus yield models* and *dynamic pool models*, have been used to predict how many fish can be caught without reducing future fish stock. Both require predictions of *recruitment*, i.e. the number of fish joining the catchable stock each year.
- Recruitment often varies widely from year to year, especially because of the variations in mortality of eggs and larvae. Attempts to predict it have met with mixed success.
- Some fish show regular population cycles, but in most species variations in mortality are likely to be caused by variations in the environment, either directly or through effects on food supply or predation. Some examples are well understood, many are not.

Further reading

Productivity and food chains:
Grahame (1987)
Longhurst & Pauly (1987)
Barnes & Hughes (1988)
Barnes & Mann (1991), Chapters 1 and 2

Fish population dynamics and management:
Pitcher & Hart (1982)
Rothschild (1986)
Laevastu & Favorite (1988)
Gulland (1988)

Chapter 5: Forests: Timber Production

Questions

- Can timber be produced at an adequate rate in a way that is sustainable long-term?
- Before human interference, were all forests mixed-age, regenerating by small gaps, or were large areas even-aged?
- Should forests be harvested by selecting individual trees, to create small gaps for regeneration? If, instead, areas are clear-felled, does the size of the felled area matter?
- Would it be better to grow all the timber we need in single-species plantations, not relying at all on existing mixed-species forests?
- After clear-felling, can we rely on natural regeneration, or should we expect to replant? Will natural regeneration happen fast enough and will it have the right species composition? If not, what can we do about it?
- Can timber be grown and harvested long-term with only natural nutrient inputs? Are there any ways, apart from adding fertilizers, that we can improve the nutrient status of forests?

Background science

- Amounts of wood used per year, worldwide.
- Rates of wood production by present forests.
- Present amount of forested area, and how it is changing.
- Tree population dynamics. Mortality. Regeneration in relation to gaps. Even-aged and mixed-age forests.
- Fires in forest: their frequency in the past, and their effects on the forests.
- Systems of forest management, i.e. of felling and regeneration.
- Yields of plantations. Importance of rotation length.
- Nutrient balance of forests. Losses at harvest.
- Trees with nitrogen-fixing symbioses. Mycorrhizas.

Forests have many functions. They provide wood that can be used for making things or for fuel; they provide fruits, tanning materials, latex and other useful products; domestic animals can be grazed in them, wild animals hunted; they influence soil erosion rates and water run-off to rivers and lakes; they provide areas for recreation; they alter the appearance of the landscape. In this chapter I concentrate on just one of these, forests as producers of wood. Forests as the habitat of wild plants and animals is a major topic of Chapter 8, on Conservation.

Production of timber and food can be seen as interacting land uses. Large parts of the world's angiosperm forests have in the past been used for shifting cultivation, which is effectively a mixture of farming and forestry. The forest fallow, as well as being a way of returning the patch to suitable conditions for another cropping period, provided the villagers with timber and other forest products. *Agroforestry*, which is today practised extensively in the tropics and a little in temperate regions, has some of the same features. Trees are planted widely spaced, with crops or grazing land between. As the trees grow larger, crop growth or grazing is reduced and finally abandoned. A further development is *alley cropping*, in which parallel rows of trees and crops are grown. The trees are cut back frequently, so allowing crop growth to continue for many years. Although I say no more about these ways of growing trees, we should bear in mind that there are viable systems of land use where timber is only one of the useful products. For further information on agroforestry see Nair (1989).

Sustainable production of wood

The key questions for this chapter are: can the world's population obtain the wood it needs on a sustainable basis? and if so, how is this best done? Sustainable wood production means that, at least when viewed over a large area, wood can continue to be extracted at the same rate year after year, and the forests can continue to grow and produce wood at the same rate for the future. This requires that after trees are harvested new trees replace them, and by having trees of different ages growing up, there are always some approaching a suitable age for felling. This is not a new idea. The Romans had management systems for forests that probably succeeded in achieving sustainable production. There have been continuous traditions of sustainable forestry in parts of Europe since the Middle Ages, e.g. in parts of Germany (Heske 1938). But in other parts of the world at other times standing forests have been treated as a non-renewable resource: people or organizations have extracted timber and then moved on. Williams (1989) provides maps which show graphically how timber felling moved across the United States: during the early part of the 19th century it was concentrated in the northeast, but as those forests became worked out it spread to the Lake States, the southeast, and by the early 20th century to the Pacific Northwest. Clear-felling as a one-off activity happened in other countries, too, for example in Australia. In tropical regions there has been widespread selective removal of particular tree species that are commercially valuable. Some people assume that once trees have been cut down that is the end of forest on that site for ever. But new trees can establish and grow. We should not assume that because the timber extractors took no interest in what happened to the cleared-felled areas after they left, therefore those areas were never forests again. Forest regeneration is considered in detail later in this chapter. But first, let us consider whether sustainable forestry is a realistic possibility, given the present population of the world and its rate of timber use. The way I approach this initially is to calculate whether the amount of additional tree-trunk growth each year in the world's forests is greater than the amount harvested each year. If it is, then there is at least a possibility of obtaining our requirements in a sustainable manner.

Rates of forest growth and timber use

The amount of wood used in the world

Table 5.1 shows that in 1990 the amount of wood harvested throughout the world totalled more than 3 km^3, an imposing volume if it had been a single cube. About half of this was used for fuel; the remainder is called 'industrial roundwood', some of which was made into paper, some into chip-board and other composites, and some was used as intact wood for making furniture, wood-framed houses and many other things. The figures for industrial roundwood should be reasonably accurate, those for fuelwood may not be. Much of the fuelwood is collected by individuals in less-developed countries for use in their own homes. Surveys have been made to estimate the average amount collected per person in various countries. Ives & Messerli (1989) point out difficulties in achieving accuracy in such surveys; often they are based on interviews, which provide scope for misunderstanding, misremembering, or deliberate lying if (for example) the person questioned is suspicious of the motives of the interviewer. In years following a survey, as the population has increased the figures in the *FAO Yearbook* have been adjusted upwards on the assumption that the fuelwood use per person remains constant. That may not be true, for

Table 5.1 Wood harvested in 1970 and 1990, in the whole world, measured in billion cubic metres

	1970	1990	% increase 1970–90
Fuelwood + charcoal	1.35	1.80	33
Industrial roundwood	1.27	1.65	30
Total	2.62	3.45	32

Data from *FAO Forest Products Yearbooks*.

example, if the total amount of wood available in the country is limited and has to be shared among the increasing population.

Some people have suggested that use of wood will decrease, as it is replaced by other materials such as concrete, metals or plastics. So far there is no sign of that. Table 5.1 shows that worldwide use of wood increased by about 1.5% per year between 1970 and 1990. You may think that use of paper will decrease, because the printed word will be made obsolete by electronics—telephones, electronic mail, compact disks and so on. So far that has not happened: world paper production almost doubled between 1970 and 1990. The fact that you are reading this book at this moment shows that little black marks on pieces of paper are not yet entirely outmoded as a way of conveying information.

The area of the world's forests

The next question is, how much new wood grows each year in the world's forests? In order to answer this we need to know the area of forested land, and how fast the trees are growing. The *FAO Production Yearbook* publishes each year figures for the area of forest + woodland in each country, and this was used as the basis for Table 1.1, which shows that forest + woodland occupies nearly one-third of the world's land surface. However 'woodland', in the sense used by ecologists, means vegetation where the trees are far enough apart to form only an open canopy. This includes much of the tropical savanna, which has little potential for production of useful wood. 'Forest' is woody vegetation with a closed canopy, and it is the area of forest alone that concerns us here. Remote sensing from satellites can distinguish forest from other ground cover (see Box 2.2), and the necessary measurements have been made by satellites, but unfortunately recent measurements have not been processed and published on a global basis. Table 5.2 provides the best data on forested area that I know of, none of it based

Table 5.2 Area of forested land in each continent

	Forest area (million km^2)	% of world forest area	% of land area of continent	Source of data
North America	4.6	16	25	1
Europe*	1.5	5	31	1
U.S.S.R.	7.9	27	36	1
Asia*	4.0	14	15	2
Oceania†	0.8	3	9	2
Africa	2.7	9	9	3
South and Central America	8.0	27	39	3
World	29.5	(100)	23	

* Excluding U.S.S.R.
† Australia + New Zealand + South Pacific islands.
Sources: 1, FAO (1985); data collected 1981–82; 2, data quoted by Kallio *et al.* (1987, Table 1.1), collected early 1970s; 3, FAO data quoted by Sayer & Whitmore (1991), collected 1980.

on very recent measurements. According to these figures, just under a quarter of the world's land surface is forested. South America and countries in the former Soviet Union (particularly Russia) are the most forested, and between them contain about half the world's forest.

Comparison of figures in Tables 5.1 and 5.2 shows that if harvesting of wood at the 1990 rate were spread exactly evenly over the world's forests, we should be extracting about $117\,m^3\,km^{-2}$, or $1.2\,m^3\,ha^{-1}$. In practice we are not exploiting all the world's forests, but $1\,m^3\,ha^{-1}$ is a useful number to bear in mind as a global average. If the world's forests are not growing at that rate then sustainable forestry will be difficult to achieve, though not necessarily impossible.

A summary table in Whittaker (1975) gives the net primary productivity of the world's major forest types, in tons $ha^{-1}\,yr^{-1}$, as:

Tropical forest	10–35
Temperate forest	6–25
Boreal forest	4–20

These figures were based mainly on work during the International Biological Program (see Chapter 2). The figures show a wide range, with substantial overlap between forest types. Since $1\,m^3$ of wood usually weighs less than 1 ton, the primary productivity appears easily enough to provide the $1.2\,m^3\,ha^{-1}$ that we need. However, these figures are for productivity of all parts of the plants. Here we are interested in the growth rate of the tree trunks alone. This is difficult to measure from satellites, and no satisfactory remote sensing measurements have yet been made. On the other hand, wood growth should be easier to measure from the ground than overall primary productivity: all we need to measure is the increase in volume of the tree trunks over a period of years. In temperate regions the rate of increase in trunk diameter can be determined from the thickness of annual growth rings in the wood; but in most tropical rainforest trees there are no annual rings, so accurate measurements of girth increase are necessary. The basic difficulty in estimating forest growth over large areas is that measurement sites have been very unevenly distributed. Data from tropical forest are particularly sparse, but large areas of boreal forest, too, have never been visited. Measurements that have been made show wide variations in growth rate. Table 5.3 shows that a more than 10-fold range was found among natural and semi-natural forests in the Great Smoky Mountains, which is admittedly a region of great variety in altitude, exposure and soil, where some of the forests are on former farmland. The figures for tropical forest wood production vary more than 30-fold. These figures are from forests in which trees established only by natural regeneration. Table 5.6 gives examples of the range of tree trunk growth in planted, even-aged forests; again there is much variation. In spite of these wide variations in growth rate, it is possible to get estimates of mean wood growth for regions of the world where

Wood production per hectare

Table 5.3 Rates of tree trunk volume growth, including bark

	Rate ($m^3 ha^{-1} yr^{-1}$)	Source of data
Ranges of values among sites		
Great Smoky Mountains		
(Tennessee and North Carolina, U.S.A.)		
Temperate deciduous forest	1.2–14.4	1
Tropics, various parts		
Forest managed, though not planted	0.5–17.0	2
Mean values: Coniferous and temperate deciduous forests		
U.S.S.R.	1.4	3
Europe (excluding U.S.S.R.)	3.7	3
U.S.A.	3.6	3

Sources: 1, Whittaker (1966); 2, Bowen & Nambiar (1984); 3, FAO (1985); data mostly from national forest inventories and other 'equally authoritative sources'.

there are well-organized forestry systems and extensive measurements of tree trunk volume have been made. Many of these measurements end up in files in offices and are never seen again, but FAO (1985) summarized a distillation of some of them, in response to questionnaires sent out in 1981. The regions covered are only in the northern temperate and boreal zone. The lower part of Table 5.3 gives mean values for major regions.

The question posed was: is the average annual tree trunk growth per hectare of forest worldwide greater than the timber harvested, which is $1.2\,m^3\,ha^{-1}\,yr^{-1}$? The data in Table 5.3, and other similar data, indicate that the answer is *yes*. This encourages us to think that sustainable wood production may be possible, though it certainly does not prove that it is. One question that arises is, if tree growth exceeds tree harvest at present, why is the world's forested area decreasing year by year? Rates of deforestation are not easy to determine on a world or even a national scale. Some widely differing figures for rates of forest loss in the tropics have been published; Sayer & Whitmore (1991) discussed some of them. They pointed out that one major reason for disagreements is that some people have counted every forest that is cut down as deforestation, whereas other people have taken into account whether the area is then reforested or not. Table 5.4 shows for some tropical and temperate countries the mean rate at which the forested area has decreased or increased; so this takes into account whether new trees grew after felling, and also any new areas of forest. Most of these figures (even those for U.S.A.) are of uncertain accuracy, but some patterns show up. In the tropical countries shown (and in all other tropical forest countries listed by Sayer & Whitmore) there was a net loss of forest. In some tropical countries the loss of forests is so fast that if this rate is continued the country will be completely

Deforestation

Table 5.4 Mean rates of timber harvest and of change in forest area in some countries

	Period	Wood harvest* (million m^3 yr^{-1})	Change in forest area: Thousand km^2 yr^{-1}	Change in forest area: % yr^{-1}
Tropical				
Brazil	1981–85	227	−14	−0.4
Indonesia	1981–85	152	−6	−0.5
Mexico	1981–85	20	−5	−2
Côte d'Ivoire	1981–85	12	−3	−7
Congo	1981–85	2.3	−0.2	−0.1
Temperate				
U.S.S.R.	1973–83	370	+4	+0.5
U.S.A	1970–77	330	−8	−0.4
Sweden	1960–80	53	+0.03	+0.02
Finland	1971–81	40	+0.15	+0.08
France	1973–81	35	+0.5	+0.3
West Germany	1971–81	30	+0.05	+0.06

* Wood removed for use. Does not include trees felled but then left or burnt on the site.
Sources: FAO (1985); Barr & Braden (1988); Sayer & Whitmore (1991); *FAO Forest Products Yearbooks*.

deforested within a matter of decades. Among the five tropical countries in Table 5.4 there was a tendency for countries with greater timber harvest to lose more forest area. However, in some tropical countries, including Brazil, much of the loss of forest is because the land is wanted for other purposes, e.g. as grazing land; in such areas the timber is often not removed to be used but is burnt on the site.

Forested area has increased in some countries

Among temperate countries, U.S.A.'s forest area decreased during the 1970s. In contrast, U.S.S.R., which produced more harvested timber than Brazil, reported a modest increase in forested area, and so did the four largest timber producers in Europe: Sweden, Finland, France and Germany (Table 5.4). Canada's forested area probably increased during the 1980s, too, though the figures are unreliable. This is proof that substantial wood production does not have to be associated with loss of forest. But it does not yet prove that the world as a whole can produce, and continue to produce, the timber it needs on a sustainable basis. For one thing, I have so far provided no evidence on whether tropical forests can be exploited for timber on a sustainable basis. However, the rest of this chapter assumes that sustainable exploitation of the world's forests is at least a reasonable aim. It considers ecological problems that have to be solved if forests, temperate and tropical, are to be managed for sustainable use, and some ways that these problems may be overcome. As a basis for this, we need first to consider some fundamental ecology, about tree population dynamics, forest regeneration and resulting forest structure.

Tree population dynamics

Imagine a large area of forest dominated by a single species of tree, then suddenly all the trees are killed by some major event, for example a hurricane or clear-felling by people. Suppose there are many viable seeds in the soil which germinate and soon give rise to a dense stand of seedlings of the dominant tree species. What will happen during the succeeding years? You might perhaps think that the trees will all grow up together as an even-aged stand until they all reach the end of their normal life-span; then they all die simultaneously (approximately) and are replaced by another even-aged stand. In reality that is not what happens, because nearly always the trees do not all die at the same time. One reason for this is that very often the number of seedlings that establishes after a major disturbance is far greater than the number of mature trees that can survive in the area. As the young trees grow larger, competition between them intensifies and some of them die. However, even in a steady-state, mixed-age forest trees do not have a 'normal life-span'. Box 4.2 explained how mortality risk is related to age in different species. In tree species there is often high mortality of seedlings, but after that trees follow the Type II survivorship curve, i.e. the mortality risk does not change with increasing age.

Tree mortality in relation to age

Figure 5.1 shows survivorship data for trees and shrubs in a mixed-age, mixed-species temperate deciduous forest. The lines show some wiggles, as one would expect with real data, but the slopes do not get consistently steeper or less steep with time. The slope of the line differs between species: *Juniperus* had a half-life of about 20 yrs, *Cornus* and *Quercus* about 30 yrs and *Carya* about 40 yrs. Studies of tree mortality in lowland tropical rainforest in West Africa, Central America and Malaysia have found mortalities within the range 1–2% yr^{-1}, corresponding to half-lives in the range 30–70 yrs (Lieberman *et al.* 1985, Manokaran & LaFrankie 1990).

Formation of gaps

This lack of relation of mortality to age results in an even-aged forest developing into a mixed-age forest with time. Because the trees die at different times gaps are formed, by death of a single tree or several neighbours that die at about the same time. If a single small tree dies the neighbours can grow to fill the gap, but larger gaps will require new individuals to colonize them. Because these gaps do not all appear at once, the new stand that develops will be mixed-age. Seeds are not the only means by which trees can regenerate. If most of the above-ground parts have died but the stump or roots remain alive, new vegetative growth may occur from them. Regrowth from stumps is called coppice growth. Many angiosperm trees have this ability, but few conifers do.

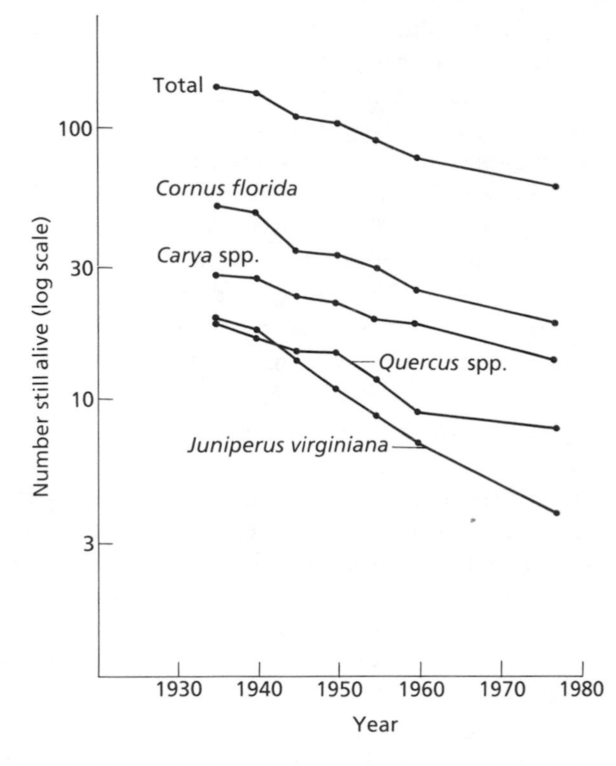

Fig. 5.1 Records of survival of marked trees and shrubs in a 0.1-ha plot in old deciduous forest in North Carolina, U.S.A. Individuals were marked in 1935; the graph shows the number still alive over a subsequent 42-yr period. *Cornus* = dogwood (shrub), *Carya* = hickory (tree), *Quercus* = oak (tree), *Juniperus* = juniper (shrub). From Peet & Christensen (1980).

Regeneration in relation to gap size

If the forest has more than one tree species the regeneration pattern will be more complex. It has been traditional in forest ecology to distinguish two groups of species: *pioneers* or *secondary forest species* can establish only in large cleared areas; *non-pioneers* or *primary forest species* can establish in small gaps, e.g. formed by the death of a single tree. Seedlings of many primary forest species can survive for a long time under the canopy of undisturbed forest, though with little growth; this constitutes the *advance regeneration*, young trees that are already present when a gap forms and thus have an advantage over pioneers which could only then germinate. A third group of species, which establish best in gaps of intermediate size, has been proposed for some areas, e.g. eastern North America (Bormann & Likens 1979). Although these different responses of species to disturbance and gap size have been widely accepted for decades, they are based on general field observations by foresters and forest ecologists, and firm published evidence to support them has been sparse. When, in the 1980s, evidence has been collected and analysed more rigorously, the behaviour of species and the differences between them have often proved to be less consistent than expected (e.g. Runkle 1982). Part of this variability may arise because when natural gaps are observed they differ in age as well as in size. This could be avoided by creating gaps of different sizes experimentally; so far only a few such experiments have been carried out. Kennedy & Swaine (1992) and Brown & Whitmore (1992) cut experimental gaps of different sizes in previously undisturbed lowland tropical rainforest in Sabah, Malaysia. To the authors' surprise, they found no consistent difference between species, in either germination or survival, in their response to gap size. The main positive result to emerge was that seedlings that were taller when the gap was formed survived better. So the ability of species to establish and survive in deep shade (advance regeneration) seemed crucial.

Clearer responses to gap size were found by Phillips & Shure (1990) in mixed-species deciduous forest in the southern Appalachians in North Carolina, U.S.A. They felled areas of four different sizes, from 160 m^2 to 2 ha, and measured the regrowth over the first 2 yrs after the cutting. In all gap sizes most of the regrowth was sprouts from stumps and roots remaining from the previous trees. Averaged over all gap sizes, growth in the second year (measured as dry weight increase) was 92% from sprouts, 6% from advance regeneration and only 2% from newly germinated seeds. Figure 5.2 shows results for the six species that formed most of the tree biomass before cutting or after 2 yrs of regrowth. The original forest was dominated by two oak species (*Quercus* spp.), hickories (*Carya* spp.) and yellow poplar (*Liriodendron tulipifera*). Of these *Quercus prinus* and *Carya* spp. regrew poorly in all gaps; *Q. rubra* regrew from sprouts only in smaller gaps, but *Liriodendron* regrew fairly well in all sizes of gap. Among species rare in the previous forest, *Robinia pseudoacacia* (black locust) and *Acer rubrum* (red maple) grew well in large gaps. Thus the gap size influenced the composition of the forest that developed: in the smallest gaps *Quercus rubra* and *Liriodendron* predominated but in large gaps it was *Robinia*, *Acer* and *Liriodendron*. Not all of these early colonizers would be expected to dominate indefinitely; *Robinia* in particular is a successional species which would in due course be replaced by others.

Disturbance to forests

The preceding section explains why the structure and species composition of a particular area of forest will depend on how recently it was subject to a large-scale disturbance, i.e. to the killing of the trees over a large area. If the major disturbance was recent enough the forest will be essentially even-aged over that area; if there has not been a large-scale disturbance for a long time the forest will be mixed-age and probably of a different species composition. Our attitude to disturbance and its effects is likely to have a major influence on decisions about

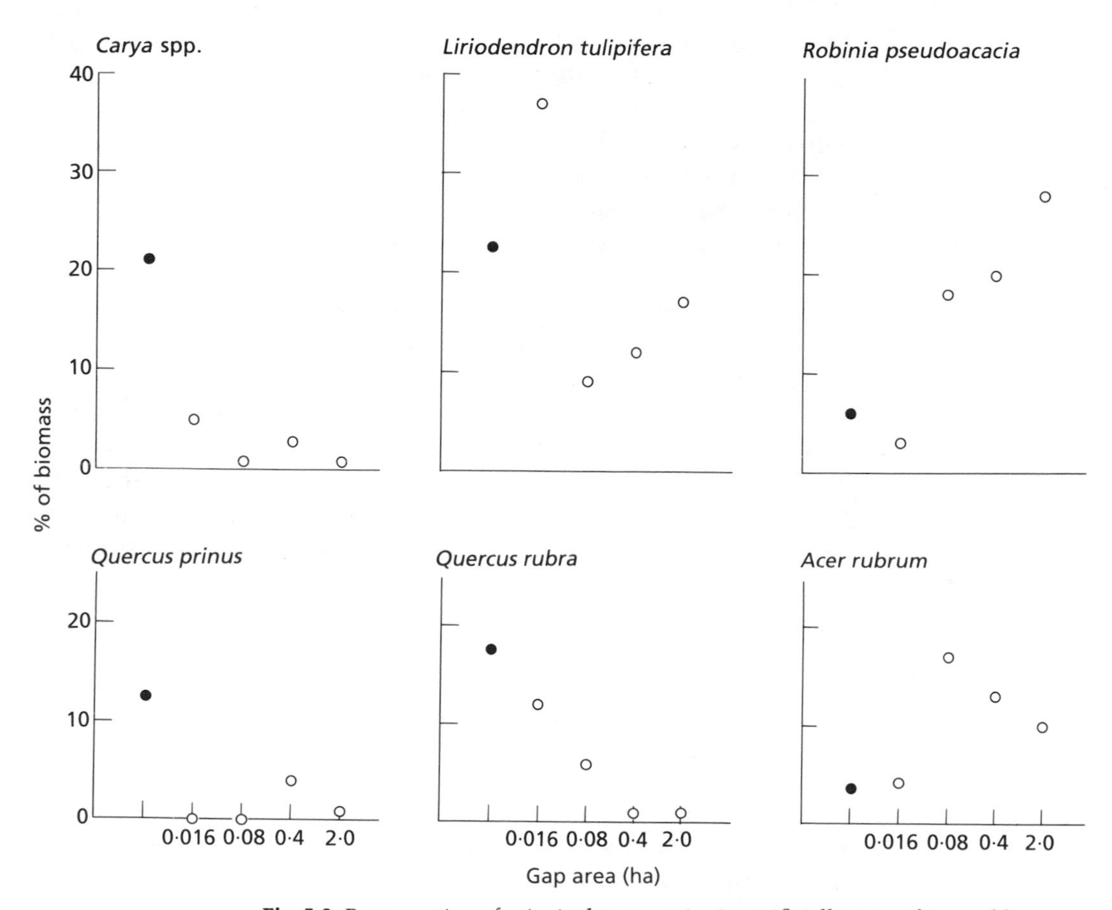

Fig. 5.2 Regeneration of principal tree species in artificially created gaps of four sizes in mixed-species deciduous forest in North Carolina, U.S.A. The abundance of each species is expressed as a percentage of the total biomass. ●, mean value before cutting; ○, second year after cutting. Data of Phillips & Shure (1990).

the best way to manage and harvest forests. If mixed-age forests are in some sense the norm, then we should consider whether to aim for systems of harvesting and regeneration that create mixed-age stands. If large-scale disturbances have been frequent enough to make even-aged stands common, this may give us an indication of the tree species, geographical areas and environmental conditions under which clear-felling is favourable.

A crucial question, then, is: how often are forests disturbed? To be more precise, how often are trees killed over large areas, hectares or more, by a single event? People have been cutting down trees for centuries, and in many areas for several thousand years. There is probably no forest in western Europe in which trees have not been felled; though there are some where regrowth after felling has always been allowed immediately, so the areas have been permanently under tree cover since the first colonization by trees after the Ice Age. In many temperate and tropical angiosperm forests there has been shifting cultivation at some time. The area of each field in shifting cultivation has usually been small, commonly less than 1 ha. One question for forest management is whether the size of the area felled at one time is important; if it is too large will that hamper subsequent regrowth?

The question I want to consider is how often the trees in a substantial area of forest are killed through causes other than being cut down by people. In particular, we want to know whether the 'return time' (i.e. the mean time between such disturbances at a site) can be of the same order of magnitude as the half-life of the dominant trees, since this would result in even-aged stands. There are some parts of the world where hurricanes are frequent enough to leave bands of regrowth forest. But the major candidate as a cause for widespread disturbance of forests is fire.

Disturbance by fire

Forest fires can be started naturally, by lightning. But people in most parts of the world have used fire for tens of thousands of years, and in some parts for over 100 000 (Russell 1967). Hunters have in the recent past, and probably for much longer, used fire to manage vegetation, to keep it more open in order to make hunting easier. It is therefore very difficult to tell whether or not past fires were caused by people. For our purposes it may not matter: the main point is that fire has been influencing vegetation so long that it has not only affected distribution of species but given time for some of them to evolve fire-tolerance characters.

The Yellowstone fires of 1988

A dramatic example of forest fire was the Yellowstone fires of 1988. Yellowstone National Park is an area of forests (mostly coniferous), rivers and lakes, about 80 km × 100 km in size, in the northwest corner of Wyoming. It was the world's first national park. In 1988 much of the United States had an unusually dry summer. Several fires that started in or near Yellowstone burnt with such ferocity and spread so rapidly that they proved impossible to control. By the time the first snows of autumn finally quenched them, about 40% of the national park area had been burnt. When I visited Yellowstone briefly 2 yrs later there were still vast areas that were entirely black, thousands of charred dead trees still standing above the blackened ground. In other burnt areas, however, herbaceous vegetation had established or small aspen seedlings were growing. The Yellowstone fires were described in more detail by Romme & Despain (1989). These fires raised questions. Were they made possible by things people have done in the 20th century, e.g. by building roads or the way the forests have been managed?

Or is such burning normal and natural? How often have forest fires occurred in the past?

Types of forest fire

Before trying to answer these questions, it is important to make clear that fires can differ greatly in severity. *Surface fires* burn litter on the forest floor. Low-growing plants, including tree seedlings, may be killed but often larger trees survive. *Ground fires* burn deeper in the soil, e.g. in peaty areas, and can kill large trees by killing their roots. The most damaging type, *crown fires*, spread through the tree crowns killing the foliage and terminal buds and therefore the trees. This was the major type in Yellowstone in 1988. For more information on these three types of fire see Spurr & Barnes (1980).

Sources of evidence on past fires

Evidence on past fires can sometimes be obtained from written records. Two other sources of evidence are *fire scars* on trees and *charcoal* in lake mud. If a fire is severe enough to damage the bark on one side of the trunk but not to kill the tree, the tree will continue to grow but the fire scar will remain in the tree trunk and can be dated by counting annual rings in the xylem. Deposition of mud at the bottom of lakes often varies through the year, so that annual layers can be recognized. If there is a fire nearby, charcoal will be incorporated into that year's layer of mud, so it can be dated. Charcoal found in soil can be radiocarbon dated (see Box 2.6) and can provide useful evidence of fire, though the dating is less precise than for the lake sediments.

Past fires in northern Minnesota

One of the most extensive studies of past fires was carried out by Heinselman (1973) in the Boundary Waters Canoe Area, 4170 km^2 of forests and lakes in northeastern Minnesota. The forests are mostly coniferous (pine, spruce, fir and larch), and about half the area has never been logged. Fire has occurred here for a long time: charcoal in the mud of one small lake in the area showed a continuous input (i.e. during every century) over the last 10 000 yrs (Swain 1973). Heinselman (1973) determined the dates of fires since 1610 over the whole unlogged area, from fire scars. He also determined the age of standing trees, from annual ring counts. The paper provides maps of the areas burnt in particular years, and figures for the land area involved. During the 19th century the largest area burnt was 44% of the forest in 1863–64; 22% was burnt in 1875 and 17% in 1894. In each of these years there were several widely separated fires which must have started independently. Of the unlogged forest standing in 1973, 65% was even-aged stands that had established in one of those three major fire years of the 19th century. Most of the rest of the forest was made up of smaller even-aged stands attributable to fires in other years. After 1910 the U.S. Forest Service operated a fire control programme and there were no very large fires.

Were these fires started by people or by lightning? There were people in the area during this time, first Chippewa and Sioux, later fur trappers. However, fires can start naturally in the Boundary Waters

Canoe Area: during the 15 yrs 1956–70, 113 fires caused by lightning were reported within the area.

Clark (1990) made a more detailed study of 1 km^2 of unlogged forest, mainly pines, in another part of northern Minnesota. Fires were dated from fire scars and from charcoal in the mud of three small lakes. He was able to map the extent of fire in each year from 1650 onwards, and Fig. 5.3 covers part of this period. The pattern is irregular. Some fires were less than 100 m across, and smaller ones could have remained undetected. Some areas were burnt several times during the 100-yr period, whereas an area at the bottom centre of the maps was not burnt once.

Arno (1980) has summarized evidence on fire frequency in the past in conifer forests of the Northern Rocky Mountain area (Alberta, Montana, Idaho, Wyoming), based on fire scar data. The mean time between one fire and the next ranged from 6 yrs to 300 yrs. This refers to occurrence of fire within a small stand (e.g. 40 ha); not all the stand was necessarily burnt each time. In any case, for the fire scar method to work, some trees have to survive the fire to show the scar.

Fires have also occurred in the past in northern conifer forest in

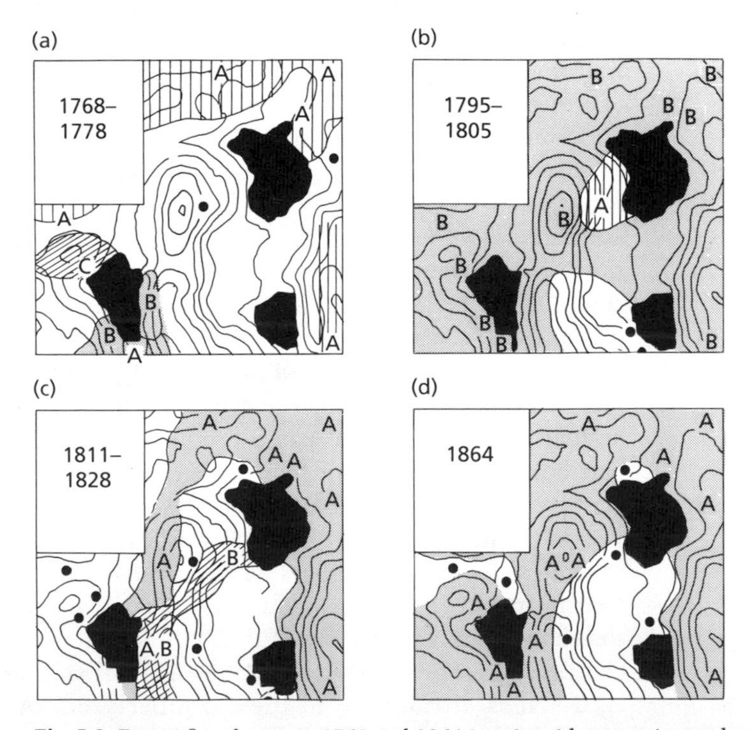

Fig. 5.3 Forest fires between 1768 and 1864 in a 1 × 1 km area in northern Minnesota, U.S.A. The three black areas are lakes. Shading shows the approximate extent of burns in each period of one or a few years. (a) A, 1768–72; B, 1775; C, 1778. (b) A, 1795; B, 1802–05. (c) A, 1811–14, B, 1828. Letters show where fire scars were recorded, ● indicates trees unscarred for that period. Other lines are contours. From Clark (1990).

Europe. For example, fire scars in pine in northern Sweden show that fires have occurred there at intervals over the last 600 yrs (Engelmark 1984). In Australia some eucalypt species normally regenerate as even-aged stands after fire. *Eucalyptus regnans* in Victoria will regenerate only after fire, and some present even-aged stands established after fires that occurred before the first European settlers arrived (Ashton & Willis 1982). However, not all eucalypt species have such an obligate relationship to fire. In some forests, e.g. jarrah (*Eucalyptus marginata*), the undergrowth burns frequently but the trees are not killed and so are mixed-age.

Fire in tropical forests

Fires can occur in temperate deciduous forests of the northern hemisphere, and even in tropical rainforest. For example, the tropical rainforest of the Gogol Valley, Papua New Guinea was extensively burnt in 1890 and again in 1930 (Saulei & Swaine 1988). At those times there were no roads into the area; the only inhabitants were living in small, widely separated villages and practising shifting cultivation (Lamb 1990). Sanford *et al.* (1985) collected soil cores from 96 locations near San Carlos, Venezuela, in tropical rainforest, and examined them for charcoal. They found charcoal in 63 (two-thirds) of them. The oldest known evidence of people living in the area is pottery remains dated 3750 BP. Of 10 charcoal samples that were radiocarbon dated, nine had dates ranging from 3080 to 250 BP, and so could be the result of slash-and-burn farming. The tenth, however, dated from 6260 BP. Overall, the evidence points to most fire in tropical forests in the past being associated with shifting agriculture, probably often localized fires, but perhaps in occasional dry years spreading substantially.

Was fire rarer in angiosperm forests than in conifer forests?

There is a widespread impression that fire is much less prevalent in temperate deciduous forest than in conifer forests, but we would like firm evidence on whether this was true in the past. Whitney (1986) estimated the frequency of fires in an area of Michigan before it was colonized by Europeans, from the records of the first surveyors in 1836–39; they recorded tree species and also burnt areas along their survey lines. Much of the area was dominated by pines, but on some richer soils the forest was a mixture of hardwoods and hemlock with pines. Whitney estimated the mean return time of fire in the pine forests to be about 100–200 yrs, but in the hemlock–pine–hardwoods to be over 1000 yrs. Taking into account the half-lives of the trees, this would probably have resulted in the pine forests being even-aged, but the hemlock–pine–hardwoods being mixed-age. Unfortunately other information on past fire frequencies in angiosperm-dominated forests is lacking, as far as I know.

The conclusion about fire in forests is that it did occur before the 20th century in all major types of forest, but not with equal frequency. It has been very widespread in northern conifer forests, to the extent that vast areas of North America were when Europeans first arrived, and still are, even-aged successional forests, not climax. The forests that burnt in Yellowstone National Park in 1988 were not primeval forests undisturbed since the Ice Age; they were even-aged stands that established after previous fires. Even the enormous, much-admired Douglas fir forests of the Pacific northwest are successional: ancient stands over a wide area appear to have established after some catastrophic event 450–500 yrs ago (Franklin & Hemstrom 1981); whether this was a hurricane or a period of drought and exceptional fires is not known. In most of this area Douglas fir, if left undisturbed, is eventually succeeded by other conifer species such as hemlock (*Tsuga heterophylla*) and fir (*Abies amabilis*), trees of smaller stature.

Key differences between conifer and angiosperm forests

In spite of the sparsity of evidence on fire frequency in the past in angiosperm forests, temperate and tropical, it seems safe to conclude that there are key differences between (a) conifer (and some eucalypt) forests and (b) most angiosperm-dominated forests. Because of the frequency of fire, most conifer forests are even-aged, regrowing after the last disturbance. And because they are not capable of coppice regrowth they have to regenerate by seed. In contrast, most angiosperm forests before the coming of farming suffered disturbance infrequently enough to allow them to be predominantly mixed-age, with small-gap regeneration; but vegetative regrowth from stumps and roots often plays an important part, as well as seeds, in regeneration.

Messages for management

This section on tree population dynamics, forest disturbance and regeneration has been provided to form a background for what follows, on ways of managing forests for sustainable timber production. The first message is that disturbance and subsequent regeneration play a key role in the structure and species composition of forests. All forests are dynamic systems. But, as this section has shown, forests differ in their disturbance and regeneration patterns, and we need to consider how far our systems of exploitation should take this into account. The widespread occurrence of even-aged, single-species or species-poor stands in boreal conifer forests, caused by fire, suggests that even-aged plantations would work well for these species. We can ask then whether the size of each even-aged stand is important, and whether natural regeneration would be just as effective as planting out trees for establishing new stands (as well as a lot cheaper). Since most angiosperm forests, temperate and tropical, are characterized by small-gap regeneration, should any system of exploitation aim for that? Should regeneration be from seed or by coppice growth? For sustainable use, do we want to maintain the same species composition as in the natural forest, or to increase the proportion of certain species? If we want to change the species composition, how is this best done?

Systems of forest management

When we consider the whole world, a wide variety of management systems has been practised; others have been tried out experimentally. Particularly important features are how trees are selected for felling and how establishment of new trees (regeneration) is brought about. Box 5.1 lists the main alternatives.

Selective harvesting of individual trees

For most 'industrial' uses (i.e. for making things) the best stand of timber is uniform in size and in wood properties. Mixed-age, mixed-species stands, on the contrary, usually vary greatly in the size of the individual tree trunks, and the wood is likely to vary in density, strength, colour, frequency of knots and other features of its appearance. If the mortality of the trees in a mixed-age stand has conformed to type II in Fig. 4.4, the diameter will also vary in approximately that way. If plotted on an arithmetic scale (not a log scale as in Fig. 4.4), that results in a 'reverse J-shaped' frequency curve; an example is in Fig. 5.4. In other words, small trees predominate, large trees are rare. For example, if we want boards 20 cm wide that requires trees of at least 63 cm girth, which in the undisturbed rainforest of Fig. 5.4 were quite rare. The combination of few large trees and many species has led to widespread selective felling in tropical forests ('selection' systems). Timber extractors have removed the largest trees of a few preferred species and then moved on. The forests here have been treated as a non-renewable resource: the effect of this selective felling on the future of the forest was not an issue. Later other species may become acceptable, either because of changes in technical ability to

Box 5.1 Systems of forest felling and regeneration

Felling

Selection systems. Fell and remove individual trees, selected by species, large size or both.
Shelterwood systems. Fell strips or areas, leaving other strips or areas in between uncut.
Clear-cutting systems. Fell large areas.

Regeneration

Vegetative regrowth ('sprouting'), from remaining stumps ('coppicing') or roots.
Advance regeneration. Seedlings and young trees that had already established under the large trees are left to grow on after the large trees are felled.
From seed: already in the soil ('seed bank');
from living trees nearby, which can be individual 'mother' trees, strips or patches;
or sown.
Young trees grown in nurseries and planted out.

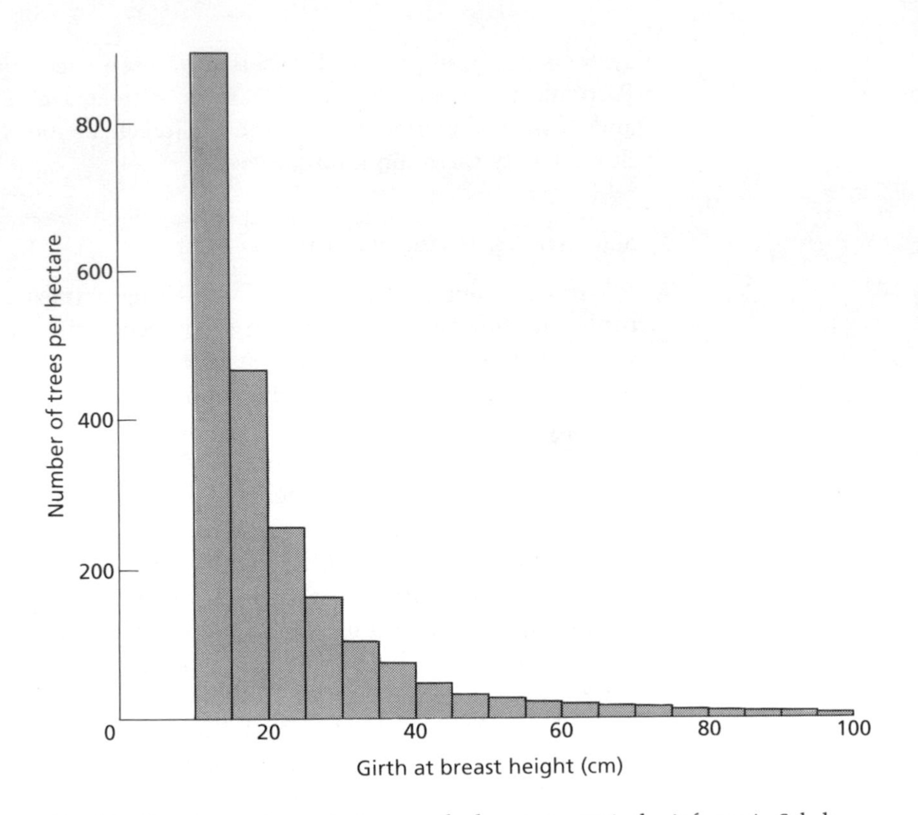

Fig. 5.4 Abundance of trees in 5-cm girth classes, in tropical rainforest in Sabah, Malaysia. The forest had been little disturbed by people. Based on measurements of all trees of girth at breast height (gbh) 10 cm or larger in two 4-ha plots. There were 63 trees ha^{-1} with gbh ≥ 100 cm. Data of Newbery *et al.* (1992), replotted from data supplied by D. M. Newbery.

use them, or because they are now the best available. So extractors may return to take some more trees.

Selective felling in tropical forests: some problems

This system does usually leave a fairly intact forest, so the question is, can this system be sustainable? If small trees of the desired species are left they will presumably grow, so in due course people should be able to return to extract some more. Several contributors to the volume edited by Gomez-Pompa *et al.* (1991) describe and discuss operation of selection systems in tropical forests; some have been operated for a time, others have been tried on an experimental basis. They usually involve rules about what size of trees can be cut, how many and how often. One problem is that during removal of the few selected trees many others may be damaged. In one Malayan selective logging area 3.3% of the trees were extracted for timber but a further 48% were irrevocably damaged (Johns 1985). Thang (1987) indicates 20–50% of trees damaged as the normal range in Malaya. To foresters experienced in thinning temperate plantations these figures may suggest gross carelessness. But in tropical rainforest the trees do not grow evenly spaced in straight lines, many of them are small and so more easily damaged, and large trees are often connected to one another by vines high up, so that the fall of one may pull others after it. Another potential problem is that selective removal of the most valuable species might in the long term result in them becoming less frequent. There is no adequate information on whether this does happen. A more immediate problem is that building roads into previously inaccessible forests to extract timber encourages other people to come in later and clear patches of forest for farming. The overall situation in the tropics is that some selection systems have been run in some areas for short periods, but none long enough to show conclusively that they can work as a sustainable system on a long-term basis. Many have been abandoned because they were too difficult to organize.

Examples of selective felling in temperate forests

Selection systems have been practised over long periods in single-species forests in Europe, removing the largest trees at intervals. The system is still practised in conifer forests in some steep, mountainous areas in France and Switzerland, where clear-felling is not favoured because of concern about soil erosion (Peterken 1981). An example of selection harvesting over more than a century is provided by the jarrah (*Eucalyptus marginata*) forest of Western Australia (Abbott & Loneragan 1986). Features that have favoured success of the system there are: (1) most of the forest is dominated by that one species; (2) the open forest structure makes it easier to remove large trees without damaging smaller ones; (3) because of the stony, infertile soil there has been little pressure to clear forest for other uses; (4) careful management, since 1919 by the Forests Department of Western Australia.

The conclusion about selective felling seems to be that in forests dominated by a single tree species it is possible, though not easy, to operate selective felling in such a way that good regeneration occurs. In mixed-species forests, especially species-rich forests, it becomes far more difficult to predict what species will regenerate in a particular gap, and the situation is ecologically so complex that it may never be realistically practical to have a selection system that works sustainably in the long term. What is 'realistically practical' clearly depends on organization, politics, economics, social systems; these are outside the scope of this book.

Clear-felling followed by natural regeneration

Clear-felling followed by natural regeneration has been practised in various ways in various parts of the world for thousands of years. In tropical shifting cultivation systems regrowth of trees is often allowed to start immediately, the crops being grown among them. At this stage the farmers may choose which species of tree are allowed or encouraged to regrow, and hence may influence the composition of the following forest. I have not found any published details of such practices, but

Dubois (in Anderson 1990) describes examples in various parts of the Amazon forest region where regrowth after farming is dominated by a single species of tree which is valuable for its timber, fruit or some other product. For example, in the southern Brazilian region of Rondonia, the regrowth after farming can be dominated by a leguminous tree *Schizolobium amazonicum*, which grows very fast and produces timber of high commercial value.

Natural regeneration systems in Europe

Clear-felling followed by natural regeneration has been practised in Europe for several thousand years. Allowing angiosperm trees to regrow vegetatively from stumps (coppice) was widespread. A 6000-yr-old trackway found preserved in peat in Somerset, England, appears to have been made from poles from coppice forest (Rackham 1976). A widespread variant was *coppice-with-standards*, in which one species was cut for coppice on a short rotation of 4–10 yrs, and widely spaced trees of a different species grown for much longer. In England hazel for coppice was often grown among oak standards; the small-diametered coppice harvest was used for such things as fuel and fencing, while the oak standards provided larger timber for making buildings and ships. During the 19th century the switch to using coal for fuel, and changes in building methods, greatly reduced the demand for coppice, and in Europe many coppice forests were converted to *high forest*, meaning large trees grown from seed. Shelterwood systems involving natural regeneration from seed were successful with some angiosperm species, for example beech in Germany (Heske 1938). France still has large areas of oak and beech high forest, much of it naturally regenerating, operating on rotations anywhere from 120 to 210 yrs (Oswald in Malcolm *et al.* 1982). Conifers can also regenerate well naturally in some parts of Europe and North America, but poorly in others. Barrett (1980) gives information about different parts of the United States, ranging from 'an embarrassing abundance of reproduction' in some parts of the Pacific Northwest, to very poor regeneration in others, e.g. where rainfall is less reliable and there are other problems such as browsing mammals.

In mixed-species forests the way felling is carried out might affect which species regenerate most abundantly. Figure 5.2 shows how the predominant species regenerating in gaps can be strongly influenced by the size of the gap. This suggests the possibility of influencing the species composition of the regrowth forest by the size of the patches felled. Cutting patches of only 160 m^2 (if we wanted to encourage *Quercus rubra*) would have many of the disadvantages of selective harvesting described earlier, but harvesting narrow strips would be easier to carry out and might have a similar effect on regeneration. The more serious limitation is that, as explained earlier, response of different species to gap size is often not consistent, in either temperate or tropical forest.

Regeneration after a tropical forest was clear-felled

It is possible to make paper pulp from wood of a mixture of tropical tree species, and this has led to a few major clear-felling operations in tropical forests. This raises the question whether tropical forests can regenerate satisfactorily if large areas are felled. One large felling operation started in 1973 in the Gogol Valley in Papua New Guinea. A description of the area and the organization of the timber project is given by Lamb (1990). Before logging started the 880 km^2 valley was almost entirely covered by tropical rainforest. About 5000 people, who among them had 21 languages, lived in small isolated communities and practised shifting cultivation. A Japanese logging company was given the right to fell and remove timber; it did not buy the land. Two restrictions placed on the clear-felling were: (1) 30 m wide strips were left on either side of streams to reduce erosion; (2) a forest area was left near each village for the use of the villagers. In the felled areas scattered trees used as food sources by villagers were left, and also some species that could not be used for pulp; but most trees were cut down. Various plans were made for what would happen after the logging, but in practice most of the land was left to natural regeneration.

Saulei carried out research in the area from 1982 to 1985 (Saulei 1984, 1985; Saulei & Swaine 1988; Lamb 1990). Saulei determined the number of seeds present in the soil immediately after the felled trees had been removed, and recorded tree establishment and growth over 2 yrs in marked plots. He also recorded trees present in areas that had been felled earlier, forming a time sequence, and in remaining unfelled forest. New trees established very rapidly: a year after felling the average density of young trees was several per square metre, and some were more than 3 m tall. Table 5.5 shows the numbers of seeds and young trees belonging to: (a) pioneer species (secondary forest species), i.e. species that require open conditions for germination and establishment; and (b) non-pioneer or primary forest species. Table 5.5 shows that there were very large numbers of seeds in the soil at the time of felling; nearly all of them were from pioneer species. The young trees present after 6 months correspond to only a tiny proportion of the seeds, but they were predominantly pioneer species that had arisen from seed. Among the few non-pioneer species the majority were not from seed but coppice growth from stumps and standing trees left after the felling operations. The remaining columns give some indication of the subsequent time-course of development, although the last two columns are not from the same plots and do not include smaller trees. As time went on pioneer species made up a decreasing percentage of the total, as individuals died. After 10 yrs about one-sixth of the individuals were non-pioneer species, so the composition was moving towards that of old-established forest, though still a long way from it. It may seem surprising that so few seeds of non-pioneer species germinated from the Gogol Forest soil, when these species predominated

Table 5.5 Abundance of seeds and of regrowth trees (number per 100 m²) following clear-felling of tropical rainforest in the Gogol Valley, Papua New Guinea. Data from Saulei (1985) and Saulei & Swaine (1988)

	Time (yrs) after clearance*					
	0	0.5	1	2	7–11	Old forest
Pioneer species						
Seeds	73440					
From seed		408	257	192	14	3.2
Vegetative regrowth		3	7	17	1.0	
Non-pioneer species						
Seeds	880					
From seed		0.4	2	5	1.4	5.5
Vegetative regrowth		8	15	24	1.7	
% non-pioneer	1	2	6	12	17	63

* At time 0: number is of seeds in soil, determined by number that germinated from moistened soil samples. Years 0.5–2: number is of trees, all sizes. Years 7 and later: number is of trees >10 cm girth at breast height.

in the preceding forest canopy. Similar results have been reported from other tropical rainforests, e.g. in Queensland (Hopkins & Graham 1984). Often seeds of pioneer species can remain dormant in soil for some time, whereas the non-pioneer seeds either germinate within a few weeks of their release from the tree or become inviable (Ng 1978).

The felled areas in the Gogol Valley had in common with the North Carolina experiment of Fig. 5.2 that regeneration was allowed to commence immediately. I have purposely not included here results from sites which were used for crop growth or grazing and then abandoned; that is a different story. The North Carolina and Gogol results have two things in common: (1) regeneration was rapid and abundant; (2) vegetative regrowth from stumps was important. In Gogol the forest left along streams and near villages (see above) provided potential seed sources, but sites in the felled areas were up to 500 m away from any remaining forest (Lamb 1990). It seems, however, that such seed sources and their distance apart were not important, since most pioneer species re-established from seed that was already in the soil seed bank, and most non-pioneers from vegetative regrowth. The most important determinant of how well non-pioneer, primary forest species re-establish could be how many stumps have survived the timber extraction processes.

Do pioneer species grow faster?

One ecological classification of species is into *r*-species, which are adapted to colonizing recently-disturbed sites, and *K*-species, which are adapted to stable situations where competition may be intense. In reality there are not two distinct classes but an *r*–*K* spectrum. A common generalization is that *r*-species grow faster than *K*-species. If this applies to trees, then the pioneer species will grow faster, and these may be the ones we want to encourage. It is certainly true that many of the conifers and eucalypts that are chosen for plantations because of their fast growth are in their natural ranges found as colonizers after fire. Examples include many of the pine species, also sitka spruce and Douglas fir (Table 5.6). Whitmore (1984) generalized that among tropical rainforest trees, pioneers characteristically have rapid growth in height and girth, at least in their early years, whereas non-pioneers are slower growing. Among the tropical species listed in Table 5.6, *Gmelina* and *Albizia* are pioneers, the slower-growing *Tectona* (teak) is a non-pioneer. The 'trade-off' is that when wood grows rapidly it is usually less dense and less strong. Among temperate trees the contrast is most often between the conifers ('softwoods') and the non-pioneer angiosperms such as oaks and beeches ('hardwoods'). Among tropical trees there are some well-known examples of light, weak woods, such as balsa (a pioneer species) and dense strong ones such as teak (a non-pioneer). Thus there are markets for pioneers and for non-pioneers, and at each site a decision must be made which species are to be encouraged. In the Gogol Valley there were plenty of seeds of pioneer species in the soil after felling, and abundant rapid regrowth by pioneer species, which needed no special encouragement. But Gogol is only one area; it is not a sufficient basis for generalization about tropical rainforests, which vary greatly.

Selecting for slower-growing non-pioneer species

If it is the slow-growing, strong, non-pioneer species that are required, allowing natural regeneration after clear-felling is probably a slow way to achieve their dominance (Table 5.5). These are the species that form the advance regeneration, the seedlings surviving under the intact canopy. This was the basis for the *Malayan Uniform System*, a method of converting species-rich, mixed-age tropical lowland forest into more uniform-aged forest in which desired species predominated (Thang 1987). In this system, which was formulated in 1948, all trees with diameter at breast height more than 45 cm were felled at one cutting and removed. Among the smaller trees, those of non-commercial species or with poor form were killed by poisoning. The remaining small trees were left to grow on to form the large trees of the more even-aged forest. In lowland Malaya members of the family Dipterocarpaceae are prominent in the natural tropical rainforest, and these are mostly valuable timber trees. This was an important contributing factor to the system working well in the lowland forest of Malaya, and would not necessarily apply in other continents. Much of lowland Malaya has by now been converted to other land uses such as rubber plantations, and the Malayan Uniform System seems to work less well in the remaining hill forests, for a variety of reasons including the steeper and more irregular terrain and hence greater damage to the small trees during removal of the large ones.

Plantations

Commonly plantations are created by growing seedlings in nurseries and then planting them out. This takes time and effort and therefore costs money, but it has clear advantages: the site can be prepared, selected races of the chosen species used, and the trees planted at a preferred spacing in straight lines that will make harvesting easy. The uniform age can lead to an even-sized crop. To replace a species-rich, mixed-age forest with a single-species, even-aged plantation, perhaps of a species not native to the area, may have disadvantages for conservation of native animals and plants, for the appearance of the countryside, and for recreation. We shall consider conservation and wildlife in forests in Chapter 8. A choice may have to be made between (1) managing forests in a way that fulfils simultaneously all our different aims, to at least some extent, or (2) retaining some forests primarily for wildlife, recreation and other amenity purposes, while felling other areas of old mixed-age forest and replacing them with plantations designed purely as efficient timber producers.

When a plantation is established, during the early years growth is slow, primarily because much of the solar radiation falling on each hectare is not intercepted by the foliage of the small and widely spaced young trees. The wood production per hectare per year, called the *current annual increment*, thus at first increases as the trees get larger (Fig. 5.5). However, as the trees get larger the percentage of their gross photosynthesis that goes to xylem production decreases, and in due course the plantation reaches an age when the current annual increment peaks and then starts to fall. You might think that this is the best moment to harvest the trees, but not so. If the aim is to grow a sequence of tree crops on the site, and to maximize the timber yield

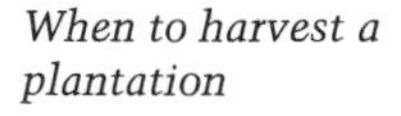
When to harvest a plantation

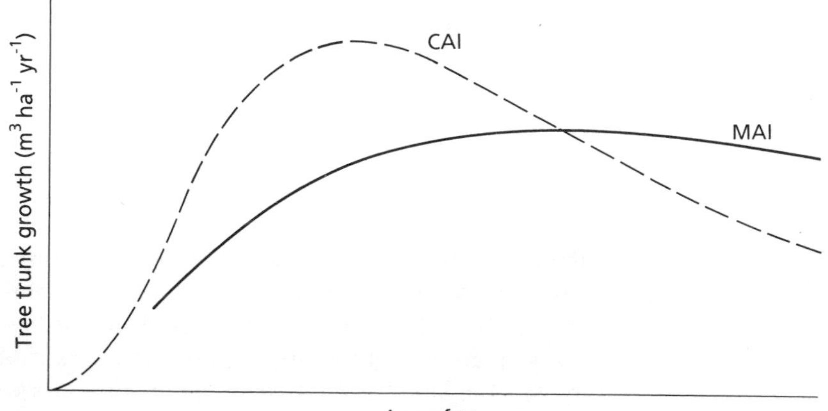

Fig. 5.5 Relation of tree trunk growth to plantation age. CAI, current annual increment; MAI, mean annual increment.

over several rotations, then harvesting should be carried out when the *mean annual increment* (MAI) is at a maximum.

$$\text{Mean annual increment} = \frac{\text{volume of timber in stand}}{\text{age of stand}}$$

This is what determines the timber yield per 1000 yrs if there is a continuous series of rotations on a site. Like the CAI, the MAI also rises to a maximum as the stand ages and then starts to decline, but the maximum occurs later than for the CAI (Fig. 5.5).

Table 5.6 shows MAI values that have been achieved by harvesting plantations at the age when the MAI has reached its maximum. Even within the same geographic region the MAI can vary greatly depending on how favourable the individual site is. The age at which maximum MAI is reached can also vary; often faster-growing stands reach their peak MAI earlier. These figures mostly come from *yield tables* produced by repeated measurements on plantation forests. For our purposes they have two serious limitations:

1 Within the range of MAIs, they do not indicate which values are more common; e.g. is the top half of the range achieved only on sites so fertile that they rarely occur or are usually reserved for farming?

2 The figures apply to stands that did not suffer any catastrophe such as fire or pest epidemic. Such catastrophes do occur; the question is,

Table 5.6 Timber yields of selected plantation species in various parts of the world

		Age at harvest (yrs)*	Mean annual increment	
			$m^3 ha^{-1} yr^{-1}$	$tons\ ha^{-1} yr^{-1}$
Fagus sylvatica (beech)	Britain	120–75	4–10	(3–7)
Picea sitchensis (sitka spruce)	Britain	70–45	6–24	(2–10)
Pinus sylvestris (Scots pine)	Central Spain	100–80	5–10	(2–4)
Pseudotsuga menziesii (Douglas fir)	Northwestern U.S.A.	130–60	2–18	(1–7)
Pinus radiata (Monterey pine)	Chile	30–25	(18–38)	7–15
Eucalyptus globulus	Portugal	20–12	5–40	(3–26)
Eucalyptus camaldulensis	Morocco	16–9	3–11	(2–9)
Eucalyptus grandis	Transvaal	12–9	13–46	(7–25)
	Uganda	7–4	24–52	(14–29)
Gmelina arborea	Tropics	17–7	19–39	10–20
Albizia falcata	Southeast Asia	?	20–60	8–23
Tectona grandis (teak)	Tropics	61–14	6–18	3–12
Coppice:				
Populus trichocarpa (poplar)	Britain	5	15	7

Figures in parentheses are approximate.

* Age to give maximum mean annual increment (except for coppice, where MAI was still increasing at final harvest).

Sources of data: Hamilton & Christie (1971), FAO (1979), Cannell (1980), Curtis *et al.* (1982), Golley (1983), Whitmore (1984), Stage *et al.* (1988).

Could we grow all our timber in plantations?

how often? (Forest pests and diseases are a topic in Chapter 6.) Nevertheless, we can take the MAI figures in Table 5.6 as an indication of what can be attained in plantations. Suppose it was proposed to grow all the world's wood requirements in plantations, how much land would this require? Table 5.6 suggests that with suitable choice of species and sites, and good management, an average MAI of $10\,m^3\,ha^{-1}\,yr^{-1}$ should be attainable. Then to grow $4 \times 10^9\,m^3$ of wood per yr, the world's likely requirement quite soon (see Table 5.1), would require 4 million km^2 of land. This is 14% of the world's present forested area (see Table 5.2), some of which is already plantations. Plantations of some species can also be successfully grown on land that is at present not forested, e.g. savanna. So it is not a ridiculous suggestion to grow all our wood in plantations, and to leave about seven-eighths of the world's forests for other purposes. Plantation forestry has been carried out in various regions of the tropics, with varying degrees of success. Major problems can arise from pests and diseases, inadequate nutrients and 'organizational and social factors'. A large and well-publicized scheme is the Jari project in Brazilian Amazonia (Eden 1990). Since 1967 about 100 000 ha of *Gmelina arborea*, *Pinus caribaea* and several eucalypt species have been planted. Although yields up to $38\,m^3\,ha^{-1}\,yr^{-1}$ were obtained, the average was only $13\,m^3\,ha^{-1}\,yr^{-1}$, which is low for any of these species in the tropics (Table 5.6), though it would be considered quite acceptable in most temperate regions. Economically the project was considered disappointing by its initiator. This serves to illustrate the point I made earlier, that yields such as those in Table 5.6 ignore many of the real problems of growing trees. That is true of temperate as well as tropical plantations.

Importance of the rotation length

Table 5.6 shows that the rotation length, i.e. the time from planting to harvest, varies greatly. In the tropics and sub-tropics some tree species grown from seed can be harvested in less than 10 years. In temperate regions coppice growth can sometimes be harvested that quickly, because of the rapid start to regrowth from the stumps; but trees grown from seed always require longer, some taking a century or more for their MAI to reach its maximum. The attitude of a landowner to a crop will be markedly influenced by such differences: it requires a special attitude of mind to take the trouble to plant trees which will not provide timber until after the person planting them is dead. Time delays are common in life: there is a delay in many activities between starting to invest effort and finally reaping the reward, whether it is growing a wheat crop, making a violin or building a dam for hydroelectricity. But delay is particularly extreme in forestry, because much of the investment is near the start (obtaining the land, fencing it off, making roads, preparing the ground, growing the seedlings and planting them), followed by a long period when there is little or no reward until the final harvest. Economists have given much attention to time delays. They generally assume that money has to be borrowed to finance an

The economic cost of delayed harvest

activity and interest has to be paid while the money is on loan. One way of applying this to forest plantations is *discounting*. This says, essentially: if that forest will be worth £1000 in 10 yrs' time, how much is it worth now (what is its *net present value*)? To calculate the net present value an interest rate is assumed. For example, if the interest rate is 5% and the value in 10 yrs' time will be £1000, the net present value is £614. In other words, if instead of buying the plantation, you invested the £614 in some other way at 5% compound interest, after 10 yrs you would have £1000. Discounting is one way to measure the 'cost' of rotation length. Sticking with the example of a final value of £1000 and interest at 5%, the net present value is: for rotation length 5 yrs, £784; 10 yrs, £614; 40 yrs, £142; 100 yrs, £7.60. The greater financial rewards of fast-growing trees arise not only from their greater timber production per year but also from their being ready for harvest sooner. For example, in Table 5.6 compare beech with sitka spruce, or *Gmelina* with *Tectona*. Although coppice has in Europe been viewed during the 20th century as an outmoded form of forestry, it has the attraction of short rotation times. Economics now suggests that where narrow stems are acceptable, e.g. for making paper or for biomass energy (see Chapter 2), coppice may have a future.

Nutrient balance of forests

Mineral nutrient supply to crop plants was considered in Chapter 3. Trees are in many ways similar: they need the same mineral nutrients and have to get them from the same sources. But the attitude of forest managers to mineral nutrition has always been different from that of farmers. As Chapter 3 explained, farmers have for thousands of years aimed to add nutrients to their fields; in contrast, nutrients have rarely been added to forests, and often the forests have been used as a source of nutrients for the farms. Today many farms receive their nutrients as artificial fertilizer, but fertilizers are only rarely added to commercial forests. This has much to do with the economics of delays, explained in the previous section. If fertilizer is to be applied this year but the trees not harvested for 50 yrs, the fertilizer needs to produce a final increase in yield worth many times the cost of buying the fertilizer and applying it, otherwise it is not financially worthwhile. As ecologists we need to ask how timber can be grown and harvested on a long-term basis without depleting the system of nutrients.

To answer that question we need to know how much of the essential mineral nutrients are removed at harvest. Table 5.7 shows the concentration of N and P in parts from trees of conifer and angiosperm species, growing in temperate and tropical regions. The variation between species is large, but it is clear that the concentration of N and P in the leaves is much higher—often an order of magnitude higher—than in the wood. Wood (xylem) contains some living parenchyma

Xylem has low N and P concentration

cells, but much of it is just cell walls of lignocellulose, which contain scarcely any N or P. Bark, which contains substantial amounts of phloem and other living cells, has intermediate N and P concentrations, as do roots and branches. The implications of this for nutrient removal at harvesting are shown, for one species, in Fig. 5.6. Although the wood in the trunk made up more than half the total weight of the trees, it contained only one-fifth of the N and one-tenth of the P. To deplete the system as little as possible of nutrients, the bark could be stripped from the trunks on the site and left, along with the branches

Table 5.7 Nitrogen and phosphorus concentrations (mg g^{-1} dry weight) in trees growing in well-established forests or plantations

	N	P
Leaves	8–28	0.4–5.7
Stem bark	2–17	0.2–1.0
Stem wood, of all species	0.5–5.7	0.005–0.8
Temperate conifers	0.5–0.9	0.01–0.06
Temperate deciduous angiosperms	0.7–2.4	0.005–0.1
Tropical angiosperms	3.1–5.7	0.1–0.8

Data from: Gordon (1964), Whittaker *et al.* (1979), Miller *et al.* (1980), Chapin & Kedrowski (1983), Bowen & Nambiar (1984), Tanner (1985), Evans (1986), Furch & Klinge (1989).

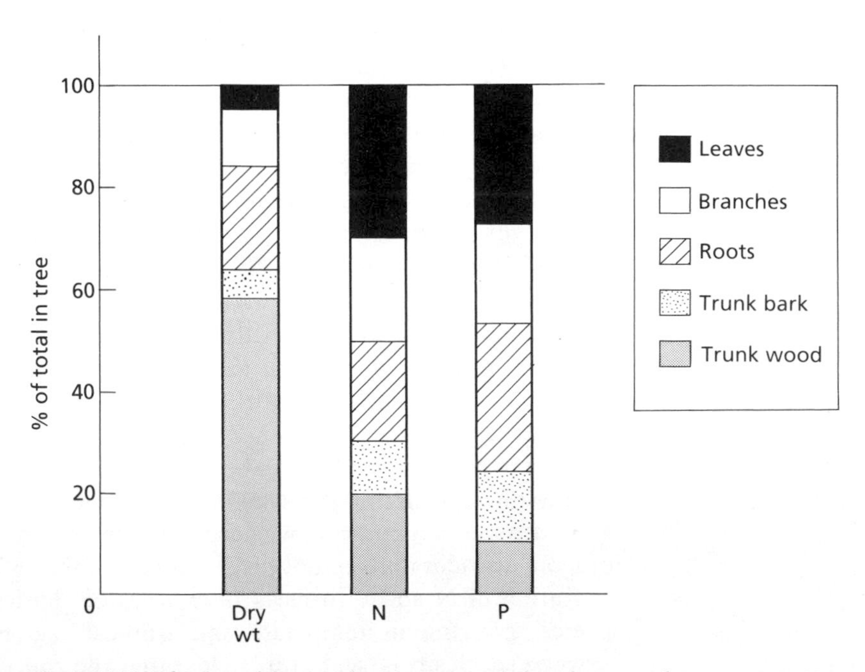

Fig. 5.6 Distribution of dry weight, nitrogen and phosphorus within trees in a 45-yr-old plantation of Scots pine (*Pinus sylvestris*) in southern Finland. Data of Mälkönen (1974).

and roots, to decompose. Using the more conventional method of harvesting the trunks with their bark on, the nutrient removal is substantially increased, because the bark, although small in weight, has higher nutrient concentrations than the wood (Table 5.7). 'Whole-tree harvesting' has been suggested, in which roots and branches are removed as well as trunks, to obtain extra material for pulp or biomass energy. Clearly this would substantially increase the nutrient removal.

Can nutrients in rain supply enough for forest growth?

Taking average concentrations in wood of temperate trees as $1\,mg\,g^{-1}$ for N and $0.05\,mg\,g^{-1}$ for P, a timber yield of $10\,tons\,ha^{-1}\,yr^{-1}$ (high for temperate forests) would result in a removal rate, averaged over the growth cycle, of 10 kg N and $0.5\,kg\,P\,ha^{-1}\,yr^{-1}$. These figures are of the same order of magnitude as inputs in rain (see Chapter 3). However, if the bark was taken away too the losses would be about 50–100% higher. Tropical trees can have higher N and P concentrations in their xylem (Table 5.7). One of the highest yielders in Table 5.6, the sub-tropical species *Eucalyptus grandis*, can have $2.4\,mg\,N\,g^{-1}$ (van den Driessche in Bowen & Nambiar 1984), so its highest yield of 29 $tons\,ha^{-1}\,yr^{-1}$ would remove $70\,kg\,N\,ha^{-1}\,yr^{-1}$, more than rain can supply in any but very polluted areas.

N and P balance of a 50-yr-old forest

The amount of nutrients in the standing biomass and the amounts gained and lost by the ecosystem were measured in the detailed study of a northern hardwood forest at Hubbard Brook, New Hampshire, U.S.A., which has already been mentioned in Chapter 3 (Likens *et al.* 1977, Whittaker *et al.* 1979). Six adjacent valleys (watersheds) were studied. Because of the impermeable rock, the only way that dissolved substances could leave each watershed was in the stream water, whose flow was measured and composition regularly analysed over some years. The original forest had been clear-felled in 1910–19, so the forest studied was regrowth. The principal tree species were sugar maple (*Acer saccharum*), beech (*Fagus grandifolia*) and birch (*Betula alleghaniensis*), with some conifers in exposed positions. Measurements made in 1965, when the regrowth forest was about 50 yrs old, allow us to estimate what would have happened if the forest had been felled then. Fifty-six per cent of the total standing biomass was in the trunks of the trees, a similar percentage to the pine stand of Fig. 5.6; and like that stand, the N and P in the trunks were much less than half the total N and P in the biomass (Table 5.8). The table shows the mean N and P gains and losses per year in rain and streamflow (including suspended organic matter), expressed as amounts per hectare averaged over the whole watershed. They have also been multiplied by 50 to give an estimate of the gains and losses during the whole life of the forest. The net gain of N approximately equals the loss in the tree trunks if they were harvested, though not if the whole trees were removed. (There may have been other gains and losses of N, e.g. by fixation and denitrification, so the net gain of N shown in the table may not be the whole story.) The gain of P was far too small to balance

Table 5.8 Data on nutrient contents, inputs and losses, from Hubbard Brook Forest, New Hampshire, U.S.A. The forest growth was about 50 yrs old at the time of the measurements

(a) *Amounts in trees*

	$kg\,ha^{-1}$	
	N	P
Trunk (wood + bark)	135	11.0
Whole trees (including leaves and roots)	552	88.0

(b) *Gains and losses by undisturbed forest*

	$kg\,ha^{-1}\,yr^{-1}$		Those figures ×50	
	N	P	N	P
Gain in rainfall	6.5	0.04	325	2
Loss in steam water	4.0	0.02	200	1
Net gain	2.5	0.02	125	1

(c) *Loss of N* ($kg\,ha^{-1}\,yr^{-1}$) *in stream water following clear-felling*

	Clear-cut watershed		Uncut control watershed	
	NO_3^--N	NH_4^+-N	NO_3^--N	NH_4^+-N
First year after felling	104	0.7	1.3	0.3
Second year after felling	147	0.5	2.8	0.2

Data from Likens *et al.* (1970, 1977), Bormann & Likens (1979), Whittaker *et al.* (1979).

the removal at harvest. However, the measured P concentration in rain falling at Hubbard Brook was much lower than has been reported from various other sites. Dissolved P input by rain was discussed in Chapter 3.

For deciduous trees the time of year when harvest takes place may influence the amount of nutrients removed. For deciduous high forest there may be some advantage in harvesting while the leaves are on the tree, because the trunk is likely to contain less nutrients at that time. During leaf senescence a proportion of the N, P and some other mineral nutrients are translocated back into the stems before the leaves fall. Probably much of this is stored overwinter in the stems, which will thus be higher in mineral nutrient content during that time. For example, in one study the stems of apple trees were shown to lose 18% of their N between February and May, as the new leaves expanded (van den Driessche in Bowen & Nambiar 1984). When coppice is harvested it is not possible to leave the leafy branches behind on the site, because they are the crop. It is therefore important to harvest during the winter season after the leaves have fallen.

Nitrate leaching after clear-felling

At the Hubbard Brook Forest an experiment was carried out which showed another way that nutrients can be lost at the time of forest harvest (Likens *et al.* 1970, Bormann & Likens 1979). After the flow rates and solute concentrations of the streams in two watersheds had been monitored for some time, the forest covering one watershed was clear-felled. This was done in late autumn, after the leaves had fallen from the deciduous trees. The cut trees were left on the site, and for 3 yrs herbicide was applied to prevent any regrowth occurring. During the first winter and spring following the cutting the concentrations of ions in the stream remained similar to those in the control watershed's stream. But from June onwards there were dramatic changes, especially in nitrate. Whereas nitrate in streams from undisturbed forest was less than 1 $mg\,l^{-1}$ for much of the year, after cutting it rose to 40–80 $mg\,l^{-1}$, and remained that high as long as regrowth was prevented. Ca^{++}, Mg^{++} and K^{+} concentrations also rose, though not by such a great proportion. Dissolved phosphorus losses were unfortunately not reported. Table 5.8 shows the N losses, expressed as kilograms per hectare, during the first 2 yrs, while regrowth was prevented. The enormous increase in N loss was nearly all as nitrate: cutting increased ammonium loss only slightly. The total N loss in the stream water during each of these 2 yrs was similar to the amount that would have been removed by harvesting the tree trunks. It is therefore potentially a major contributor to the long-term N balance of a forested site, and we need to consider how such losses can be minimized. After felling, the rate at which ammonium was produced from decomposing plant material and also the rate at which nitrate was formed from ammonium were both faster than in uncut forest (Bormann & Likens 1979). Nitrate formation is important, because it leaches much more readily than ammonium. The causes of its faster production after felling are uncertain. Knowing more about this might lead to ways of reducing nitrate loss after felling, which is important for long-term forest management.

This experiment purposely enhanced the effects of clear-felling by preventing regrowth for 3 yrs. After that the herbicide application was stopped, and vegetation rapidly started to grow. Nitrate loss decreased markedly in the second year of regrowth and was close to that of the control forest by the third year. Losses of Ca^{+} and K^{+} were also back to near-normal by then. Thus the rate at which revegetation can occur is likely to have an important influence on nutrient losses following felling.

Ways to increase nutrient supplies to trees

When fertilizers have been applied experimentally to established plantations the growth responses have sometimes been large (e.g. Miller & Miller 1976), so there is no doubt that some forests are deficient in nutrients; but of course this will vary depending on soil conditions. Experiments with tree seedlings have sometimes obtained very high rates of growth by supplying ample nutrients. Ingestad has developed a system for supplying nutrients very frequently at a rate related to the size of the plant, which has demonstrated that tree seedlings can show rates of growth far more rapid than normally observed in the field (Ingestad 1982). Cannell *et al.* (1987) grew willow (*Salix viminalis*) outdoors in pots in Scotland and supplied NPK fertilizer dissolved in irrigation water. For much of the growing season the nitrogen concentration in the leaves was $50–65\,mg\,g^{-1}$, much higher than concentrations normally found in the field (Table 5.7), and this was associated with high rates of photosynthesis and growth. In spite of the unfavourable Scottish climate, the stem production was 10 $tons\,ha^{-1}\,yr^{-1}$, a rate rarely achieved by any tree species in any temperate region (Table 5.6).

These results should make us ask whether we need to accept the current rates of timber production (e.g. Table 5.6), as normal, or whether forest management should in future aim for much higher growth rates and yields. Agricultural research has produced enormous increases in crop yield per hectare; foresters seem to be lagging behind. This difference in attitude relates to time and economics. Most things that could be done to increase growth rates and yields of forests turn out to be not economic. The long delay before you can harvest the increased yield is just as much of a deterrent to breeding improved tree varieties as it is to adding fertilizer. If fertilizers are likely to be used rarely, can biologists suggest any other way of increasing nutrient supply to trees?

Trees with N-fixing symbionts

There are trees that have symbiotic relationships with N-fixing microorganisms. Many species of the Leguminosae are trees, some temperate and some tropical. In addition, two tree genera, *Alnus* (alder) and *Casuarina*, have nodules inhabited by the N-fixing actinomycete *Frankia*. *Alnus rubra* and members of a few tropical legume genera, e.g. *Dalbergia* and *Pterocarpus*, provide some useful wood; and *Leucaena leucocephala* is used in alley cropping and provides fuelwood and small poles. So there may be some scope for increased use of such trees. However, none of the species mentioned is a major timber producer. Another possibility is to grow two-species mixtures of a N-fixing species with a more valuable timber tree. Table 5.9 shows results from one experiment. The wood production by a mixed stand of Douglas fir and N-fixing alder was much higher than Douglas fir alone. Growth of Douglas fir was not significantly altered, so the gain was the large yield of alder wood, a less valuable wood than the

Table 5.9 Trunk weight of 23-yr-old trees in plots where Douglas fir (*Pseudotsuga menziesii*) was grown on its own or with alder (*Alnus rubra*) in Washington state, U.S.A.

	Trees in plot		
	Douglas fir only	Douglas fir + alder	
Trees measured	Douglas fir	Douglas fir	Alder
Number of trees per hectare	650	540	2200
Trunk dry weight (tons ha^{-1})			
Individual species	35.0	41.2	53.5
Whole stand	35.0	94.7	

Data from Binkley (1983).

Douglas fir. Also the high N concentration in the alder leaves would probably improve the soil N status for the next rotation.

Mycorrhizal associations

Most trees form mycorrhizas. Many important timber trees form the ectomycorrhizal type, e.g. pines, spruces, firs, Douglas firs, oaks, beeches, most eucalypts, dipterocarps. There are other timber trees that are VA-mycorrhizal; these include many tropical species, such as *Gmelina arborea* and the two species listed after it in Table 5.6, but also temperate species such as the redwoods (*Sequoia sempervirens* and *Sequoiadendron gigantea*) and the ashes (*Fraxinus* spp.). Since mycorrhizas often improve the nutrient-uptake ability of plants, we can ask whether there is potential to manage the mycorrhizas in forests to increase their beneficial effects. In some special situations mycorrhizal inoculum can be extremely sparse, e.g. after open-cast mining, and here inoculation can enhance growth of woody plants (Jasper *et al.* 1989). But in most forests mycorrhizal fungi are naturally abundant, and the relevant questions are: would it be beneficial to alter the mycorrhizal species composition? and if so, could we do it on a field scale? We have no answers to either question at present. In fact, we do not have any direct evidence on how mycorrhizal fungi affect large trees; all experiments comparing mycorrhizal with uninfected trees have been carried out with seedlings or saplings. Among ectomycorrhizal fungi, some species infect only young trees, others only old trees (Ingleby *et al.* 1990), so experiments on seedlings are bound to miss some important information. One piece of evidence that not all mycorrhizal fungi are equally beneficial comes from research on the ability of ectomycorrhizal species to obtain N from different sources. Abuzinadah & Read (1986) showed that in culture some of the species of fungus could obtain N and grow well when a protein was the only available N source, whereas others required NH_4^+. In other words, some had proteinase activity. When some of these 'protein fungi' were used to establish mycorrhizal association with pine seedlings, the pine could grow and obtain N when protein was the only N source, which it could not do when non-mycorrhizal (Abuzinadah *et al.* 1986). These results suggest that in the field some mycorrhizal fungi (but not others) enable their associated trees to obtain N from organic N sources in litter without depending on free-living decomposer organisms, which could put them at an advantage where decomposition rates are slow. So further research on mycorrhizas may open up opportunities to alter the mycorrhizal species composition in forests in ways that would benefit the trees' mineral nutrition.

Conclusions

- The present tree trunk growth of the world's forests is greater than the amount of timber harvested. This suggests that it should be possible for the world's timber requirements to be obtained in a sustainable way.
- The total forest area is decreasing in many tropical countries, but in many of the major timber-producing countries of the north temperate zone it is not.
- In the coniferous forest area of North America fires have been frequent enough in the past to result in much of the forest being even-aged patches. In contrast, much temperate deciduous and tropical forest was, before farming, disturbed infrequently enough for it to be mixed-age.
- Selective removal of individual large trees can work well in some temperate forests. Whether it can be a satisfactory method long-term in tropical forests is doubtful.
- After clear-felling natural regeneration is often rapid. However, the species that predominate in early regeneration may not be the most desirable for timber.
- Under some circumstances input of N in rain can be sufficient to balance removal in trees at harvest, provided only the trunks are removed. However, clear-felling can increase N loss by leaching.

Further reading

Forest ecology:
- Temperate: Waring & Schlesinger (1985)
- Tropical: Whitmore (1984)

Forest nutrition:
- Bowen & Nambiar (1984)

History—people and forests in the past:
- U.S.A.: Williams (1989)
- U.K.: Rackham (1976)
- Germany: Heske (1938)

Management and exploitation of tropical forest:
Eden (1990)
Gomez-Pompa *et al.* (1991)

Plantation forestry:
Temperate: Savill & Evans (1986)
Tropical: Evans (1992)

Economics of forestry:
Price (1989)

Chapter 6: Invaders and Pests

Questions

- Can we tell in advance whether a species will survive and multiply in a new region, if it arrives there?
- What are the special risks from releasing genetically engineered organisms? How can they be minimized?
- Can we predict how fast a species will spread?
- Do pests and diseases ever have major effects in natural ecosystems? Are the host and its attacker always in a stable relationship, or are there sometimes outbursts of damage to the host? How does the host species manage to survive?
- Can we predict conditions under which host and attacker will reach a stable equilibrium, and other conditions under which they will cycle widely in abundance?
- Can pests in arable crops and forest plantations be controlled by altering the management? Can study of pests in natural ecosystems suggest ways of doing this?
- In biological control, should we expect to introduce the control species just once or many times?
- Can we tell in advance whether a species will be effective in biological control?
- Is an initially effective control species likely to evolve to become less effective?

Background science

- What proportion of the species that have arrived in a new area have become established there.
- Attempts to identify characteristics common to all successful invaders.
- Characteristics of bacteria and of their environment that increase their chances of survival and of exchanging genes.
- Rates at which invader species have spread in the past.
- Mathematical models of host–parasite population dynamics.
- Lessons from past successes and failures in biological control. Where the successful species have come from.
- Predictions from models of characteristics likely to make species effective in biological control.

As its title says, this chapter is about *invaders* and *pests*. What do those words mean? A *pest* is a species we do not want. It may be a parasite—such as a virus, bacterium, fungus or protozoan—that causes a disease; it may be a free-living nematode, insect or other animal that attacks a useful plant, for example in a farm crop or a forest; or it may be a weed, an unwanted plant in cropland, grazing land or forest. Our main interest in pests is how to get rid of them or to reduce their harmful effects.

An *invader* is a new arrival in an area, which we may view as either good or bad. Species have invaded new areas naturally in the past, and some continue to do so. Others have been introduced by people, accidentally or on purpose; some of these have been beneficial, others have become pests.

Thus this chapter is concerned with a wide range of species—animal, plant and microbe—which we may regard as anything from very dangerous, e.g. rabies virus, to very useful, e.g. successful biological control species. They are all in this one chapter because the study of them is linked by fundamental features, as the chapter will show.

Predicting whether an invading species will survive

There are many occasions when we want to know, before a species arrives in an area, whether it will survive and multiply there. It may be a plant or animal which could affect native species, by competing with them or eating them, and we want to know whether action should be taken to prevent it reaching the new region. If it is a proposed biological control agent we want to know whether it will survive well enough to be effective against the species we want to control, but also whether it will spread and harm other, non-target species. The only sure way to find out whether a species will survive in a new region is to introduce it experimentally. But sometimes we cannot take such a risk, because once introduced the species, if it proves to establish well and to be harmful, may be difficult or impossible to eradicate.

To provide a basis for answering in advance the question 'If that species is introduced, will it survive?' we need to know characteristics that favour establishment; that is characters of species that make them good at establishing in new areas, and characters of areas that make them prone to invasion. To try to find such characters we must study successful and unsuccessful invasions of the past. A major problem is that, although we have substantial information about successful invasions, species that failed to establish may often go unrecorded.

Introductions: importance of the initial population size

Another problem is that if the number of individuals arriving initially is small, chance may play a large part in whether the species establishes or not. An example is provided by introductions of red deer into New Zealand. Between 1851 and 1892 six small groups of red deer, each including at least one male and one female, were released in South Island. Of these only one established a permanent population (Clarke 1971). Crowell (1973) experimentally introduced a mouse species and a vole species to small islands off the coast of Maine; both were native on a nearby larger island. If only one or two pairs were introduced they usually failed to establish, whereas introducing more than two pairs often was successful. However, the number of replicate introductions carried out was too small for statistical significance. One message for people carrying out deliberate introductions is that the number of individuals introduced may be important. The topic of the risk to small populations of becoming extinct is dealt with in more detail in Chapter 8.

Percentage of invaders that survive and become pests

Table 6.1 summarizes figures, drawn together by Williamson & Brown (1986) from various sources, showing the fate of species that are known to have invaded Britain at some time in the past. The percentages from the three major groups are sufficiently similar to provide a clear message. A high percentage of the invader species survived, at least in a restricted area, but only about one-fifth of them have become widespread. These figures in the top two rows of the table may be spuriously high, if there are many species that have invaded briefly but died out without being recorded. Of the species that survived about 10% have become pests. A similar figure has been calculated for the United States (Waage & Greathead 1988): of 837 insect species that became established between 1920 and 1980, 10% became pests. It may seem comforting that 90% of the species that established were not harmful, but the remaining 10% that were classed as pests could between them do a great deal of damage. Even one pest species can have a very serious effect.

Are there consistent characters of successful invaders?

Are there key characteristics of species that would allow us to predict whether they will be successful invaders or not? Scientists who have considered this question carefully have often come to the

Table 6.1 Data on fate of non-native animal and plant species that have invaded Britain, i.e. that are known to have been present in a free state at some time

	Vertebrates	Insects	Vascular plants
% of invading species that:			
(a) survived (remained localized or became widespread)	87	41	66
(b) became widespread	19	22	18
Of the species that survived (i.e. line (a)), % that became pests	10	7	12

Data from Williamson & Brown (1986).

conclusion that there are no such characters that apply consistently (see symposia edited by Kornberg & Williamson 1986, Mooney & Drake 1986). Generalizations suggested by some writers have often later been found not to apply to other groups of species or to other areas. For example, the features of the classic *r*-species, such as rapid growth, early reproduction, mechanisms for rapid spread, have often been suggested as the key to successful invasion of new areas. But there are clear examples where this does not fit the facts (Lawton & Brown 1986). These *r*-characters help a species to be an early colonizer of a newly available habitat, but ability to invade an area already occupied by thousands of species may require different characters. Again, one might guess that a species whose native range is very small would be unlikely to invade and spread widely in another continent. A conspicuous example not fitting that hypothesis is *Pinus radiata*, whose native habitat is 40 km^2 on the Monterey Peninsula in California. In Australia it not only grows well in plantations, but can invade native eucalypt forests (Burdon & Chilvers 1977).

It has been suggested that herbivores are more frequently limited by parasites and predators than are carnivores, which are more frequently limited by food. Crawley (1986) argued that if this is true then it would probably be easier for herbivores than for carnivores to invade a new area: carnivores would almost certainly have to compete for food, herbivores might leave their parasites behind. In support of this, he showed that although about three-quarters of native British mammal and bird species are carnivores, the majority of introduced species that have established are herbivores (Table 6.2).

Are some areas more easily invaded?

In trying to predict which areas or ecosystem types are more prone to invasion we are again short of consistent indicator features. It is often suggested that plant invasion is aided by frequent disturbance, though the evidence is often anecdotal. One example was the spread of European annuals into western prairie and sagebrush in North America, which was almost certainly hastened by trampling and heavy grazing by cattle, in areas where native herbivores had been sparse (Mack 1981). Crawley (1986) ranked 30 British habitat types according to the percentage of their total flora made up by alien species, and concluded that a high percentage of aliens occurs when there is a combination of much disturbance and proximity to a source of alien seed. For example,

Table 6.2 Percentage of mammal and bird species in Britain that are carnivorous (as opposed to herbivorous)

	Mammal	Bird
Native	71 (58*)	77
Introduced, established	12	26

* If bats excluded.
Data from Crawley (1986).

open banks and shingle by lowland rivers had a high percentage of aliens, whereas upland screes and rock ledges had none, perhaps because of lack of suitable seed. Invasion of islands, especially remote islands with many endemic species such as Hawaii and New Zealand, has aroused much interest. The arrival of competitor or predator species has often led to the extinction of endemics (Elton 1958). But it is still not clear that islands are more prone to invasion than mainland areas.

There seem to be few reliable generalizations about why species succeed or fail in becoming established in a new area. It may be that there are many different reasons why a potential invasion fails, some of them subtle and not easy to spot in advance. One example is provided by fig trees (Kjellberg & Valdeyron 1990). In this genus, *Ficus*, each species is pollinated by a different species of wasp. In its native range, around the Mediterranean, *F. carica* is efficiently pollinated by its wasp, and this depends on the timing of its flower development and of the wasp's life cycle. When *F. carica* has been introduced into northern France it grows well and flowers, but it fails to set seed because the cooler climate does not allow the wasp to complete its life cycle. Without a detailed study of the intimate relationship between this plant and insect, it would have been impossible to predict that *Ficus carica* would fail to maintain itself in northern France.

Release of genetically engineered organisms

Genetically engineered organisms: some advantages from releasing them into the open

Genetic engineering has been developed to the stage where it is now possible to insert a gene into many species of animal, plant and microorganism. During the early development of the techniques in the 1970s there was much concern about possible dangers in the laboratory from genetic engineering experiments, but today safe methods for growing and handling genetically engineered organisms (GEOs) within physically contained areas are thoroughly developed and widely practised. GEOs, mainly bacteria, are now in use commercially to produce various substances, e.g. for medical treatments and diagnosis, but these species are kept within confined systems, so they are not released into the open environment. There are important potential uses of GEOs that would require releasing them into the open (Fincham & Ravetz 1991, Lindow *et al.* 1989). One example is biological control: it is possible to produce strains of microorganism that are more effective and more specific at control of insects, fungi and weeds. Another example is in pollution control, to produce species that can break down particular toxic organic chemicals or take up heavy metal elements efficiently. Genetic engineering provides opportunities to produce improved strains of crop plants and farm animals (Gasser & Fraley 1989), e.g. with increased disease resistance, and improved bacteria, e.g. for the rumen and for silage. These species will be little use unless they can be released into the open on the farm.

If a new GEO is to be released, we want to know in advance whether it will survive long enough to be useful, but also whether it will spread and have undesired side-effects. These are the questions we have already considered in the previous section, for existing species arriving in a new region. But with GEOs there are additional difficulties.

1 The organism has never previously existed in the open, so we cannot learn from studying it in its native habitat.

2 The novel genetic features of the new GEO are likely to be a single gene or linked genes (e.g. a plasmid), so we need to know whether this could be transferred to other species, with possible undesirable effects. For example, it is possible to produce by genetic engineering crop plants that are tolerant of a particular herbicide. This would allow the herbicide to be used to control weeds without harming the crop. But we need to know whether there is a risk that the gene for herbicide resistance could be passed on to weed species, for example by occasional cross-pollination.

In most developed countries the release of GEOs into the open is strictly controlled and each new GEO is allowed to be released only after the risks have been assessed. This section considers ways of predicting the outcome of releasing a particular GEO, assessing the risks involved and also of reducing those risks.

Predicting whether a strain of bacterium will survive

Whether a bacterial strain will survive can sometimes be usefully studied in *microcosms*. For example, Dwyer *et al.* (1988) put 'synthetic sewage' and sewage sludge into a 2.5-litre container designed to simulate a sewage plant, and inoculated a bacterial strain genetically engineered to be able to decompose a group of toxic organic substances, substituted benzoates. The microcosm provided useful evidence that the bacteria survived and were able to break down the benzoates, but was not able to answer other questions such as whether they might be transported away from a real sewage plant into fresh water. Some guidance on the likely fate of GEOs can be gained from studying what has happened to new strains, produced by conventional means, that have been released in the past. For example, strains of *Rhizobium* have been applied to agricultural land since early in the 20th century to improve nodulation in legumes, and Beringer & Bale (1988) have summarized lessons to be drawn from this. The responses of rhizobia to physical factors such as temperature, soil moisture, pH and salinity have been extensively studied, and provide a basis for predicting whether the physical conditions in a new area will be suitable. The major difficulty is to predict whether a new strain will be able to compete effectively against existing strains. If a new legume species is introduced to an area and there are no indigenous rhizobia that can infect it, then suitable new rhizobia can be introduced without difficulty. But if rhizobia that can infect the legume are already present there is no reliable way to predict whether a new strain introduced will be successful. Strains with low nitrogen-fixing ability can be more competitive than efficient nitrogen-fixers. In north-central U.S.A. a single strain of *Bradyrhizobium japonicum*, which forms nodules in soybean, has become very widespread and prevents the establishment of most other strains that have been tried; the reason for its success is not known. Beringer & Bale comment 'We know so little about the biochemical determinants of competitiveness that we are unable to utilize modern genetic techniques to produce more competitive inoculants.'

A field test of a GEO: Ice$^-$ bacteria

For some GEOs microcosms cannot well reproduce their outdoor environment, but test releases present some risks. An example is the Ice$^-$ strain of *Pseudomonas syringae* (Lindow & Panopoulos 1988). The naturally occurring Ice$^+$ strains of this species promote ice crystal formation in plant tissues and thus increase frost damage. Deletion of a single gene results in an Ice$^-$ strain which does not promote ice formation, but can compete with the Ice$^+$ strain and so may reduce frost damage. A bacterium lacking a natural gene may not sound very dangerous, but one possibility to be considered is that if it multiplied and spread it might increase the frost resistance and hence abundance of undesirable weed species, or alter the relative abundance of wild species. The Ice$^-$ strain was inoculated on to 67 plant species under controlled, contained conditions. It grew on most of them, as well as the Ice$^+$ strain but no better; so there was no indication that it would multiply at the expense of Ice$^+$ in the field. The natural Ice$^+$ was sprayed outdoors to indicate how far Ice$^-$ was likely to spread and hence what width of bare soil would form an effective barrier; the species is known to survive only a short time in plant-free soil. These tests suggested that Ice$^-$ could be safely released outdoors; but they left some uncertainties about the survival and spread of the bacteria outdoors, since the environmental conditions in the growth rooms were found to markedly affect growth and survival of the bacteria on the plants, but growth rooms can never exactly mimic conditions outdoors. Field tests were authorized and carried out in California in 1987. The plot of test plants was surrounded by a bare buffer zone 15–30 m wide. The inoculated bacteria were applied by an aerial spray in some experiments. The number of these bacteria landing in the outer part of the bare zone was low but not zero, about 100 per square metre in one experiment. Spraying was carried out only when wind speeds were very low; to investigate spread of the bacteria under a variety of weather conditions would require a lot more research. It is probably more useful to know about survival of the bacteria than about their spread. Figure 6.1 shows that the Ice$^-$ strain survived on potato for several weeks after application, but not through the whole season. This makes it unlikely that it would multiply and spread on other, non-target plant species, though it does not entirely rule out that possibility. If this limited survival proves to be true for other target plant species also, the Ice$^-$ bacteria would need to be applied every season, or perhaps more than once per season. Such organisms

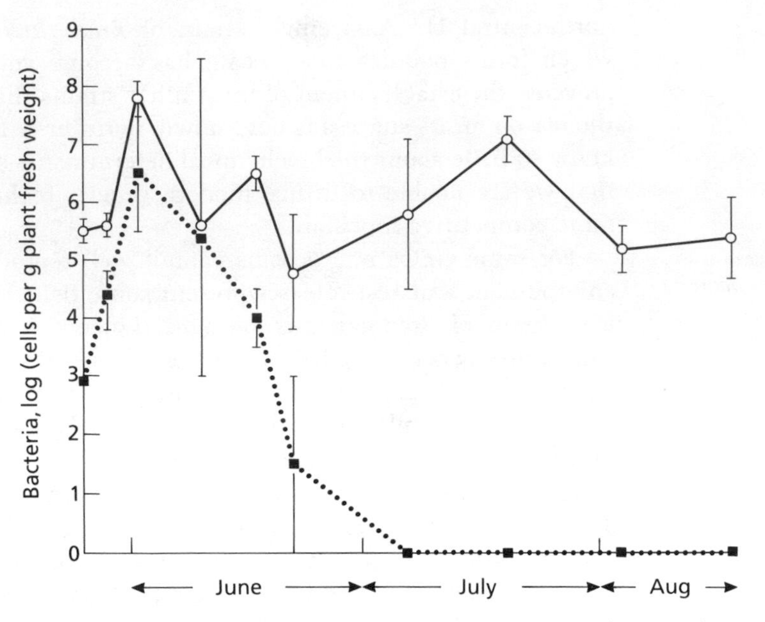

Fig. 6.1 Abundance of bacteria on the above-ground parts of potato plants in a field trial in California. ○, all bacterial species; ■, Ice⁻ strain of *Pseudomonas syringae* that had been sprayed on to the plants on 28 May. Vertical bars show standard errors. From Lindow & Panopoulos (1988).

are inherently safer than those which, once released, are expected to multiply and persist indefinitely.

A major concern in the release of genetically engineered bacteria arises because DNA can be transferred naturally between bacterial species, even species that are not closely related. For example, it has been known for some time (Sykes & Richmond 1970) that genes for antibiotic resistance can be transferred between bacterial species that are not at all closely related. This has presumably contributed to the build-up of populations of antibiotic-resistant bacteria, e.g. in hospitals. It is clearly important to know what controls the frequency of such gene transfers between species. Transfer of plasmids by conjugation is a common way for genes to pass between unrelated bacteria. Because of this, there is at present great reluctance among government regulatory committees to authorize release of bacteria carrying plasmids that have been inserted or altered by genetic engineering.

Environmental conditions favouring gene transfer

Research using microcosms has shown that environmental conditions can influence the frequency of genetic exchange between bacteria. Table 6.3 shows frequencies of transfer between two *Bacillus* spp. of a plasmid carrying resistance to the antibiotic tetracycline, in soil altered in various ways. None of the frequencies was high. In unamended soil no transfers were detected; transfers were increased by addition of nutrients which would support bacterial growth, by bentonite clay and particularly by the two together. The soil was a

Table 6.3 Frequency of transfer of a plasmid between *Bacillus cereus* and *B. subtillis* in soil during 2 days

	Transfers per 10^7 cells
Unsterilized soil	0
Unsterilized soil + bentonite clay	0.8
Sterilized soil*	0
Sterilized soil* + nutrients†	0.5
Sterilized soil* + nutrients† + bentonite clay	5

* Soil sterilized before inoculation with *Bacillus*.
† The nutrients provided soluble organic energy substrate + amino acids + mineral salts.
Data from Van Elsas *et al.* (1987).

loamy sand, and the clay added probably provided particles suitable for bacteria to adhere to. Adding the nutrients increased the number of *Bacillus* cells, and sterilizing the soil before introducing the *Bacillus* did too, by removing competitor bacteria. These results illustrate how the frequency of plasmid transfer is likely to be strongly influenced by the abundance of donor and receiver cells, and hence how often they encounter each other; and how a type of microsite where both are favoured (clay surfaces in this example) can also increase transfer frequency, presumably again by increasing the frequency of encounters. Plasmid transfer occurs only between cells that are in a metabolically active state, whereas many bacteria in soil, fresh water or the oceans are starved of organic nutrients much of the time and so are metabolically inactive. Thus in soil plasmid transfer may be concentrated in microsites such as the surface of organic particles or the rhizosphere. One of the few demonstrations of plasmid transfer in soil not sterilized or amended with nutrients was between *Klebsiella* bacteria in soil close to the roots of radish plants (Talbot *et al.* 1980). Sewage plants and animal guts are other nutrient-rich environments that may be sites for plasmid transfer.

Gene transfer can also occur by *transformation*, meaning that DNA is extruded by a living cell or released when the cell is damaged, and the free DNA is later taken up by another cell and incorporated into its genome. This depends on the survival of intact free DNA in the environment. DNA is a potential source of nitrogen, phosphorus and organic substrate to plants and heterotrophs, and enzymes to break it down (DNAases) are widespread. So the question arises how DNA can remain intact long enough for transformation to occur. Among the suggestions (Stewart & Carlson 1986) are that in soil DNA interacts with lignins in humus or is adsorbed to surfaces of mineral particles, and that these reduce its susceptibility to attack by DNAase activity. Evidence on this is limited.

These examples show that the risk of releasing a GEO into the open varies between species, and is affected by how the new genetic material is carried within the GEO and also by environmental factors. Taking such information into account, release of some GEOs has been considered sufficiently low-risk to be authorized, though so far mainly for experimental purposes rather than general use.

Rate of spread of an invader

Rate of spread by diffusion

If a small drop of coloured dye is placed in the centre of a large dish of agar, it will spread outwards slowly by diffusion. The radius (r) of the coloured circle at any time will be given approximately by

$$r = \sqrt{2Dt} \tag{6.1}$$

where t is time since the drop was put on, and D is the diffusion coefficient, which depends on properties of the dye and the agar. So the boundary advances more slowly as time goes on: if it advances 1 cm in the first day it will take 4 days to reach 2 cm and 9 days to reach 3 cm. Suppose a population of animals is introduced into a new area and they each move about randomly ('random walk'), but they do not multiply; then the outer boundary of the area they are found in would be expected to advance at a rate indicated by equation 6.1, in other words their spread should get slower and slower (if measured in kilometres, not in square kilometres). But supposing they do multiply? Skellam (1951) presented a mathematical model of a population moving randomly and increasing in numbers, which predicted that the radial extent of the area occupied would be

$$r = a\gamma t \tag{6.2}$$

Rate of spread of animal species

where a is related to the rate at which individual animals disperse, γ is related to the rate at which the number of animals increases, and t is time. This very simple model does not allow for the fact that dispersal and reproduction may vary with age of the individual: species of mammal and bird that have territories usually disperse when juvenile but then remain in their territory. The model does not include any density-dependent control of population increase, so the population would go on increasing exponentially however many animals there were. This is unlikely to be true indefinitely, but could be approximately true while the species continued to increase its area. As long as it is true, equation 6.2 predicts that the rate at which the species' boundaries extend will be constant.

The spread of many species has been monitored during the last 100 yrs. One example, shown in Fig. 6.2, is the spread of muskrat (a native of North America) in central Europe from five individuals introduced south of Prague in 1905. In Fig. 6.2(b) the linear rate of spread has been averaged by using $\sqrt{\text{area}}$; it is virtually constant.

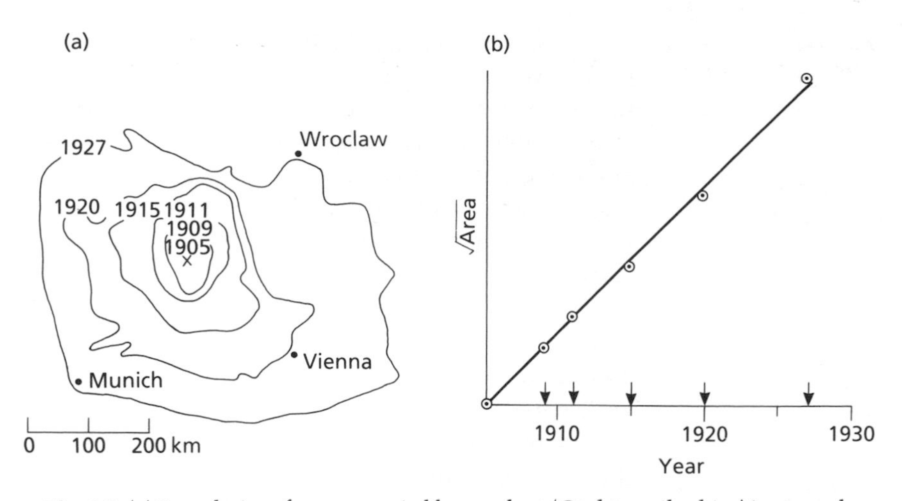

Fig. 6.2 (a) Boundaries of area occupied by muskrat (*Ondatra zibethica*) in central Europe. (b) Square-root of the area enclosed by the boundaries shown in (a). From Skellam (1951).

(After 1930 the rate of spread was slower (van den Bosch *et al.* 1992), probably because of extensive trapping to control muskrat's abundance.) Elton's (1958) book on invasions contains maps showing the spread, from initial starting points, of example mammal, bird and insect species. Elton did not analyse his examples in a quantitative way, but many of the same examples, and others, are analysed in more detail by Hengeveld (1989), who shows that many of them do show approximately constant rates of spread. There have been exceptions, however. For example, three red deer (one male and two females) were introduced near the north end of the South Island of New Zealand in 1861 (Clarke 1971). For about 30 years the population increased in numbers but scarcely spread; but then they started to spread at several kilometres per year.

This fairly steady rate of spread shown by many species allows us to predict, for a species already established and spreading, when it is likely to reach a particular place. It does not allow us to predict how fast a species not yet present in a region would spread if it became established. Van den Bosch *et al.* (1992) proposed a model which allows this to be done. The input data needed are: (1) mortality in relation to age; (2) number of offspring produced in relation to age of parents; and (3) distance individuals disperse as shown by mark and recapture (e.g. bird ringing). Van den Bosch *et al.* were able to obtain such data for muskrat and four bird species; some of the data were from the native range of the species, some from the region in which it was spreading. The predicted rate of range extension for each species agreed fairly well with the actual rate; the least good was for collared dove expanding westward in Europe, where the predicted rate was 50% higher than observed.

Rates of plant spread after the Ice Age

So far we have considered only the rate of spread of animals. For plant species we have information on the rates of spread during several thousand years at the end of the Ice Age, as the climate warmed. Pollen sequences, dated by radiocarbon (see Box 2.6), have been studied at enough sites in eastern North America and western Europe to allow maps of spread of individual species to be made. Figure 6.3 shows two examples for North American tree species. These species were responding to climatic warming, but white pine did not spread in exactly the same direction as the oaks. Table 6.4 shows mean rates of spread for members of major tree genera in Europe and North America. Two things are particularly surprising. Firstly, one might expect species with winged, wind-dispersed seeds to spread more rapidly than those with larger seeds which have no obvious means of dispersal, but this was not consistently so. Secondly, the rates are surprisingly fast, hundreds of metres or even 1–2 km yr^{-1}. If you imagine an oak forest spreading as a massed front, and you take into account its large seed, you might guess at a rate of spread of a few metres per year. Obtaining these rates from pollen data is not entirely straightforward, and Delcourt & Delcourt (1991) have suggested that oak and beech in North America spread at 0.1–0.2 km yr^{-1}, slower than the rates given in Table 6.4.

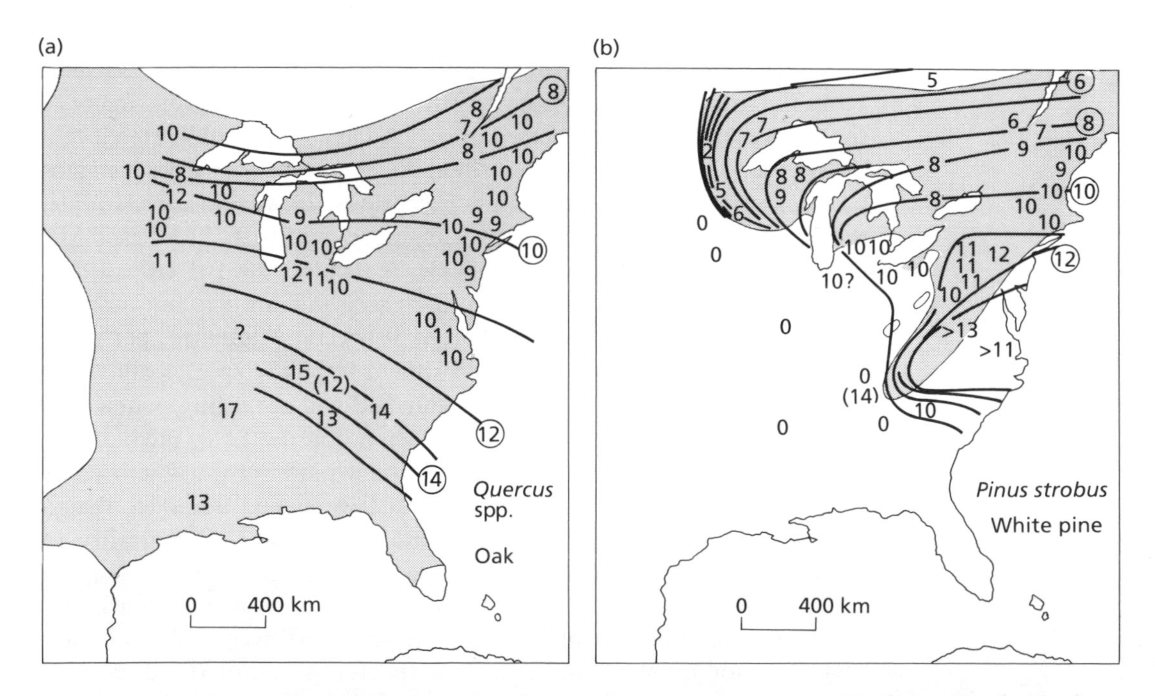

Fig. 6.3 Change in distribution of tree species in eastern North America during the last 17 000 yrs. (a) Oak (*Quercus* spp.); (b) white pine (*Pinus strobus*). Figures show the date of first arrival (in thousands of years BP); the figures within circles apply to the thick lines, which approximately join points of equal arrival date; uncircled figures apply to individual sites. 0, absent. Shaded areas are the present distribution. From Davis (1981).

Table 6.4 Rates of spread of tree genera during the post-glacial period

		Europe (mainland)	North America		
	Type of seed	Rate of spread (km yr^{-1})	Rate of spread (km yr^{-1})	Age (yr) at first seed production	Rate of spread (km per generation)
Birch (*Betula*)	Winged	>2			
Pine (*Pinus*)	Winged	1.5	0.3–0.4	3–5	1.2–1.7
Hazel (*Corylus*)	Large, hard coat	1.5			
Oak (*Quercus*)	Large, hard coat	0.15–0.5	0.35	20	7
Elm (*Ulmus*)	Winged	0.5–1.0	0.25	15	4
Spruce (*Picea*)	Winged		0.25	4	1
Beech (*Fagus*)	Large, hard coat	0.2–0.3	0.2	40	8
Chestnut (*Castanea*)	Large, hard coat	—	0.1	12	1.2

Data from Davis (1981), Webb (1986), Birks (1989).

However, that still leaves rates much higher than simple observation of present trees would suggest. Most tree species do not produce seed in their first year, so it is more informative to work out the average distance each species must have spread per 'generation', i.e. per length of time from seed germination to seed production by the resulting tree. Table 6.4 shows such calculations for North American trees. These indicate that seeds must have dispersed several kilometres; the longest distances are for the large-seeded oak and beech.

Long-distance dispersal

The spread was not necessarily as a massed front; individual seeds could have been carried long distances and have formed isolated colonies, followed by slower spread from these nuclei into the intervening areas. These contrasting modes of spread are sometimes called 'jump dispersal' and 'diffusion dispersal'. One possible means of long-distance dispersal of large seeds is in rivers. However, in eastern North America there are no major river systems that flow northwards, the principal direction of the rapid post-glacial spread. It is likely that hazel invaded Britain by seeds carried by ocean currents, since the pollen records show that it first occurred along the west coast of Wales, northern England and Scotland but along the east coast of Ireland, and then spread inland (Birks 1989). Its seeds can germinate after floating in sea water for some days. Another possible agent of long-distance dispersal is large animals. Johnson & Webb (1989) considered possible mammals and birds of North America, and pointed to bluejays as the only species known to carry large seeds more than a few hundred metres and then cache them in the ground. The jays carry several thousand seeds per bird per year to form caches in their nesting territory, which may be some distance from the seed-bearing trees. The nesting caches can be in forest edges, young woody vegetation or even grassland. The jays carry beech, oak and chestnut, and have been observed to carry beech seeds up to 4 km and acorns up to 2 km. Neither of these distances is as great as the observed rate of spread per generation given for the species in Table 6.4, but they are the right order of magnitude. Birds of another genus, nutcrackers, have been observed to carry pine seeds up to 22 km in North America and up to 15 km in Europe (Vander Wall & Balda 1977). Thus it seems quite possible that birds could play a major part in rates of spread as fast as those shown in Table 6.4.

A clear example of jump dispersal during historic times is the invasion of the American west by the grass *Bromus tectorum*, one of the European annual species that have become very abundant in rangeland formerly dominated by perennial grasses and shrubs. *B. tectorum* was first recorded in the west in 1889, and for about 30 yrs after that was present at widely separated sites, which increased in number and size (Mack 1981, 1986). By 1930 much of the area between these sites had been filled in, giving large continuous ranges. People were probably a major dispersal agent for the seeds during this time, for example in contaminated crop seed.

One might perhaps expect that birds would often spread by jump dispersal, but well-recorded examples show a very steady wave spread; e.g. the European starling from an initial population introduced to Central Park, New York (Elton 1958), and the westward spread of the collared dove across Europe (Hengeveld 1989).

Disease spread

Diseases can extend their range at a fairly steady rate, too; for example the spread of chestnut blight through the forests of U.S.A. during the first half of the 20th century, and the spread of rabies in foxes from an outbreak in Poland in 1939 (Newhouse 1990, Anderson 1982). But the science called epidemiology is much concerned not with the area occupied by a disease, but with what proportion of a particular population of animals or plants is infected. The time-course of percentage infection usually follows an S-shape, levelling off as all the population, or all the susceptible members of it, become infected. The spread of Dutch elm disease among elms in areas of Britain, following the arrival of a virulent strain of the fungus *Ophiostoma ulmi*, is one example: Fig. 6.4 shows the early and middle parts of an S-shaped curve. These time-courses often fit fairly closely to the logistic equation

$$\frac{dy}{dt} = \beta y(1 - y) \tag{6.3}$$

where y is the proportion of the population infected, β is a constant measuring the pathogen's ability to spread under these conditions, and t is time. Much attention has been given to more sophisticated and hopefully more informative equations for fitting disease progress curves (Leonard & Fry 1989, Gilligan 1990). When applied to a crop, such curves can predict how much of it will be infected by the time of harvest, and hence whether control measures are justified.

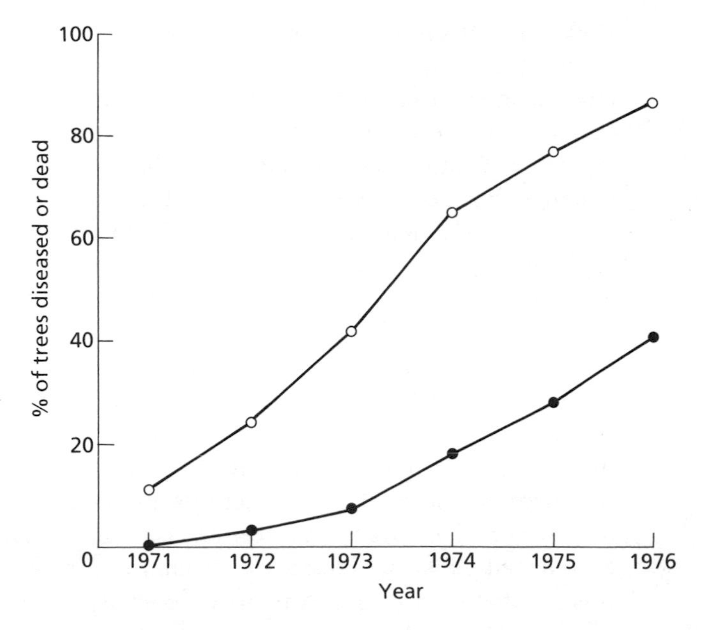

Fig. 6.4 Progress of Dutch elm disease in two regions of England. The disease is caused by a fungus *Ophiostoma ulmi*, which is spread from tree to tree by bark beetles. ○, south-eastern England; ●, East Anglia. From data of Gibbs (1978).

Pests: their abundance and their effects

In this chapter I consider under the term 'pests' all parasitic species—viruses, bacteria, fungi and animals—that cause diseases in plants and animals, and also insects that eat enough of their food plants to be considered as pests. We are now concerned not merely with whether they maintain themselves and how fast they spread, but also with their abundance in the host population, whether they show outbursts from time to time, how much effect they have on the host population. The aim is to use this understanding as a basis for planning control measures. Let us first consider some examples of diseases and pest insects in natural and semi-natural ecosystems—in other words attacking plants and animals that are not in farms or forestry plantations—to see what we can learn from them about what influences the abundance and spread of the attacker species and the damage they do to their hosts.

Diseases and pest insects in quasi-natural ecosystems

A fungal epidemic of native forest: chestnut blight

The first message is that epidemics *can* occur in near-natural ecosystems. An example is chestnut blight in North America, which is caused by the fungus *Cryphonectria parasitica* (Van Alfen 1982, Buck 1988, Newhouse 1990). The fungus probably reached U.S.A. in young chestnut trees imported from Asia. It was first noticed in 1904 in New

York, and within the next 50 yrs had spread through the whole natural range of the native chestnut (*Castanea dentata*) in eastern U.S.A. Once a major tree of mixed deciduous forests in the Appalachians, the chestnut now occurs there only as regrowth coppice shoots from surviving stumps. These shoots rarely live long enough to set seed, so whether the species will survive indefinitely in the area is uncertain. The fungus also reached Mediterranean Europe, where it killed many of the native *Castanea sativa*. There, however, after about 15 yrs less damaging 'hypovirulent' strains of the fungus began to appear, which caused some symptoms but did not kill the tree. The character for hypovirulence was shown to be carried in the cytoplasm (not the nucleus) of the fungus in double-stranded RNA similar to a virus, which could transfer to virulent strains, reducing their virulence. In this way, hypovirulence spread in Italy until after about 25 yrs the disease was no longer a serious problem. Hypovirulent strains have occurred naturally at a few sites in U.S.A., and spread has been encouraged as a possible means of controlling the disease. Although this has helped tree survival in some areas, e.g. Michigan (Fulbright *et al.* 1983), in the main eastern native range of *Castanea dentata* hypovirulence has not spread or had a major effect. The reason for this difference between Europe and U.S.A. is not clear.

A native pathogen in tropical forest

An example of severe plant mortality caused by a disease that was not introduced by people is provided by Augspurger's (1983) study of death of seedlings of the tree *Platypodium elegans* in tropical forest on Barro Colorado Island, Panama. There is a wet season, starting in May, and in 1980 germination of *P. elegans* was concentrated in the last 2 weeks of May. During the following 3 months 82% of the seedlings died and 87% of those deaths were due to damping-off. The fungus or fungi involved were not identified, but there is no serious doubt that fungi were the cause. Mortality decreased with increasing distance from the parent tree (Fig. 6.5). The mortality over 3 months was closely correlated with the rate of spread of the fungus between seedlings during their first 6 weeks (Fig. 6.5); by the age of 3 months the stems became more woody and deaths from damping-off were rare after that. The seedlings were less abundant further from the parent tree (see numbers above bars in Fig. 6.5), so the reduced mortality could be due either to reduced infection originating from the parent tree or to slower spread of the fungus caused by lower density of seedlings. In two experiments Augspurger & Kelly (1984) varied distance from parent tree, density of seedlings and shade independently of each other, and showed that all three had a significant influence on death of *Platypodium elegans* seedlings by damping-off; lower density also reduced damping-off in seedlings of several other tree species.

A fungus of wide host range

Chestnut blight provided an example of a disease caused by a parasite which can attack plants only within a single genus (*Castanea*). In contrast, some other plant pathogens have a wide host range; one

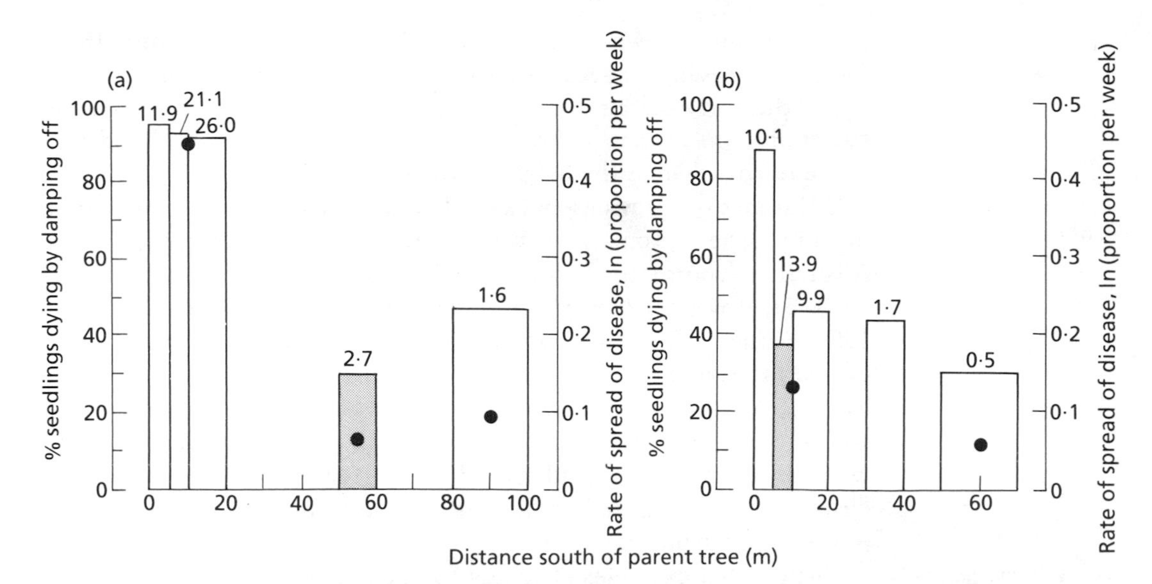

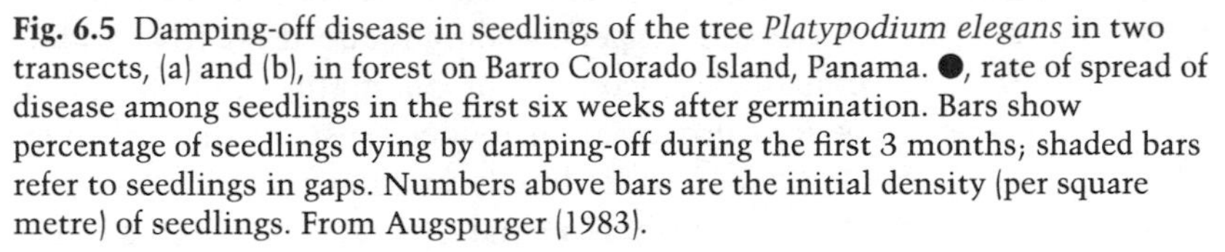
Fig. 6.5 Damping-off disease in seedlings of the tree *Platypodium elegans* in two transects, (a) and (b), in forest on Barro Colorado Island, Panama. ●, rate of spread of disease among seedlings in the first six weeks after germination. Bars show percentage of seedlings dying by damping-off during the first 3 months; shaded bars refer to seedlings in gaps. Numbers above bars are the initial density (per square metre) of seedlings. From Augspurger (1983).

example is *Phytophthora cinnamomi*. This fungus infects the roots of many woody plant species, causing root rot, which can be followed by die-back of branches and ultimately death of the whole plant. The fungus is confined to soil and roots, and has no airborne spores, so it normally spreads slowly in water or with eroding soil. Yet it occurs in every continent except Antarctica (Zentmeyer 1985). It attacks commercially important plantation and orchard trees as well as species of natural forest. It has caused conspicuous die-back in eucalypt forests in several parts of Australia; research on this has been reviewed by Weste & Marks (1987). The most detailed published study has been in the Brisbane Ranges in Victoria (Dawson *et al.* 1985, Weste 1986). There the fungus produced major changes in the vegetation: susceptible species decreased greatly in abundance, while *Phytophthora*-resistant species greatly increased. However, some of the very susceptible species did not disappear altogether from study plots: even after 20 yrs a few individuals remained. These might be genetically different individuals with higher tolerance of the fungus. Their survival could be aided by a decrease in abundance of the fungus, which happened in some areas after the main epidemic had passed. This decline in the pathogen could be a response to decreased abundance of susceptible plants; but there is also evidence that in soils in some parts of Australia microbial populations build up which suppress *P. cinnamomi* (Malajczuk 1979). This same fungus is found in apparently undisturbed forests, including tropical rainforest, where it is presumably endemic. The results from the Brisbane Ranges suggest that it may be able to reach some sort of equilibrium in Australian eucalypt forests, with a lower abundance of susceptible hosts than before the epidemic, and a lower abundance of the fungus than at the height of the epidemic.

Myxomatosis in rabbits

Myxomatosis provides an example of an epidemic disease in a wild mammal. The myxoma virus was introduced from South America into Australia to control the European rabbit, and began to spread rapidly in 1950–51. It was introduced into France in 1952 and reached Britain in 1953 (Fenner 1983, Dwyer *et al.* 1990). In each of these countries it caused massive mortality, but in none of them has the rabbit become extinct. At one site in Australia 5000 rabbits were reduced to 50 within 6 weeks of the virus being introduced; the next summer there were 60 (Fenner 1983; Anderson & May 1986, Fig. 1). The host and virus have thus reached some sort of equilibrium, at a much lower host density than before.

A beetle controlling a weed

Many other examples of successful biological control have resulted in the target species being reduced greatly in abundance but not eliminated. The control of Klamath-weed (St John's wort, *Hypericum perforatum*) by an introduced insect is described by Huffaker & Kennett (1959), and more briefly by Harper (1977) and Debach & Rosen (1991). Klamath-weed is a small herbaceous plant, a native of Europe, where it is not considered a problem. When introduced into western U.S.A. and Australia it became a serious weed in pastureland. Few animals will eat it, probably because of a polyphenolic, hypericin, which it contains. The species was successfully controlled in California: three beetle species known to feed on it elsewhere were introduced, but most of the reduction was caused by one of them, *Chrysolina quadrigemina*. Soon after the beetles arrived in an area the Klamath-weed population was almost wiped out, followed by a rapid decline in the beetle numbers (Fig. 6.6), but in most areas both species persisted at low abundance.

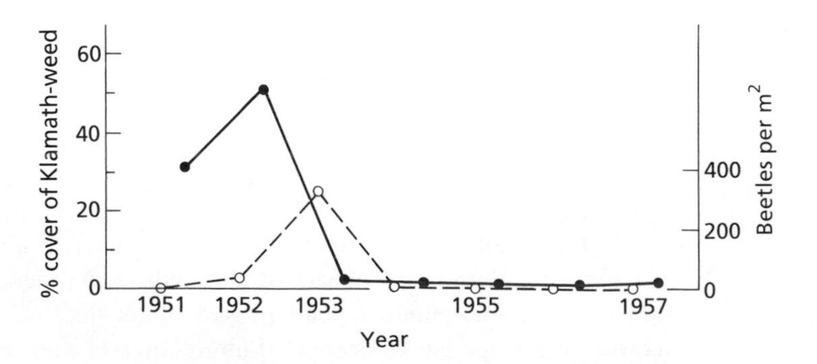

Fig. 6.6 Abundance of Klamath-weed (*Hypericum perforatum*) (●) and *Chrysolina quadrigemina* beetles (○), at a site in California about 1 km from where the beetles had previously been introduced. From Huffaker & Kennett (1959).

The beetles rarely lay eggs in shade, and *Hypericum* is now commoner in shaded habitats than in the open.

Population cycles in a forest insect

There are herbivorous insects and pathogens which show periodic outbursts: after remaining at a low abundance for some time the population increases rapidly to a high and damaging density, then after a while decreases again to a low level. An example is the spruce budworm, *Choristoneura fumiferana*, in parts of Canada (where it is native). The larva feeds on leaves and buds of spruce and fir. When it is at its most abundant the insect causes severe defoliation of trees over large areas of forest, leading to reduced stem growth, and if the outbreak persists for several years to death of trees (Crawley 1983, Speight & Wainhouse 1989). Figure 6.7(a) shows that in an area of New Brunswick

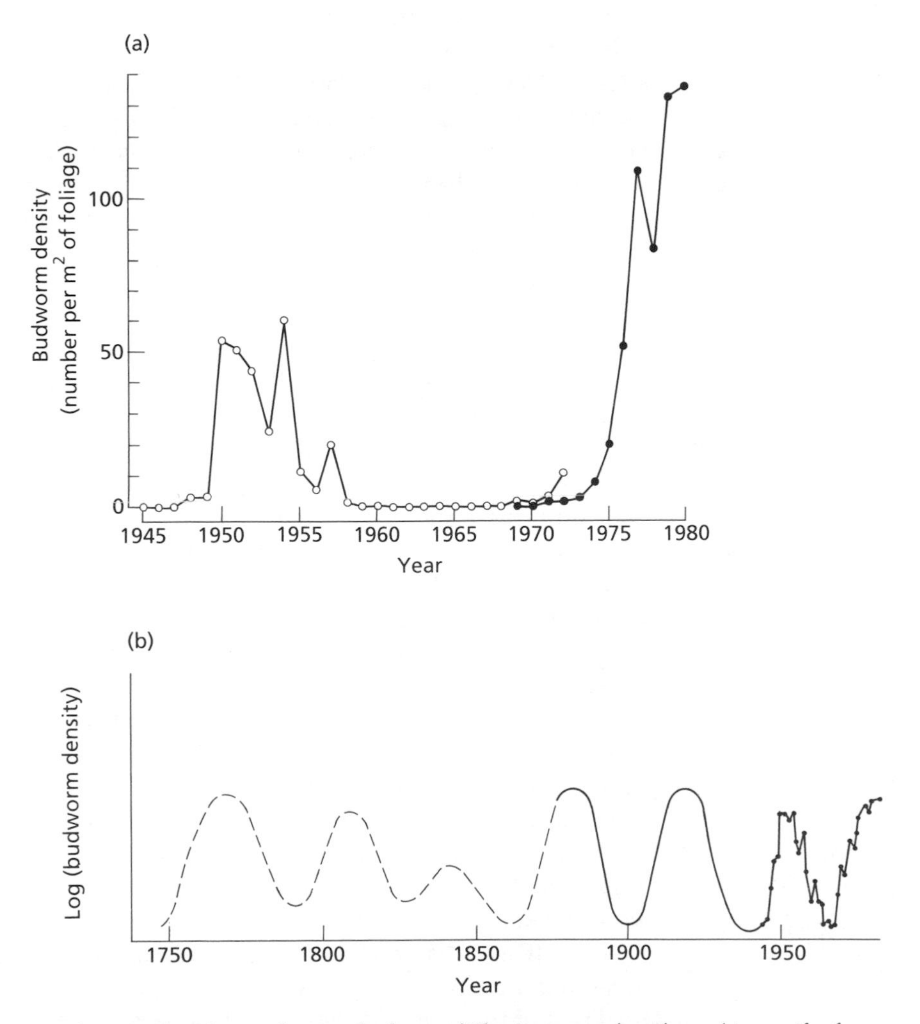

Fig. 6.7 Abundance of spruce budworm (*Choristoneura fumiferana*) in conifer forest in New Brunswick. (a) 1945–80; ○, third- to fourth-instar larvae; ●, eggs.
(b) 1750–1980; ---, estimated from xylem ring widths in trees; —, based on written records; 1945 onwards, from counts *see* part (a). From Royama (1984).

the larvae were abundant during the 1950s, then sparse (though never absent) for a decade, but then increased again in the 1970s. There are written descriptions of two earlier outbreaks, about 1880 and 1912–20. Before that other outbreaks are indicated by changes in tree-ring width. Figure 6.7(b) shows that according to these estimates outbursts of the insect have occurred at fairly regular intervals of about 35 years. Chapter 4 explained how cycles in animal numbers can be caused by interactions between species at two or three trophic levels, even if the physical environment remains constant. In spite of extensive research, the cause of the cycles in spruce budworm is uncertain. Contributory factors that have been proposed include: (1) weather; (2) food abundance and quality, influenced by the frequency and age of different tree species and their nutrient status; (3) predation by birds; and (4) parasitoids and microbial diseases (Crawley 1983, Royama 1984). At four sites in New Brunswick, during the years 1951–58 when budworm numbers were high and then falling (Fig. 6.7) the percentage of budworm larvae containing parasitoids was high, usually 20–50% (Royama 1984), so this was probably a major cause of death; but it is unlikely that parasitism on its own could account for the population cycles. Spruce budworm cycles may not have been as frequent or as regular in all parts of its range in Canada as those in Fig. 6.7. Blais (1983) has argued from tree-ring data that in some areas outbreaks were less frequent and less severe in the 19th than the 20th century. This could be because human activities in this century, such as felling or fire control, have favoured the spruce budworm's preferred food plant, balsam fir.

Speight & Wainhouse (1989) give information on six other species of forest insect that have undergone major outbreaks. Like spruce budworm, these species live in temperate or boreal forests where there are only one or a few abundant tree species. However, outbreaks can also occur in much more species-rich communities, including tropical forest. Wong *et al.* (1990) describe conspicuous defoliation in 1985 of one tree species, *Quararibea asterolepis* in Barro Colorado forest, Panama, by larvae of two moth species, *Eulepidotis* spp. Among 460 *Q. asterolepis* trees in a 50-ha plot, defoliation ranged from 0% to 100%, with about 100 of them suffering more than 90% defoliation. *Eulepidotis* larvae had not been abundant in the previous year, nor were they in several years following the outbreak.

Spread of a parasite from deer to moose

A parasite can be ecologically important even if it is carried by a species that is little harmed by it. An example is the meningeal worm *Parelaphostrongylus tenuis* in white-tailed deer in eastern North America (Davidson *et al.* 1981). The parasite has a complex life cycle involving slugs or snails as an alternate host. The larvae end up in the brain of the deer, but they do not cause any marked symptoms and are not a cause of death. Often over 50% of deer in a population are infected. During the 20th century the range of white-tailed deer has extended. Where they have moved into the ranges of moose, the moose have become infected with *P. tenuis*; in them it is a harmful parasite, often causing paralysis and death. The worms seem not to spread among moose in the absence of the deer, but have almost certainly been the cause of decline or disappearance of moose from areas into which the white-tailed deer has spread.

Some messages for pest control

These examples of pests in natural and semi-natural ecosystems have been purposely chosen to range widely among different types of host and attacker, different sorts of ecosystem and areas of the world. It is now time to consider whether any key messages or questions emerge. One message is that diversity is in itself no protection: the examples show that pests can have a major effect in species-rich temperate forest of the Appalachians and tropical forest in Panama, as well as in less species-rich communities. Another important lesson is that often the host species was not eliminated; it might be reduced to much lower density (e.g. rabbit, Klamath-weed) or suffer occasional major damage (e.g. the attacks by spruce budworm and by *Eulepidotis* moths), but the host–attacker system persisted. Two examples involved local extinction of a species: moose due to the meningeal worm, and some plant species due to *Phytophthora* root rot. In both these examples the parasite had an alternative host on which it could survive—deer for the worm, less susceptible plants for the fungus. These examples point to several questions that need to be considered when planning biological control.

1 Are we aiming for complete extinction of the pest or only to reduce its abundance markedly? The answer to this question may affect our choice of control species.

2 A control species that allowed occasional massive outbreaks of the pest would not be satisfactory. Can we predict in advance which potential control species would lead to this?

3 What are the advantages and disadvantages of a control species with high specificity?

Study of natural communities may also suggest to us ways of reducing damage by pests to farm and forest species. Plants and animals survive successfully in nature in spite of their pests and diseases, so can we arrange for this to happen in farms and plantations, too? When the attacked species survives, what allows this to happen? Why does not the pest wipe it out? The answers to these last two questions fall into three classes, which are summarized in Box 6.1. Examples of all three mechanisms occur among the host–pest relationships described in the previous few pages. Low density of the host (mechanism 3) was important in the survival of damping-off by some seedlings of *Platypodium* in tropical forest, and allowed rabbits and Klamath-weed to reach a stable relationship with the myxoma virus or *Chrysolina* beetles after successful biological control. It is difficult to determine the relative importance of these three types of control mechanism in natural ecosystems. Often they act together. For example, the continued

Box 6.1 Mechanisms by which a host species can coexist with a pest species in natural ecosystems.

1 Resistance characters in the host:
(a) Immune systems in vertebrates.
(b) Toxic chemicals in plants and invertebrates.
(c) Physical barriers, e.g. insect cuticle.
(d) Low resource quality, e.g. low protein concentration.

2 The pest species is itself attacked by a parasite or predator, which controls its abundance.

3 Low abundance and hence wide spacing of the host species interferes with spread of the pest, sufficiently to keep its abundance low. This only works if the pest is specific to the host, so there are no alternative hosts to support it.

existence of Klamath-weed in California requires not only its low abundance, to reduce spread of *Chrysolina*, but the polyphenolic in its tissues which prevents most other herbivores from eating it. Sometimes two mechanisms can be in conflict: for example, if plants of a particular species are widely spaced this may reduce the proportion that are found by a herbivorous insect, but it could also reduce the percentage of those insects that are found by a parasitoid.

Mathematical models of host–parasite population dynamics

These examples of parasites and herbivorous insects in quasi-natural ecosystems have led to useful questions and suggestions, but they have also indicated how difficult it is to understand fully what is going on. One way forward is to apply mathematical modelling. I start by outlining briefly a model devised by Anderson and May which has substantially advanced our understanding of the relationship between animals and their parasites (Box 6.2). Equation 6.3, given previously, predicted that the rate of spread of a disease through a population is dependent on the proportion of infected host individuals and of uninfected individuals, and on a 'transmission coefficient' β. The basic model of Anderson & May (1979, 1986) modifies this by allowing that infected individuals may recover and become immune. Thus at any time the total population of N host individuals consists of S uninfected susceptibles, I infected and M immune. The rate of transmission of the infection through the population is βSI. A crucial difference between Anderson & May's model for animals and the sort of disease spread typified by Fig. 6.4 is that their model allowed for birth and death of host animals. The key equations are set out in Box 6.2. According to equation 6.4, in the absence of disease the population could grow exponentially without limit: clearly this is not realistic. Later models did include density-dependence at this point. The presence of disease imposes an additional death rate on the infected individuals; see equation 6.5. The number of infected individuals *increases* by spread

Box 6.2 The Anderson–May model of the population biology of infectious diseases of mammals.

This box summarizes only a few of the basic points of the model. The letters are defined below. From Anderson & May (1979, 1986).

In the absence of disease the rate of change in host population size is

$$\frac{dN}{dt} = (a - b)N \qquad (6.4)$$

In the presence of disease

$$\frac{dN}{dt} = (a - b)N - \theta I \qquad (6.5)$$

The rate of increase in the number of infected individuals is

$$\frac{dI}{dt} = \beta SI - cI = I(\beta S - c) \qquad (6.6)$$

where:

$$c = b + \theta + \nu \qquad (6.7)$$

The disease will maintain itself when

$dI/dt \geqslant 0$

i.e. when

$\beta S \geqslant c$

i.e. when

$S \geqslant c/\beta$

Therefore, the critical number of susceptible host individuals above which disease will establish, is

$$S_T = c/\beta = (b + \theta + \nu)/\beta \qquad (6.8)$$

Where the host–parasite population reaches stability the host population size is

$$N^* = \frac{\theta c}{\beta[\theta - (a - b)(1 + \nu/b)]} \qquad (6.9)$$

Meaning of letters

a = birth rate
b = death rate from non-disease causes
c = rate at which infected individuals cease to be infected, by recovery or death
I = number of infected host individuals
M = number of immune host individuals
N = total number of individuals in the host population; $N = S + I + M$
N^* = total number of host individuals in a stable host–parasite relationship
S = number of uninfected, susceptible host individuals
S_T = critical number of susceptible individuals, below which the disease fails to persist
t = time
θ = death rate of infected individuals caused by disease (= virulence)
β = transmission coefficient of disease in the population
ν = rate at which infected individuals recover and become immune.

of the parasite to new individuals, but it *decreases* by death of infected individuals and by recovery of others; it is the balance between these increases and decreases that determines whether the parasite survives in the population. Equations 6.6 and 6.7 set this out, and lead on, by simple algebra, to show that there is a critical population size of uninfected susceptible individuals, S_T, below which the parasite will not survive, above which it will survive; this is defined by equation 6.8. S_T can alternatively be expressed as a critical population *density*, if N, I and S in the preceding equations are densities, i.e. numbers of individuals per unit area. This idea of a critical population density of the host, below which the disease will die out, is a very important one. Equation 6.8 shows that a parasite can maintain itself in a sparser population if it is more effective at spreading (larger β), which is perhaps not surprising. Less obvious, perhaps, is that a disease that kills its host more quickly (higher virulence, larger θ) needs a denser population to survive: the quicker the animals die, the fewer infectives there are to spread the disease. This may be the reason why the meningeal worm fails to maintain itself in moose populations unless white-tailed deer are present (see above): the worm is too effective at killing the moose.

The critical population density of the host, below which the disease cannot establish

When a parasite reaches a potential host population for the first time all the animals will be uninfected, so equation 6.8 defines the density of the population above which the disease will establish. In a population in which the disease is endemic the total population density of the host will be higher, because it has infected and immune individuals in addition to at least S_T susceptibles. The Anderson–May model predicts conditions under which the host and parasite populations will reach a stable equilibrium, i.e. the disease is endemic, with a stable host density defined by equation 6.9. I return to this rather complicated equation later.

The host species and the disease do not necessarily reach a stable equilibrium. An alternative possibility is that so many individuals die or become immune that the number of susceptibles falls below the critical level (equation 6.8), so that the disease starts to decrease (equation 6.6). The disease could then disappear altogether, to perhaps reinvade from a neighbouring area at a later date, after the disease-free host population had increased again; or the declining parasite abundance might allow the host to recover and reach S_T before the parasite had completely died out. In either case, what we have is cycles of host and disease abundance, i.e. repeated epidemics.

The Anderson–May model, as presented here, applies to parasites that are directly transmitted from one living host animal to another. It needs to be modified if the parasite can survive for a long time away from the host (e.g. as eggs), if there is a vector or alternate host within which the parasite multiplies, or if there is direct 'vertical transmission' from a mammalian mother to the unborn foetus.

Application of the model to rabies in foxes

The model can be made clearer, and its importance illustrated, by applying it to a real example, rabies in foxes in Europe (Anderson *et al.* 1981, Anderson 1982). The disease is caused by a virus, which spreads between animals by direct contact. It occurs in fox populations in much of central Europe. The primary interest in rabies arises because it can be spread to humans, usually via domestic dogs and cats. In humans the disease has very unpleasant symptoms and is usually fatal. The model of Box 6.2 was modified in a few ways so that it more closely applies to foxes. There is no recovered-immune class because the disease is almost always fatal in foxes; there is, however, a latent period while a fox is infected but not yet infectious. The model was also modified to make the disease-free population's rate of increase density-dependent, so that in the absence of disease it will reach a plateau density. From research on foxes and rabies it is possible to derive values for the key variables in the model. Figure 6.8 shows predictions by the model of the percentage of foxes infected and the effect of the disease on fox numbers, for a range of initial (pre-rabies) fox densities. Since most foxes at any time are uninfected susceptibles, the model predicts that for any initial (disease-free) fox density within the range 1–9 per km^2, rabies will maintain the population near the critical density (S_T), which is predicted to be about 1 per km^2. Above a disease-free density of about 9 per km^2 the model predicts cycles: rabies reduces the fox population, rabies then becomes less prevalent, so the fox population increases again, followed by another outbreak of

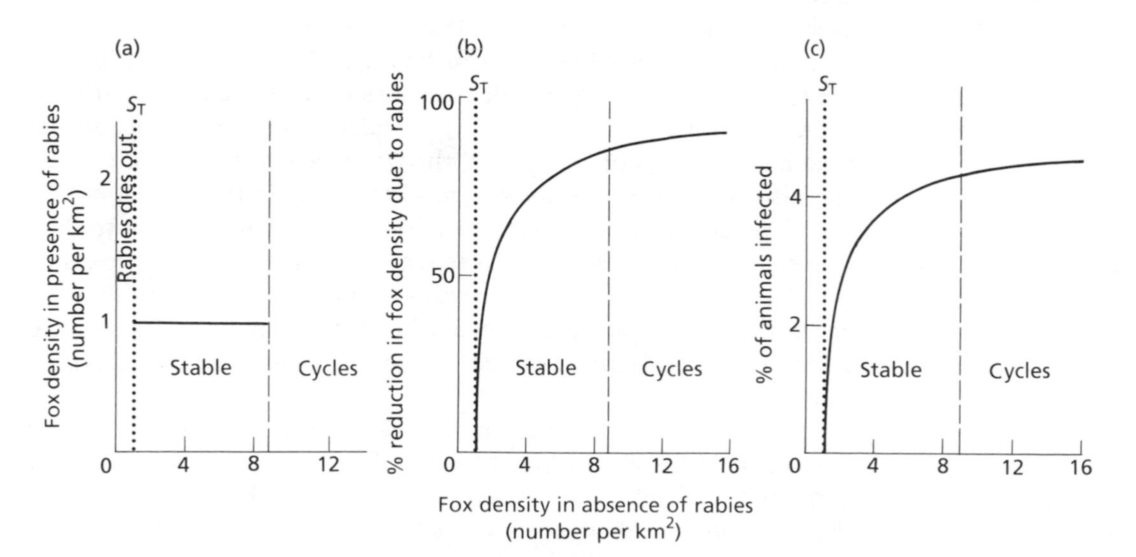

Fig. 6.8 Predictions by model of Anderson *et al.* (1981) of population dynamics of foxes and rabies. (a) Abundance of foxes when rabies present; (b) percentage reduction in fox abundance caused by rabies; (c) percentage of foxes infected at any time. S_T is the critical fox density, below which rabies does not persist.

rabies. Outbursts of rabies and consequent cycles of fox abundance, with peaks every 3–5 yrs, have been recorded in various parts of Europe and North America. Within the range where rabies and fox densities are stable, the proportion of foxes infected at any one time is predicted to be usually only 2–4% (Fig. 6.8(c)), in reasonable agreement with field observations which have shown the figure to be usually 3–7% where the disease is stable and endemic. In spite of this the reduction in the fox population can be large (Fig. 6.8(b)). Rabies shows that a disease can be the most important single factor controlling the population density of its host and yet only a small percentage of the population may be infected at any time. This could make us wonder whether in natural ecosystems there are many species whose abundance is controlled by a parasite, parasitoid, herbivore or predator but we have not guessed because the attacker species is rare.

The critical fox density, S_T, below which rabies will not establish is of interest to people in areas where rabies has not yet arrived, but might arrive in the future, for example Britain. Foxes have become common in some British cities, living especially in gardens; the fox density varies among towns and cities, from zero to more than 4 per km^2 (Harris & Smith 1987). If rabies was introduced to a city, e.g. by illegal importation of an infected domestic animal, whether rabies survived and spread or died out would depend on whether the fox population was above or below the critical density, about 1 per km^2 according to the Anderson model. The model is, however, clearly too simple to be used on its own as a basis for deciding which British cities are at risk and preparing contingency plans for controlling rabies in each one if it arrives. Smith & Harris (1991) have modelled in much more detail the spread of rabies in four actual urban areas in Britain, taking into account how fox density varies locally, and how far individual foxes move. The aim was to predict how rabies would spread from a single point of infection, if there was no control measure taken and also if some of the foxes were killed as a control measure. The model predicted that in order to have a high chance of controlling rabies it would be necessary to reduce the density of adult foxes to 0.3–0.4 per km^2 throughout an area extending about 19 km from the point of infection. Since this would involve killing up to 90% of the foxes, Smith & Harris doubted whether it would be a practicable control strategy.

Host–parasitoid dynamics

Parasitoids are promising for biological control of insect pests, for example because of their high host specificity and hence low danger of harming beneficial species. A key question is whether they can reduce the abundance of their host substantially and then maintain it at a stable low level. If, instead, they result in large cycles of pest abundance that would not be acceptable control of the pest. Beddington *et al.* (1978) discussed various host–parasitoid models and how much reduction in host abundance they predict. This reduction in host abundance can be expressed by q, where

$$q = N^*/K \quad (6.10)$$

where K is the number of host individuals in the absence of the parasitoid, and N^* is the number of host individuals in a stable host–parasitoid relationship. For the simplest model, and several variants of it that they tried, q was never below 0.3. In other words, the parasitoid could not reduce the host population by more than two-thirds if it were to maintain that level stably. The basic reason for this is that stability of the host population was dependent in these models on interactions between host individuals such as competition for food; if the host population became very sparse this density-dependent control became very weak. Host–parasitoid interaction does not involve a density-dependent control of this sort: if the host population grows rapidly it can outstrip its parasitoid population so that the control becomes less effective (at least for a time) just when a more effective control is needed. In four laboratory experiments parasitoids did, as predicted, reduce the host population by about two-thirds, but in six field studies cited by Beddington *et al.* the reduction was much greater, by one or two orders of magnitude. A family of models that predict a much lower stable q are those that introduce patchiness. For example, there may be a 'refuge' within the habitat where the host can escape attack by the parasitoid. It may seem surprising that such a refuge should lead to a lower predicted host population, but that is so. An analogous situation may occur with Klamath-weed and *Chrysolina* beetles (see earlier): the ability of some Klamath-weed plants to escape the beetles by growing in shade may be crucial for allowing the two species to coexist at low abundance. Another possible type of patchiness is if the parasitoid searches for new host individuals in a non-random way, for example searching preferentially in patches of high host density. These and later models of host–parasitoid population dynamics have an important message for biological control: choice of a parasitoid as an effective biological control agent will need to take into account 'patchy' behaviour of the host and the parasitoid.

Models of relations between herbivorous insects and plants

Crawley (1983) presented a family of models of interactions between a plant species and its insect herbivore, which have some basic similarities to the Anderson–May model of Box 6.2. The models provide estimates of the abundance of plant and insect, and whether or not these will be stable. Two of the predictions are similar to those from the fox model (Fig. 6.8): (1) that plant abundance in the presence of the insect is not related to its abundance in the absence of the insect, i.e. if the initial density of the plant is higher then the insect gives a greater percentage reduction of it; but (2) the more abundant the plant when insect-free the more likely that the insect will cause cycles rather than stable populations. Another prediction is that the lowest stable plant population will be achieved when the insects have an intermediate rate of eating and killing each individual plant; if the insect kills the plant very rapidly the insects then become less abundant because many die without finding another plant.

In this section I have summarized some key points of important models, but I have by no means done full justice to them. These models have provided important advances in our fundamental understanding of host–parasite and plant–herbivore relations. So far their importance remains at the strategic level: it has rarely been possible to apply them quantitatively to particular species, so it has not been possible to validate them in a rigorous manner. As we shall see later, their contribution to pest control has usually been at the level of basic understanding rather than in predicting whether a particular species will be an effective biological control agent.

Control of pests

This section considers how ecology can be applied to help in the control of pests. The aim is to reduce the abundance of the pest to an acceptably low level, but at the same time to avoid undesirable side-effects of the control measures such as damage to other, beneficial species.

Box 6.3 summarizes the main methods of control that have been used. I shall say only a little about the first three, and concentrate mainly on management and biological control, since ecology can make the most contribution to those. These five methods need not be seen as alternatives: it is often most effective to use two or more of them in combination, sometimes called integrated pest management.

Chemical pesticides: successes and problems

Chemical pesticides have achieved major successes in the control of some fungal pathogens, insect pests and weeds. Some arable weeds have become so rare that people are now concerned about their preservation (Potts 1991). Kogan (1986, p. 256) gives a list of insect pest species that have been virtually eliminated by use of insecticides. However, there have also been failures. Some pesticides have become ineffective because the pest species evolved resistance to them. Pesticides can damage non-target species, i.e. beneficial species or harmless wild species. The chemical may kill a species that was providing a natural control of the pest: this can happen when a fungicide inhibits a mycoparasite, i.e. a fungus that was attacking the fungus pest (Cook & Baker 1983), or when an insecticide kills a parasitoid of an insect pest. This can also result in previously sparse fungi or insects increasing to pest levels of abundance. An example of this is given by Waage (1989): when parathion was sprayed on apple trees in Pakistan to control a moth pest, one result was a marked reduction of two parasitoid species of a scale insect and a 20–50-fold increase in abundance of the scale. However, insecticide can also enhance the effect of a parasitoid: if the insecticide reduces the insect population before the

Box 6.3 Methods that can be used for controlling pests and diseases.

1 *Chemical treatments*. Many are aimed at killing the pest (fungicides, insecticides, herbicides). Chemicals produced by insects which elicit a response in other insects can sometimes be used; e.g. sex pheromones can be used to disrupt mating or to attract insects to traps.
2 *Immunization* of mammals and birds.
3 *Genetic alterations* to animals and plants to make them more resistant to the disease or more tolerant of it. This can involve:
(a) selecting individuals with desired characteristics;
(b) conventional breeding;
(c) treatments to promote mutation;
(d) genetic engineering.
4 *Management*. Possibilities include:
(a) altering the density of the animal or plant population;
(b) mixtures, strips or patches of different plant species or varieties;
(c) rotation.
5 *Biological control*. Using another species to control the pest. Principal methods:
(a) *Classical*. Introduce the control species once only. It is then expected to multiply and spread.
(b) *Augmentation*. Increase the abundance of an already existing species.
(c) *Repeated introduction* every few generations, of a species that cannot persist permanently.
(d) *Inundation ('biological pesticides')*. Release large numbers of a control species every year, not expecting it to persist.

time for parasitoid attack, then the number of parasitoids per host individual can be greater and the percentage mortality increased (Waage 1989).

Protection of mammals from disease by immunization is mostly too medical a subject for this book, but one point is relevant. Models of the population dynamics of disease can predict what proportion of the animal population needs to be immunized in order to prevent spread of the disease. The fox–rabies model of Anderson *et al.* (1981) predicts that the proportion immunized (P) needs to be

$$P \geqslant 1 - S_T/K \qquad (6.11)$$

where K is the stable fox population density in the absence of rabies, and S_T is the critical density defined in Box 6.2. For example, among possible methods to prevent the spread of rabies one is to treat foxes in a neighbouring zone (as yet disease-free) with oral vaccine. If the critical density, S_T, is calculated to be one-quarter of the actual disease-free density, the prediction is that at least three-quarters of the foxes alive at any one time would need to be immunized to prevent the disease spreading.

Management

Control of a pest by management often depends on detailed and specific knowledge about the pest and its relation to the host. I summarize briefly some examples.

Control of a pest based on specific ecological knowledge

If a pathogen also infects other, wild hosts, or requires a vector or alternate host this can indicate possible methods of management. The liver fluke *Fasciola hepatica*, which infects sheep and cattle, requires as alternate host a small snail, *Lymnaea truncatula*, which lives in wet areas on farms. This knowledge suggests several possible management methods: drain the wetter areas, keep the stock away from them, or keep ducks since they eat the snails (Wilson *et al.* 1982).

In forests and plantations tree stumps are essential for the survival or establishment of some insects and pathogenic fungi, which can then move on to attack young trees. Examples are the pine weevil, *Hylobius abietis*, and the fungus *Heterobasidion annosum* (formerly *Fomes annosus*), which are both serious forest pests, and also the fungus *Fomes lignosus* in rubber plantations (Speight & Wainhouse 1989, Rishbeth 1988, Fox 1965). Damage by the pest to the living trees can be greatly reduced by ensuring that there are few stumps, either by removing stumps physically or hastening their decay. Another possibility is to prevent the pest colonizing stumps. Successful control of the fungal pathogen *Heterobasidion annosum* has been achieved by treating stumps with another, antagonistic fungus (Rishbeth 1988).

Because these methods of control require specific knowledge they have to be based on research on each pest organism, probably in each separate host and quite possibly in several or many areas of differing habitat. Such research may be justified by the importance of the problem, but it takes time, effort and money. The question to consider here is whether more fundamental ecological knowledge and understanding can help to save time and effort and achieve satisfactory control more rapidly.

Plant spacing as a means of pest control?

One message that came from examples of diseases and herbivorous insects attacking wild species, and from models, was that the density of the host often plays a part in the natural control, by influencing the spread of the attacker. Can this be applied to farming or forestry? Can altering the spacing of crop plants or forest trees help to control pest species that attack them? Risch *et al.* (1983) drew together results from a large number of investigations of abundance of herbivorous insects in crops, in which there was comparison between pure stands of crops and mixtures of species; the mixtures involved different species intermingling or in alternating rows, or there were weedy and weed-free plots. Table 6.5 shows that 53% of the insect species were less abundant in the mixed-species plots and only 18% were more abundant. The difference, 18% vs 53%, is statistically significant ($P \ll 0.001$). However, it was only the monophagous insects that showed this dif-

Table 6.5 Summary of experiments in which the abundance of herbivorous insects was compared in 'more diverse' crops (i.e. mixed species, alternating rows of different species, or crops + weeds), and less diverse crops. Data from 198 species of insect in all. The columns do not add up to 100% because some species had varying responses and have been omitted

	Monophagous species*	Polyphagous species	All species
% of species that were:			
More abundant in more diverse crops	10	44	18
Little different	11	4	9
Less abundant in more diverse crops	61	27	53

* i.e. insect ate only one of the plant species present.
Data from Risch *et al.* (1983).

ference. This suggests that abundance of insects was reduced if they had difficulty in spreading from one food plant to another, because their food plants were more widely spaced or were obscured by other plants. Burdon & Chilvers (1982) reviewed research on plant diseases in relation to host density. Results on fungal diseases mostly showed more disease infection when the host plant density was greater. In contrast, the amount of infection of crops by aphid-borne virus diseases was often greater at lower crop density; this could be either because the number of aphids per hectare was unchanged so there were more aphids per plant, or because aphids are actually attracted when bare ground is visible between the plants.

These surveys suggest that distance between crop plants can influence the amount of attack by insects or microbial pathogens. However, most of the experiments suffer from limitations. Firstly, it is often not clear whether the pest is really reduced sufficiently to contribute usefully to its control, or whether there is merely a slowing of its spread. Burdon & Whitbread (1979) sowed plots of mixed barley and wheat in different proportions, and of pure barley. They monitored the rate of increase of powdery mildew of barley (*Erysiphe graminis* f.sp. *hordei*); this does not infect wheat. Its rate of increase on the barley leaves was slower if the barley was growing among wheat, but not vastly slower: when barley formed only 10% of the mixture the disease increase averaged 58% of the rate in pure barley. Figure 6.9 gives an indication of the spacing needed to actually prevent disease spread. It suggests that to give effective control of these rusts oat plants would need to be spaced at least 0.5 m apart and maize plants more than 1 m. One example of response of a herbivorous insect species to plant spacing is provided by the prickly pear, *Opuntia inermis*, in regions of eastern Australia (Munro 1967). After *Opuntia* had been successfully controlled by the moth *Cactoblastis cactorum*, the plant and insect species appeared to be approximately in balance, and the mean distance

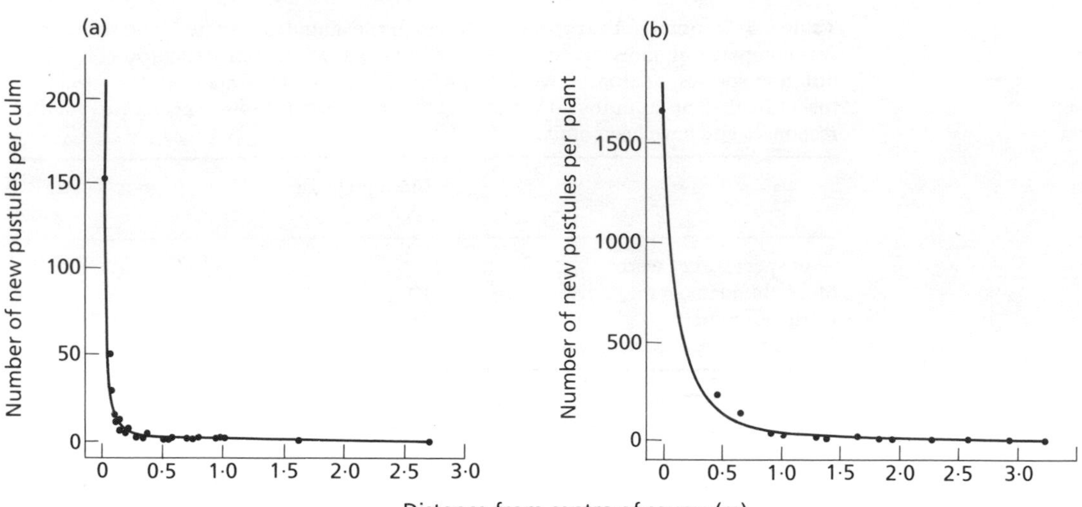

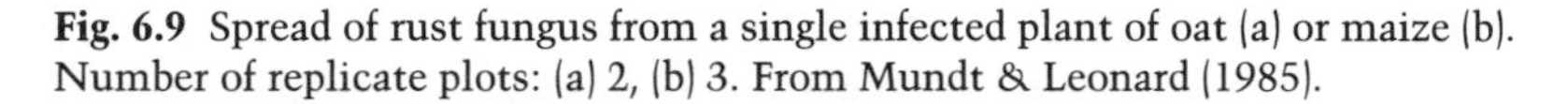

Fig. 6.9 Spread of rust fungus from a single infected plant of oat (a) or maize (b). Number of replicate plots: (a) 2, (b) 3. From Mundt & Leonard (1985).

between surviving prickly pear plants ranged from 5 to 20 m. These are only a few examples, and the spacing needed to control other diseases and insects will be different. Although it is rash to generalize, these and other examples suggest that, to provide control, spacings would usually need to be far wider than are currently normally used in agriculture or forestry. They could be considered only if the space in between can be used for something else. Use of widely spaced trees may be practical, if the space in between can be cultivated or grazed. This is agroforestry, which was discussed briefly in Chapter 5. I am not aware of any reports that trees are less prone to pests when grown in agroforestry than in normal plantations. One possibility for arable crops is use of multilines, a mixture of genotypes of a crop that differ only in their resistance to particular races of a parasite. From the point of view of cultivation and harvest the crop behaves as a single species, but for each race of the parasite the potential host plants are widely spaced.

Intervening plants can reduce insect attack on crop plants

In most experiments which compare mixtures and monocultures it is not clear whether distance between host plants is the key factor; alternatively the other plants growing between the host plants might influence the pest in some other way than merely by being non-hosts. Bach (1980) carried out an experiment in Michigan which aimed to distinguish between these two possibilities. Table 6.6 summarizes the treatments, in which cucumbers were grown at different spacings, with or without intervening maize and broccoli. The larvae of the cucumber beetle feed on the roots of cucumber, and the adults attack the above-ground parts, but the species does not attack maize or broccoli. In Table 6.6, if we compare plots with the same total number

Table 6.6 Abundance of the striped cucumber beetle (*Acalymma vittata*) on cucumber plants grown at two densities, with or without maize and broccoli. Figures in parentheses are standard errors

Number of plants per 100 m²				Number of beetles (*A. vittata*)	
Cucumber	Maize	Broccoli	Total	per 100 m²	per cucumber plant
72	36	36	144	137 (±60)	1.9 (±1.4)
145	72	72	289	81 (±8)	0.7 (±0.6)
144	0	0	144	1062 (±248)	7.8 (±3.4)
289	0	0	289	2092 (±618)	7.2 (±3.8)

Data from Bach (1980).

of plants (e.g. line 1 vs line 3) the beetles were much less abundant on the cucumber plants if they were in a mixture with maize and broccoli than in monoculture. Merely comparing lines 1 and 3 (or 2 and 4) would leave undecided whether the difference in beetle abundance was due to wider spacing of cucumber plants or to the intervening maize and broccoli. But comparing all four treatments allows us to see clearly that most of the effect was due to the intervening plants, not the cucumber spacing.

How the maize or broccoli (or both) reduced cucumber beetles in the previous experiment was not clear, but some experiments have suggested how other plants can be beneficial. Dempster (1969) determined the mortality of larvae of cabbage white butterfly on Brussels sprouts plants grown at 90-cm spacing. If weeds were allowed to grow unchecked between the sprouts larva mortality was twice as high as in plots where the weeds were removed (Table 6.7). This increased mortality was probably caused mainly by a ground-living predatory beetle which climbed up the sprouts plants at night. It was abundant among the weeds (Table 6.7), which provided a habitat for it to shelter in by day. In spite of the reduced caterpillar attack, the Brussels sprouts plants were smaller in the weedy plots, no doubt due to direct competition with the weeds. Possibly some other plant species, or a non-living ground cover, could encourage the beetles without competing against the crop plants.

Table 6.7 Results from an experiment in which plots of Brussels sprouts were either kept weed-free by hoeing or were left unweeded

	Unweeded	Weed-free
% mortality of cabbage white caterpillars	70.3	34.8
Number of *Harpalus rufipes* (beetles)*	69	13
Weight of Brussels sprouts produced (kg per plant)	0.41	0.64

*Caught in pit-fall traps.
Data from Dempster (1969).

Hedges and field margins have sometimes been suggested as a source of natural enemies of pests. For example, nettles can harbour a nettle aphid which in turn can support a parasitoid, *Aphidius ervi*, which also attacks the grain aphid *Sitobion avenae* (Wratten & Powell 1991). Thus patches of nettles in field margins might lead to reduced abundance of the grain aphid in cereal crops, though this has not yet been demonstrated. However, hedges can also be a source of pests. For example in northern Europe the disease fireblight, caused by the bacterium *Erwinia amylovora*, can spread from the common hedge shrub hawthorn to apple and pear trees (Billing 1981). As well as harbouring control species, weeds can also increase disease in crops. Cucumber mosaic virus infects lettuce and can cause discoloration of the leaves that makes the lettuces unmarketable. The virus is also carried by some British weed species, though without causing symptoms. The virus can be carried over from one year to the next in weeds that survive the winter, for example in the field margin, and in a few weed species the virus passes from one generation to the next in seeds. It can be transmitted from weeds to lettuce by aphids, and infected weeds are thus a potential source of infection (Tomlinson & Carter 1970, Tomlinson *et al.* 1970). Some fungi and insects have a specific requirement for two host plant species to complete their life cycle. For example, in western Canada mixtures of spruce and Douglas fir are prone to outbreaks of a woolly aphid, *Adelges cooleyi*, which requires both these tree species to complete its full life cycle (Speight & Wainhouse 1989).

But other plants can also increase insect attack on the crop

Crop rotation is a method of pest control that has been successfully used for a long time, primarily against insects, nematodes and fungi in soil that attack roots and that survive in the soil, or on crop residues, from one season to the next. It involves growing alternating crops, or a sequence of several crops, each of which is susceptible to different soil pests. A single year's break before a crop species returns is not necessarily enough to reduce its pests to acceptably low levels. In northern Germany at least 2 yrs of cereal is needed between each year of sugar-beet to control beet nematodes (Heitefuss 1989); Baker & Cook (1974) give examples where an even longer break is recommended. There are some examples where continued growth of the same crop year after year can lead to a decrease in disease (Campbell 1989), the best-known example being take-all decline. Take-all disease of wheat is caused by the soil-borne fungus *Gaeumannomyces graminis*. It has been observed in many countries that if wheat is grown every year on the same field take-all increases in severity for about 3 yrs, then declines again. This decline appears to be at least partly caused by living species in the soil, since fumigation, irradiation or heat treatment of the soil results in the disease increasing again (Baker & Cook 1974). It is thus similar to the decline in *Phytophthora cinnamomi* found in some Australian eucalypt forests, as mentioned earlier. Other examples of disease decline

Crop rotation can reduce soil-borne pests

But many years of the same crop sometimes leads to a decline in disease

are shown by potato scab (*Streptomyces scabies*), *Fusarium oxysporum* wilt and *Phymatotrichum* in alfalfa and cotton (Campbell 1989).

This section has failed to provide any suggestions of management systems that will reduce pests in every crop. Mixtures of species within the field or between field and margin sometimes reduce pests but in other cases increase them. Rotation often reduces soil-borne diseases, but some decline in prolonged monoculture. Management for pest control has to be based on knowledge of the individual species.

Biological control

How often has biological control been successful?

Biological control has had many successes. Two of them (rabbits by myxomatosis, Klamath-weed by beetles) have already been summarized in this chapter, others will be described soon. But not all attempts have been successful. Waage & Greathead (1988) summarize the success rate of insect species that were introduced for biological control of insect pests and weeds. Of the species that became established, only 40% of those aimed at insect control and 31% of those aimed at weed control were 'substantially successful'. This does not take account of species that were introduced but did not establish, for which records may not always be kept. The question to consider here is whether there are any ways of spotting in advance a species that is likely to be effective, so that we have a better chance of being successful first time.

One decision required is whether to aim for classical biological control, which requires only a single introduction of the control species, or for one of the other methods which require repeated introductions (see Box 6.3). In general, classical biological control has worked less well for arable crops than in systems not cultivated each year, i.e. forests, orchards and pastureland. For example, there are some well-known success stories of control of weeds in pastureland (Debach & Rosen 1991). Control of Klamath-weed and prickly pear by insects has already been described; another example is control of *Chondrilla juncea* in Australia by the rust fungus *Puccinia chondrillina*. But for control of weeds in arable crops the best hopes lie in 'mycoherbicides', which involve application of spores of the control fungus every year. Several mycoherbicides are commercially available (Cullen & Hasan 1988).

Biopesticides

Bacillus thuringiensis shows considerable promise as a 'bio-insecticide'. It produces a protein which, when ingested by insects, is broken down in their guts to toxic polypeptides. Only insects are killed. Many strains of the bacterium are known, some of which kill only a limited range of insect species (Payne 1988). The bacterium is a slow multiplier and spreader under field conditions, but it is easy to grow in artificial media, and is marketed commercially as the bacterial spores or as the toxin itself in crystalline form, to be sprayed on crops. Limitations to this application method are: (1) the toxin remains active on plant surfaces for only a few days; and (2) since the insects have to ingest the spores or crystals the toxin is effective only against insects that feed on the outside of the above-ground parts of plants. The protein toxin is controlled by a single gene, which has been successfully transferred to other species, where it is expressed. When the gene was inserted into the soil bacterium *Pseudomonas fluorescens* the bacterium showed some toxicity to a root-attacking insect, root cutworm, though not to the more serious pest corn rootworm (Lindow *et al.* 1989). The gene has also been transferred into several crop plant species; it reduced caterpillar attack on them in field experiments (Gasser & Fraley 1989).

Among viruses, baculoviruses show promise for the control of various insect pests of arable crops, orchards and forests (Payne 1988, Wood & Granados 1991). They are usually specific to a particular host species. Although a few are commercially available, their effectiveness is limited by low virulence and inability to maintain a population for long after release. Attempts are being made to improve the effectiveness, using genetic engineering.

If a natural enemy of the pest already occurs in the area but at low abundance, it might seem obvious to try augmentation, i.e. releasing a large number of additional individuals to increase its abundance greatly. This can work: control of root pathogens is an example discussed later. But there are limitations (see May & Hassell 1988); one is that the factors which normally control the natural enemy may soon reduce its population towards the original level, so that repeated augmentation is needed. For the remainder of this chapter I concentrate on classical biological control.

How to choose species for classical biological control

Our main source of clues on how to choose successful biological control agents must be the successes and failures of the past. One difficulty is to obtain adequate records of failures. For example, Crawley (1983) noted that no insect successfully used for weed control has been a seed-eater. Is that a message not to bother trying seed-eaters in future, or does it merely mean that seed-eaters have not been tried in the past? Another difficulty in learning from past attempts is that often several possible control species have been introduced nearly simultaneously, in the hope that one of them will work, and it may be difficult afterwards to decide how much each of them contributed to the successful control of the pest. A third problem is that control species that failed to establish first time have sometimes later proved very effective. For example the first few releases of myxoma virus in Australia failed to establish permanently, but once established the virus gave very effective control of rabbits. Maybe some other potential successes have been missed because they were abandoned after failing to establish first time. The risk of chance extinction of small populations is discussed in Chapter 8.

Should we aim for extinction of the pest or a low, stable level?

A key decision to be made is whether it is better to aim for extinction of the pest or to reduce it to a low abundance. It might at first sight seem obvious that extinction is better, but whether this is so will depend partly on whether the control organism is host-specific.

A major advantage of it being host-specific is that this removes the danger of it attacking non-target species. A disadvantage is that if the pest species becomes extinct, so too will the control species. Then if the pest survives somewhere else there is no control against it reinvading later. Aiming for extinction can in practice result in subsequent periodic outbursts of the pest.

If it is decided to aim for a stable pest abundance, clearly this abundance must be very low, so a key question is whether the control agent will reduce the pest enough. Looking again at the equations in Box 6.2, if the virulence θ is high (i.e. infected animals die quickly) c becomes approximately equal to θ (equation 6.7), and equation 6.9 reduces approximately to

$$N^* = \theta^2/\theta\beta \quad \text{or}$$
$$N^* = \theta/\beta$$

The same equation applies approximately to pathogens of insects (May & Hassell 1988). So high virulence gives a relatively high pest abundance, which is not what we want. But if θ is low the part of equation 6.9 in square brackets becomes small, so again N^* is high. In other words, if θ is too low the parasite is ineffective at killing the pest, but if θ is too high it kills the pest individuals so quickly that it hampers its own spread. So according to the basic Anderson–May model of Box 6.2 the lowest stable population of the pest will be given by control species of *intermediate* virulence (θ) plus *high* ability to spread (β). Control species with high virulence can either give only modest reduction of the pest, or else produce cycles or local extinction of the pest.

Parasitoids are attractive as control agents for insect pests because they are host-specific, but they too may give either only a limited reduction of the pest or else give very unstable pest populations. The model of Beddington *et al.* (1978), mentioned earlier, predicts that a parasitoid cannot reduce the host abundance by more than two-thirds, if this reduced abundance is to be stable; however, patchiness in the system can allow a much greater reduction to be stable. Beddington *et al.* predicted that a parasitoid is most likely to be an effective control agent if it has high ability to search for its host, high ability to disperse between patches and a strong tendency to aggregate in host patches.

Should we look for control species among those that attack the pest in its home range?

Many species become pests only when they are introduced to a new area outside their native range. Their ability to reach pest-level abundance in their new home might be because they have left some of their natural control species behind. So a promising source of species for biological control is to look at what is attacking the pest species back in its original native area. This has been the basis of many of the success stories of classical biological control. An early example was the cottony-cushion scale insect (*Icerya purchasi*), which was accidentally introduced from Australia into California about 1868 and became a serious pest of citrus. It was successfully controlled by two insects, a predatory ladybird (beetle) and a parasitic fly, introduced from Australia in 1888 (Thorarinsson 1990, Debach & Rosen 1991). Another example is the winter moth (*Operophtera brumata*), introduced from Europe into parts of Canada and later controlled by a parasitoid. These two examples show that in practice choosing the control species may not be straightforward. When American scientists went to Australia to look for a species controlling cottony-cushion scale, they found that neither the scale nor any attackers were at all common there (Debach & Rosen 1991). And a detailed study of the winter moth in Wytham Wood, England (where it is not a pest) found that the parasitoid *Cyzenis albicans* has little effect on its numbers there; yet this parasitoid was very successful in controlling the winter moth in Nova Scotia. The difference probably arises from high mortality of larvae of the moth in winter in Wytham Wood having a major controlling influence. This is due mainly to predation by beetles, which does not occur in Nova Scotia (Hassell 1978, 1980).

It can alternatively be argued that if the pest species and its attacker have been together in the native area for a long time they may have coevolved, resulting in the attacker being less damaging; and that it will therefore be more promising to look for a control species in an area where the pest species does not occur. Some successful biological control organisms have originated from other hosts. One example is myxoma virus, which occurs naturally in forest rabbits (*Silvilagus brasiliensis*) in South America. It does not cause severe disease in these rabbits, and its lethal effect on European rabbits (*Oryctolagus cuniculus*) was discovered only by accident when some of them in a research laboratory in Uruguay became infected (Fenner 1983). Another example is that the prickly pears (*Opuntia* spp.) which became a serious pest in grazing land in Australia originated from the Gulf of Mexico; whereas the control insect *Cactoblastis cactorum* is native in South America, where it feeds on other species of cactus. Waage & Greathead (1988) analysed data from 441 introductions of insects (parasitoids or predators) aimed at control of pest insects. Their data excluded introductions that failed to establish. The percentage of cases classed as giving completely or partially successful control of the pest were: if the pest and control species had been associated before elsewhere, 40%; if they had not, 34%. These two percentages were not significantly different statistically. So it is evident that either old or new associations can result in successful biological control, and there is no strong reason to think that one is more likely to prove successful than the other.

Control species from 'suppressive' soil

When searching for potential control agents for fungi that cause plant diseases, it is promising to examine individual plants or patches that are less diseased than their neighbours, or to look in the soil beneath them. The cause of the reduced disease is sometimes a physical or chemical property of the soil, or a feature of the microclimate; however, there can be biological causes. A single plant may carry a form of the pathogen with transmissible hypovirulence, as in the chestnut blight fungus described earlier. Although transmissible hypovirulence carried by RNA is probably quite common in fungal pathogens, it has so far proved difficult to use in pathogen control. If a disease spontaneously declines over several years while the same crop is grown, this may be a promising place to look for control species. Several examples of this were mentioned earlier, including take-all disease of wheat, caused by the fungus *Gaeumannomyces graminis*. If the decline can be reversed by sterilizing the soil, and if an inoculum from the soil introduces suppression of the pathogen into another soil, then the soil is referred to as 'suppressive' and it seems very likely that a microbial species in the soil is involved in reducing the pathogen. This could be a bacterial, fungal or protozoan species, or more than one; as there are a great number of microbial species in soil, screening for effective control agents can be very time-consuming. Potential control agents against *Gaeumannomyces graminis* that have been tested include amoebae that directly attack fungi, bacteria of several genera, and fungi, including the closely related non-pathogenic *Phialophora* spp. (Campbell 1989). It may well be that the cause of suppressiveness is not the same in all areas and that a control species effective in one area may not be in another.

Mechanisms by which soil bacteria suppress fungal pathogens

If we can find out more about the mechanism by which the control species suppresses the pathogen this could indicate simpler ways of initially screening large numbers of soil microbial species. For example, if antibiotics were shown to be a common mechanism, it would be possible to start by selecting in the laboratory species that produce an antibiotic inhibitory to test fungi. It has long been known that some soil bacteria can produce antibiotics when grown on laboratory media, but there has been uncertainty about how much antibiotic they produce when they are in soil and about how much effect these have on other microorganisms. Experiments involving DNA transfer are helping to elucidate this. Tomashow & Weller (1990) isolated strains of the bacterium *Pseudomonas fluorescens* from soil suppressive to take-all, and found one strain (2–79) which in culture produced an antibiotic, phenazine-1-carboxylic acid, that was inhibitory to *Gaeumannomyces graminis*. Using genetic engineering techniques, they produced a mutant of 2–79 that did not synthesize the antibiotic, and showed that it was no longer inhibitory to *G. graminis* in culture and was much less suppressive of take-all in soil. This is strong evidence that this antibiotic was involved under field conditions in the ability of this bacterial strain to control this pathogenic fungus.

Will initially effective control species evolve to become less effective?

A major problem with chemical pesticides has been that the pest may evolve to become more tolerant of the chemical. With biological control this could also happen; and there is the additional possibility of the control species evolving too. One might expect that from the point of view of the pathogen the ideal situation is for infectious individuals not to die, so they remain to pass on the pathogen. In terms of the model in Box 6.2, if $\theta = 0$, this will make c lower and hence dI/dt higher (equations 6.6 and 6.7). So will natural selection favour the control species becoming less virulent? Some parasites, for example baculoviruses and many parasitoids, always kill infected hosts, and presumably could not complete their life cycles without doing so; here reduced virulence is not possible. It is not always true in practice that a less virulent strain of a parasite will increase at the expense of a more virulent strain. One example is Dutch elm disease: a low-virulence strain of the fungus was present in western Europe from the 1920s, but after two much more virulent strains arrived during the 1960s they replaced it, killing a large proportion of the elm trees (Brasier 1983); see Fig. 6.4.

Although the Anderson–May model (Box 6.2) appears to predict that zero virulence will be most favourable for the parasite, May & Anderson (1983) pointed out that this is true only as long as θ (virulence) can change without any accompanying change in β or ν (transmission or recovery). In practice reduced virulence is often accompanied by reduced transmission and increased rates of recovery. The importance of this is illustrated by a detailed study of genetic change in rabbits and myxoma virus in Australia (Fenner 1983, May & Anderson 1983, Dwyer *et al.* 1990). The virus was classified into five virulence grades, assessed by the percentage of infected rabbits that died in laboratory tests and how long it took them to die. The most virulent grade, I, caused more than 99% mortality, whereas the least virulent, V, caused less than 50% mortality. It is likely that the strain originally introduced into Australia was the highly virulent Grade I. Figure 6.10 shows that this quickly became less abundant, as less virulent grades increased. Within a few years Grades III and IV, of intermediate virulence, became predominant, and remained so for at least 25 yrs. To investigate whether the rabbits' susceptibility was changing, too, wild uninfected rabbits were caught and inoculated with virus of Grade III. Figure 6.11 shows results from one area in Australia where there was about one myxomatosis epizootic (outbreak) per year. After a few epizootics the susceptibility of the rabbits began to decline.

Genetic changes in rabbits and myxoma virus

Thus within a decade of the arrival of the myxoma virus it had changed to become less virulent and the rabbits had also become less susceptible. If we extrapolate that onwards we might predict that within a few decades the virus would cease to control the rabbits at all. Is that likely? Figure 6.10 shows that for the virus such extrapolation is not valid: the virus did not evolve towards eventual dominance of the least virulent strain, V, indeed that strain decreased in percentage abundance after 1958. This stabilizing of the host–parasite relationship happened because virulence, recovery and transmission interact in ways that tend to be balancing. The less virulent virus takes longer to kill its host, which favours the virus by giving it more time to spread;

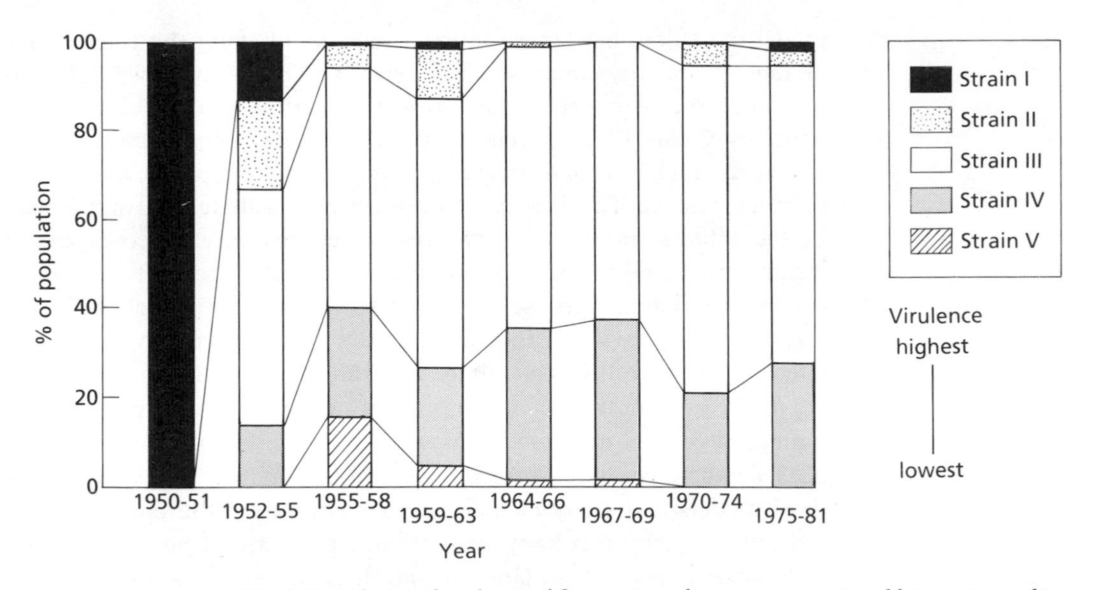

Fig. 6.10 Relative abundance of five strains of myxoma virus in rabbits in Australia. Strains graded from I (most virulent, i.e. highest percentage mortality and most rapid death) to V (least virulent). Data from Fenner (1983).

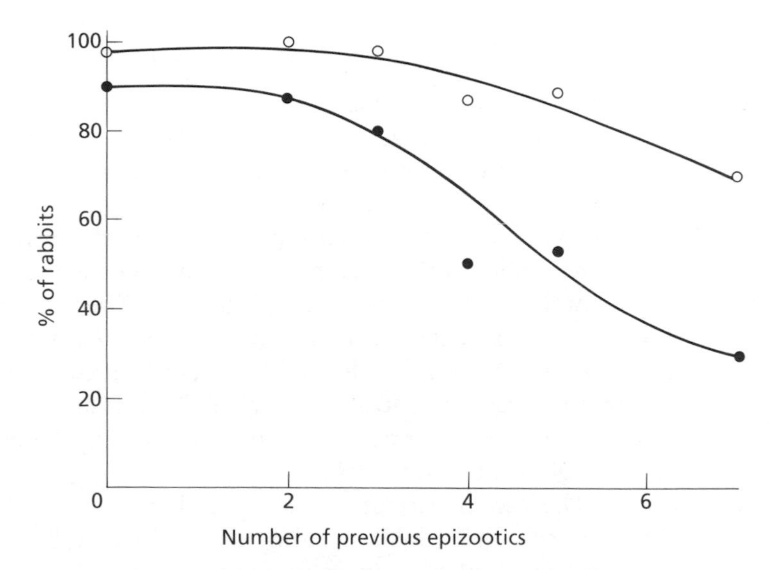

Fig. 6.11 Response of wild rabbits from Lake Urana area of Australia to inoculation with myxoma virus of virulence Grade III. The rabbits were collected after various numbers of myxomatosis epizootics had occurred in the area. ○, rabbits developed moderate to severe symptoms; ●, rabbits died. Data from Fenner (1983).

but more of the rabbits recover and become immune, which slows the spread of the virus (equations 6.6 and 6.7). Spread of the virus by fleas and mosquitoes requires open lesions in the rabbit, which are fewer if the virus is less virulent. Dwyer *et al.* (1990) developed a complex model to explore how these interactions would affect the coexistence of rabbits and myxomatosis in Australia. Their model was able to predict correctly that the rabbit–virus interaction would become dominated by Grade III virus (Fig. 6.10). Their fundamental prediction is that if the rabbits evolve greater resistance, then the virus will evolve greater virulence. This is because a certain intermediate level of response in the rabbits (in terms of survival time, recovery percentage and lesion formation) is most favourable for the virus, and if the rabbit evolves to change this, selection will favour the virus changing to restore it. Dwyer *et al.* therefore predict that the virus will remain effective in controlling the rabbits for some time to come.

At the start of this section on evolutionary change I suggested that biological control might suffer worse problems than chemical control because there are two species capable of evolving. The conclusions from models and study of rabbit myxomatosis are the opposite, that biological control can have some inbuilt protection against evolution of resistance in the pest. One limitation to this is its assumption that the parasite is always capable of matching any increase in host resistance by a further increase in virulence; the host might eventually outstrip its parasite in the evolutionary race.

Undesirable side-effects of biological control

Will the biological control agent harm non-target species?

One concern about chemical control of pests is that the control agent might kill non-target species. This is also a concern for biological control, with added potential dangers because organisms can multiply and exchange genes whereas chemicals cannot. In this respect augmentation and inundation types of biological control (see Box 6.3) are more similar to chemical pesticides in the risks they pose, because the control species will not multiply greatly. There have in practice been few serious effects on non-target species from any type of biological control. One example that did occur is that predatory snails introduced into Hawaii for control of the giant African snail were thought to be responsible also for the extinction of some native endemic snail species (Waage & Greathead 1988).

The earlier part of this chapter that discussed introductions and invasions ended by concluding that it is very difficult to predict in advance whether an introduction will establish and spread. The same applies to trying to predict whether a species introduced for biological control will spread. If the species is known to have high host-specificity in its native range this clearly reduces the danger of it spreading to non-target species in its new area. For example, an attractive feature of parasitoids for biological control is that they are often strongly host-specific. An economic conflict may arise here, because wide-spectrum pesticides (whether chemical or biological) are likely to have a wider market than highly specific ones, and so firms are more likely to decide that they are worth developing, producing and marketing. To be set against this, however, are costs of testing for undesirable side-effects, if this is required by law. Waage & Greathead (1988) reported that in Canadian programmes to develop biological control of 28 weed species, about half the total cost was screening the control species for ability to harm other plants. The question here is, how far should the screening go? If each potential control species were to be tested on every native plant species in the area the costs could be prohibitive. So far most testing has been on useful plants, especially crop plants. If we are to extend testing to wild plants, can we decide which are most likely to be at risk? One procedure has been to test species related to the target weed, e.g. in the same genus. Another is to test plants with similar secondary chemistry, i.e. similar protection systems. To provide a sound basis for testing, we need a fundamental understanding of what controls the ability of particular parasite species to attack certain potential host species and not others, the ability of certain insects to eat certain plants and not others.

Conclusions

- Of species that have invaded in the past, only a small percentage have become pests.
- There are few consistent characteristics of species that have become successful invaders, or of areas in which invaders have established.
- The physical conditions under which genetically engineered bacteria are likely to survive and to exchange genes are becoming known, but to predict whether these bacteria will compete successfully against indigenous bacteria is more difficult.
- A mathematical model has predicted that a species which is multiplying in a new area will extend its range at a uniform rate. Many recorded examples conform approximately with this prediction.
- At the end of the Ice Age tree species, even those with heavy seeds, extended their ranges at mean rates of hundreds of metres per year.
- Pests and diseases occur in species-rich as well as species-poor natural communities, sometimes as stable endemics, sometimes in intermittent outbursts.
- There is a critical population density of the host species below which a parasite cannot survive.
- A parasite that at any one time infects only a small percentage of the host population can nevertheless be a major controller of host abundance.
- Mixtures of plant species within a field sometimes suffer less pest attack than monoculture, but in other cases more attack. Management of crops for pest control needs to be based on information about each individual species.
- When choosing species as potential biological control agents, models

predict that parasites or predators of intermediate virulence will be most effective at maintaining a low, stable host abundance.
- Patchiness can allow parasitoids to reduce their host species to low, stable abundance.
- Some very effective biological control agents have been found by looking at what attacks the pest in its original range; but about equally often effective control species have been found elsewhere.
- If the pest adapts to be less susceptible to the control species, the control species may adapt to increase its virulence and so maintain the control.

Further reading

Invasion and spread:
Kornberg & Williamson (1986)
Mooney & Drake (1986)
Delcourt & Delcourt (1991)

Release of genetically engineered organisms:
Sussman *et al.* (1988)
Fincham & Ravetz (1991)

Modelling host–parasite relationships:
Hassell (1978)

Pests and their control:
General: Kogan (1986), Wood & Way (1988)
Plant diseases: Campbell (1989), Campbell & Madden (1990)
Animal diseases: Anderson (1982)
Herbivorous insects: Crawley (1983), Speight & Wainhouse (1989)

Chapter 7: Pollution

Questions

- There are thousands of potentially useful synthetic chemicals. We want to know for each of them whether it poses a risk to living things. How can this be determined, without impossible requirements of time and resources? Are there useful tests that are simple and quick?
- Which test species should we use?
- How can we scale up from simple tests to predict effects of a chemical outdoors?
- Can we predict the concentration of a pollutant that will build up within a plant or animal?
- Do pollutant concentrations increase up the food chain?
- How do interactions between species affect their response to pollutants?
- How can we find out whether decline of a species is caused by pollution?
- Do particular structures in organic molecules make them less easily degraded?
- How can we speed up the breakdown of organic pollutants?
- How can we obtain suitable plants to revegetate toxic mine waste?
- Can changed farming methods reduce nitrate leaching to safe levels?
- How can we get rid of algal blooms on nutrient-rich lakes?

Background science

- How substances enter plants and animals. Uptake by plants from soil, by aquatic animals from the surrounding water.
- Movement of substances along food chains.
- How species differ in their response to pollutants.
- How mycorrhizas affect the response of plants to heavy metals.
- Using microcosms and mesocosms to study effects of pollutants on communities.
- A case study: why numbers of peregrine falcons declined.
- Biochemical pathways by which complex organic chemicals are broken down.
- The genetic basis of heavy metal tolerance in plants.
- The interacting role of nutrients and grazers in determining amounts of phytoplankton.

Like the rest of this book, this chapter concentrates on biological aspects. Therefore it says little about sources of pollution and how pollutants are spread. The basic questions considered are how to determine whether a chemical is likely to be harmful, and if so what to do about it.

The chapter is not organized chemical-by-chemical, but in each section chooses whichever pollutant best illustrates the topic being considered. Box 7.1 lists pollutants that are mentioned in the chapter and gives a little information about each. Some important pollutants are unfortunately scarcely mentioned (I cannot cover everything). There is very little about gaseous pollutants and acid rain, though greenhouse gases were a subject of Chapter 2.

Measuring how toxic a chemical is

Box 7.2 provides an outline of this section. Because there are so many potentially harmful chemicals, and so many species that they might harm, testing a new chemical usually starts with methods that are simple and quick. Some of these are outlined first. The lower part of Box 7.2 lists five basic problems in scaling up from these simple tests to predicting whether the chemical will be harmful in the real world outdoors. Each of these is discussed in the text.

Simple tests in artificial conditions

To ask 'Is this chemical harmful?' is rarely useful. The effect of a chemical will depend on the *amount* or *concentration in the animal or plant*, or in certain sensitive parts of it. This in turn will depend on the *concentration in its surroundings* (e.g. in the water around an aquatic animal, or the soil around a plant's roots) or on the amount (the '*dose*') eaten by an animal. A very common procedure is to determine what dose or concentration is required to kill individuals of the species.

Amounts that cause death

Figure 7.1 shows an example, in which five sets of aphids were sprayed with an insecticide, each at a different concentration. From the graph it is possible to determine the concentration that would kill 50% of the individuals: it is 4.9 mg l^{-1}. This concentration is termed the LC_{50} (LC standing for lethal concentration). If the chemical is fed by mouth, e.g. to rats, the dose (amount per animal) sufficient to kill 50% of the animals is the LD_{50} (LD = lethal dose). If a fish or other aquatic species is immersed continuously in water containing the test chemical, the number dying is likely to increase with time. Figure 7.2 shows an example. We can say from these results that half the trout would survive NH_4Cl at 20 mg l^{-1} however long the treatment went on; but the experiment stopped too soon to allow a similar statement for $ZnSO_4$. We can say that for Zn the 48 h LC_{50} is 3.5 mg l^{-1}.

The LD_{50} and LC_{50} are rarely intended as direct indicators of the amount or concentration of a chemical that can safely be released. If

Box 7.1 Pollutants mentioned in this chapter, with their principal sources.

Gases. Sulphur dioxide. Mainly from burning fossil fuels.

Inorganic elements

Arsenic (As) Cadmium (Cd)* In P fertilizer Copper (Cu)* Lead (Pb)* Motor exhaust fumes Nickel (Ni)* Zinc (Zn)*	Waste and contamination from smelters, foundries, mills. Mine waste. Industrial waste in sewage sludge

*Heavy metals

Nitrate, Leached from farmland.

Phosphate. From domestic sewage, and from animal waste in intensive animal production.

Radioactive isotopes

	Half-life (yr)
^{90}Sr	28
^{134}Cs	2
^{137}Cs	30

From accidents at nuclear power stations (e.g. at Chernobyl, Ukraine in 1986) and at nuclear waste processing works. From former atmospheric testing of nuclear weapons.

Organochlorine compounds

Insecticides, including DDT, lindane, aldrin, dieldrin, endrin. Production and use severely restricted in many countries.

Herbicides, including 24D, 245T, atrazine, dichlobenil, quintozene.

Polychlorinated biphenyls (PCBs).

Have the framework

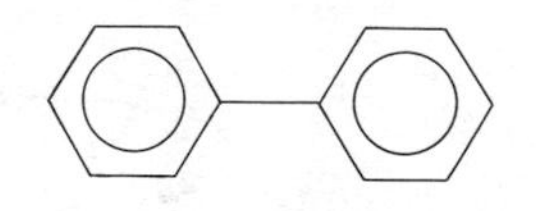

with Cl at any positions

Examples in Fig. 7.12. Used for insulating materials, paints, lubricants. Production restricted, but they are still abundant.

Chlorinated aliphatic compounds, including trichloroethylene (TCE). Produced industrially.

Organophosphorus compounds

Insecticides, e.g. malathion.

Herbicides, e.g. glyphosate.

Other organic pollutants

Natural pesticides, e.g. rotenone (Derris), a complex aromatic used as an insecticide.

Polynuclear (polycyclic) aromatic hydrocarbons (PAH). Produced by burning fossil fuels or wood.

Crude oil. From oil wells, tankers, industrial waste.

Hellawell (1986, Table 7.22) and Freedman (1989, Table 8.1) give much longer lists of synthetic pesticides, with their full chemical names.

Box 7.2 Measuring how toxic a chemical is.

Short-term tests on individual species in artificial conditions: responses by the species that are commonly measured.

1 Death.

2 Growth rate.

3 Population increase.

4 Measures of metabolic state or activity:

(a) RNA:DNA ratio.

(b) Amounts of high-energy compounds, e.g. lipids, glycogen.

(c) ^{14}C-aminoacid incorporation into proteins.

(d) Respiration rate; useful to measure microbial activity in soil, less useful for animals (see text).

(e) Plants: photosynthesis, e.g. by rate of CO_2 uptake; root extension rate (see Wilkins 1978).

(f) The Microtox test. Uses the natural luminescence of the marine bacterium *Phosphobacterium phosphoreum*.

Scaling up these results to predict responses in more natural conditions: problems to be considered.

1 The substance may act over a long time.

2 Other species may be affected, besides those tested.

3 Environmental conditions outdoors may be different from those in the test.

4 There may be a mixture of more than one chemical, whose effects could interact.

5 Interactions between species may alter their response to the chemical.

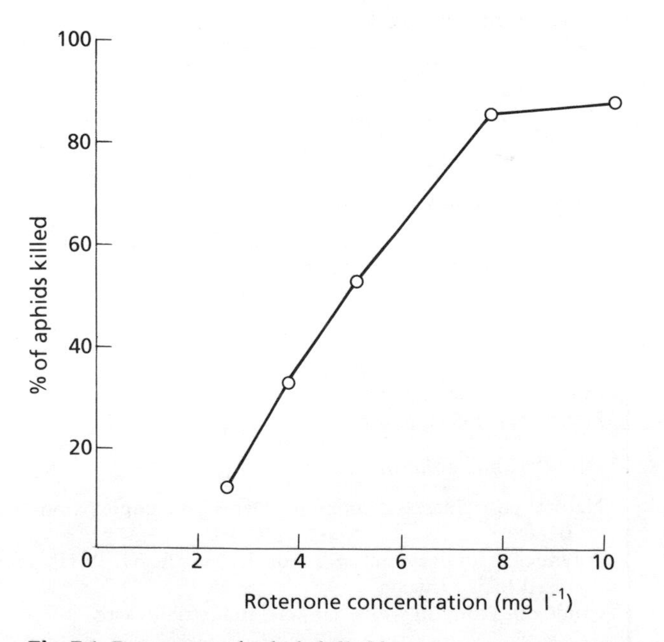

Fig. 7.1 Percentage of aphids killed by spray containing different concentrations of rotenone. Data from Butler (1978).

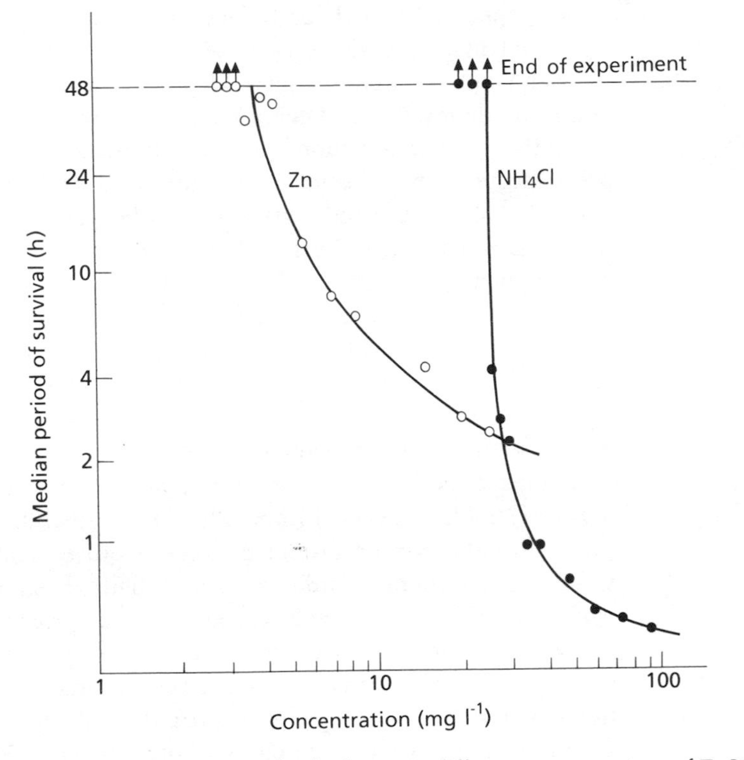

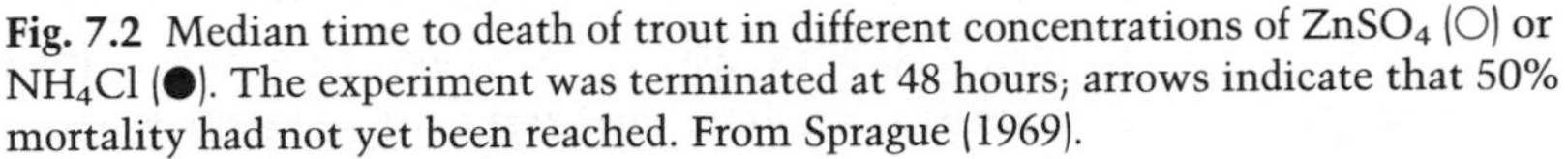
Fig. 7.2 Median time to death of trout in different concentrations of $ZnSO_4$ (○) or NH_4Cl (●). The experiment was terminated at 48 hours; arrows indicate that 50% mortality had not yet been reached. From Sprague (1969).

you read on a food packet 'Only 50% of the people who eat this will die' would you be happy to eat it? The question 'How much damage to a species should we allow?' is unlikely to have the same answer for all species. A 50% decrease in one species of soil nematode would probably pass unnoticed, but many people might be unhappy if half the fish in their local lake died suddenly. As we shall see, the LD_{50} and LC_{50} tests are useful to measure the *relative* toxicity of different chemicals and the *relative* sensitivity of different species. We shall consider later whether we can calculate from the LD_{50} and LC_{50} acceptable levels of pollutants. Another approach, using the example of Fig. 7.1, would be to try other concentrations below $2.6\,mg\,l^{-1}$ to find the highest concentration that kills no aphids (assuming that is your aim). A problem with that approach is that there may be no sharp cut-off point, i.e. the line curves towards the bottom axis, and the definition of 'no aphids killed' depends on how many aphids you have in your sample. However, in this example, and most throughout this chapter, high precision is not essential; this is because in deciding a safe or acceptable level of a pollutant chemical it is normal to allow a substantial margin of safety.

Effects on growth

We may reasonably ask whether death is the best measure of response to a chemical. It has been the most widely used response for aquatic species (animals and algae), but has less often been used for terrestrial plants, perhaps because death of a large plant is rarely a clear and sudden event. One could argue that harm may be done to an animal or plant without it being killed, and that we should be measuring other things that are more sensitive indicators that the species has been affected. An indicator commonly used for land plants is growth. Figure 7.3 shows the shoot weight of barley plants grown for 11 weeks in different concentrations of SO_2 under near-natural conditions in open-topped chambers. At $270\,nl\,l^{-1}$ SO_2 reduced growth by half. This is the EC_{50}, where EC means 'effective concentration'; EC is analogous to LC, but something other than death is the response measured. The results show that under these conditions barley growth begins to be reduced when SO_2 is above about $100\,nl\,l^{-1}$, and this is likely to be more useful information than the EC_{50} for planning controls on SO_2 emissions. Experiments on other crop and grass species have indicated SO_2 thresholds between 15 and $120\,nl\,l^{-1}$ (Roberts 1984). To compare these with SO_2 concentrations outdoors requires conversion to $\mu g\,m^{-3}$, since measurements outdoors have usually been expressed in that way. Multiplying $nl\,l^{-1}$ by 2.7 gives $\mu g\,m^{-3}$ approximately, though the conversion factor depends somewhat on temperature (Wellburn 1988). Among the mean annual SO_2 concentrations for 1988–9 given in the *UN Environmental Data Report* (UNEP 1991) the highest was $211\,\mu g\,m^{-3}$, in Tehran, which is about $80\,nl\,l^{-1}$. Many were below $50\,\mu g\,m^{-3}$ (about $20\,nl\,l^{-1}$), suggesting that direct harmful effects of gaseous SO_2 are localized rather than widespread. However, concentrations are unlikely to remain always at the mean: since most SO_2 comes from burning of fossil fuels, concentrations tend to be higher in winter, and particularly high locally on some windless days. It is not obvious, without research, whether such short-term events are more

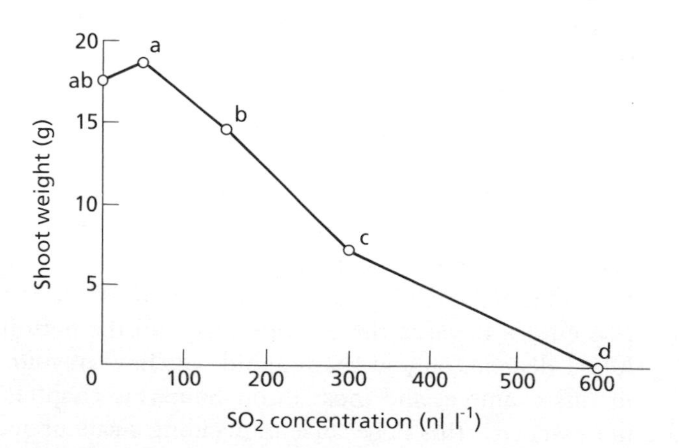

Fig. 7.3 Final shoot dry weight of barley plants grown for 79 days in an atmosphere containing different concentrations of SO_2. Points not bearing the same letter are significantly different at $P = 0.05$. From Murray & Wilson (1990).

important than the long-term average. The question of how living things respond to varying pollutant concentrations is an important one, but there is not time to consider it in this chapter.

Since sulphur is an essential element a fertilizing effect of SO_2 is possible, though the slight rise at 50 nl l^{-1} in Fig. 7.3 was not statistically significant. Zinc and copper are examples of elements that are essential to plants in small quantities but become toxic at higher concentrations. Even phosphate and nitrate can be harmful, in the wrong place at the wrong concentration, as we shall see later. This serves to emphasize the importance of studying how different concentrations and doses affect living things.

Other toxicity tests

Growth and reproduction can be measured on many invertebrates, algae and microorganisms in an acceptably short time; but for large animals or plants the time needed for direct measurements will be far too long, and other, quicker methods have been devised (see Box 7.1). Respiration rates of animals are difficult to interpret: the initial response to a harmful chemical is sometimes an increase in respiration rate, so that is no indication that the substance is harmless. Decline in high-energy compounds or in the RNA/DNA ratio are more reliable indicators of a harmful effect on metabolism. However, the respiration rate of soil samples is often used to assess effects of pesticides and other pollutants on soil microbial populations. The response is best measured over some days (Somerville & Greaves 1987), to provide an indication of changes in the size and overall metabolic activity of the microbial population, rather than a short-term response to the added chemical itself. It is also common to measure nitrification rate in soil samples; this is chosen primarily because nitrate production is easy to measure, not because nitrification is a particularly important or informative process. Rate of breakdown of a standard cotton (cellulose) cloth is easy to measure and is an indicator of decomposer activity. Cotton breakdown assays have been used successfully in other research, but rarely to assess pollution effects.

Response to short- and long-term exposure

These toxicity tests can mostly be completed in hours or days, but in the real situation the organisms may be subjected to the pollutant for much longer. Figure 7.2 has already shown that a concentration which has little effect in a few hours can nevertheless cause 50% mortality in a few days, and some substances might continue to build up an effect over much longer times. So the question arises whether responses to 'acute' (i.e. short) exposure can predict the effect of 'chronic' (i.e. long-term) exposure to the chemical. Figure 7.4 compares the acute and chronic toxicity of 50 substances, including organic pesticides and also salts of heavy metals such as lead, copper and mercury. Each chemical was tested on a fish species (either fathead minnow or rainbow trout) and on the crustacean *Daphnia magna*, and the results used from whichever species proved to be the more sensitive. The horizontal axis ('acute') in Fig. 7.4 is the 96 h LC_{50}, the vertical axis ('chronic')

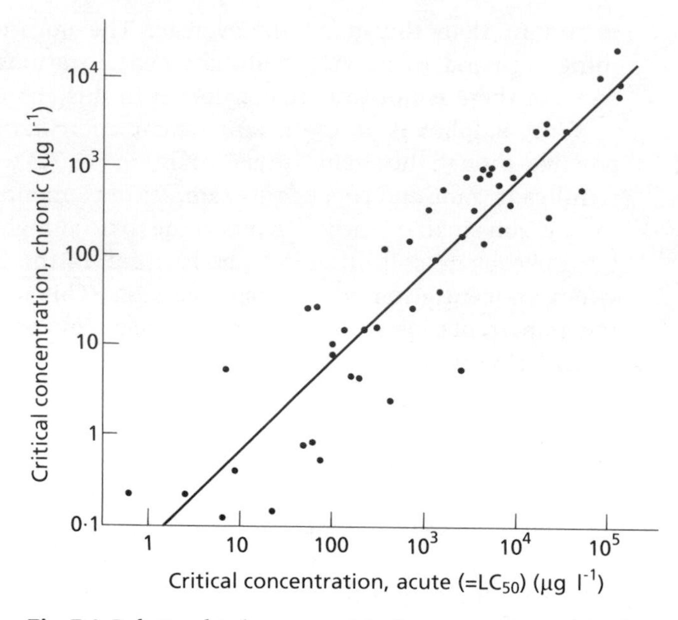

Fig. 7.4 Relationship between critical concentrations of 50 chemicals in acute (short-term) and chronic (longer) toxicity tests on fresh-water animals. The meaning of 'critical concentration' is explained in the text. Each point is for a different chemical. Both axes are log scales. The straight line is the linear regression of chronic on acute. Data from Giesy & Graney (1989).

gives the concentration to which the species could be exposed 'indefinitely' without suffering any observable adverse effect on survival, growth or reproduction. As we would expect, this chronic concentration is always lower than the acute concentration: if a concentration is just low enough not to kill any fish when acting over a long time, it will not kill half of them in 96 h. There is clearly a strong correlation between the two measures of toxicity: the correlation coefficient is 0.919 ($P \ll 0.001$). The regression line fits closely to

$$\text{Chronic level} = (\text{acute level})/14.8.$$

This suggests that a measurement of LC_{50} made in 4 days can be used to predict a safe concentration to which the species can be exposed long-term, simply by dividing by 14.8. Notice, however, that some individual points diverge widely from the regression line. For example, the equation above predicts that if the 96 h LD_{50} is 414 $\mu g\, l^{-1}$, any concentration below about 28 $\mu g\, l^{-1}$ should have no observable effect long-term. Yet one point on the graph shows that cadmium which gave an LC_{50} value of 414 $\mu g\, l^{-1}$ can have long-term effects at a concentration as low as 2.4 $\mu g\, l^{-1}$; the prediction was wrong by more than a factor of 10. So LC_{50} values determined over a few days are by no means worthless in providing an indication of the long-term toxicity of chemicals, but there could be a wide margin of error in particular cases. Relationships between acute and chronic toxicity have been determined for other chemicals and other species; the equation above is not intended to be definitive, only an example.

Which species to test

If a chemical is to be released into the air or into water it will come into contact with many species of animal, plant and microorganism. It is not feasible to test the response of every one of these species. So how many species do we need to use to get a reliable indication of toxicity, and which species should we choose? There is a tendency for the test species to be chosen at least partly for convenience, because they are easy to breed and grow in laboratory conditions. This has been a major reason why among invertebrates *Daphnia* (freshwater) and *Artemia* (marine) have been so widely used. There is a danger that species robust enough to be easily cultured may tend to be less susceptible than average to toxic chemicals. The Microtox test uses a bacterial species chosen because changes in its luminosity can be measured very quickly, not because it is an important species in key ecosystems. Obviously there are differences between species in their response to at least some chemicals: herbicides such as atrazine that inhibit photosynthesis have little direct effect on animals, though they can kill non-target plants such as algae. Hellawell (1986) provides tables of LC_{50} values for many organic and inorganic chemicals on fresh-water fish and invertebrates. Table 7.1 summarizes a few of the results. Even though only a few species have been tested, the toxicity of a chemical to fish can vary 10-fold or even 100-fold, and to invertebrates nearly as much. Invertebrates are not consistently more sensitive or less sensitive than fish, and the range of sensitivity within the two groups is at least as large as the difference between them. A common practice for potential pollutants of fresh water has been to test them on three species, of which one is a fish, one an invertebrate and one an alga; but Table 7.1 shows that this may be no more informative than testing (for example) two fish species.

Table 7.1 Range of response among species of fish and aquatic invertebrates to four potential pollutants

	Number of species tested		96 h LC_{50} ($\mu g\ l^{-1}$)	
Substance	Fish	Invertebrates	Fish	Invertebrates
Cu^{2+} in hard water	10		300–10 200	
Dichlobenil (herbicide)	4	3	4200–8000	8500–13 000
Endrin (organochlorine insecticide)	6	4	0.27–1.96	0.25–5.0
Malathion (organophosphorus insecticide)	7	3	120–20 000	0.76–50

Data collated by Hellawell (1986).

One suggestion would be to use particularly sensitive species for tests. Some experiments suggest that there is no single species that is the most sensitive to all chemicals: the species most sensitive to one chemical is not the most sensitive to another (e.g. Schafer 1972). However, field observations have often found that in very polluted sites certain groups of animal disappear while others survive or even increase, and this has led to the use of such groups as pollution monitors (see Fig. 7.5 and accompanying text).

There are special problems about choice of species when testing effects on soil microorganisms. Usually chemicals are applied to soil samples, and the response of the whole sample, e.g. its respiration or nitrate production, is measured (see above). Such tests are useful, but probably not adequate on their own, since they may fail to show a major change in species composition: if many of the microbial species died, but were replaced by other (pollutant-tolerant) species, the overall respiration rate might well remain little changed. One argument against testing pollutants on individual species is that many bacterial species present in soil will not grow on standard culture media. The species available for testing may thus not be biochemically representative of the full population. Furthermore, their metabolism may be different when they are growing on laboratory media from their state in soil, where they are usually severely starved of degradable carbon substrates. Greaves (1987) has put an alternative viewpoint, that modern microculture techniques would allow tests of pollutants on hundreds of species simultaneously, and that the results of such tests could be very useful in showing differences in response between different microbial species. Such methods were successfully tried in his laboratory, but have not so far become widely used.

Effects of pollutants outdoors

The aim of all the tests so far described is to predict responses of individual species and whole ecosystems to pollution in the real world outdoors. This presents several sorts of difficulty (Box 7.2). One is that environmental conditions may affect the toxicity of particular chemicals. For example, Table 7.1 shows the LC_{50} of Cu to fish species in hard water, meaning for those experiments $CaCO_3$ 220–360 mg l^{-1}. Two of the species were also tested in softer water ($CaCO_3$ 31 and 42 mg l^{-1}); their LC_{50} was lower by a factor of more than 10; in other words Cu was then much more toxic to them. Temperature can also affect toxicity. Crosby *et al.* (1966) measured the LD_{50} of some insecticides to *Daphnia magna* at 21°C and 25°C. Some of the insecticides were much more toxic at the lower temperature (e.g. DDT was 70 times more toxic), whereas the toxicity of others, e.g. lindane, was about the same at both temperatures. These examples illustrate the importance of conducting tests under conditions similar to those in the field. In laboratory tests it is common to have all conditions, apart from the test chemical, kept favourable for the test organism. In the field the species may well be suffering some environmental stress, which could alter its response to the test chemical.

Effects on animal behaviour

Another type of problem in scaling up from a laboratory test to the field is that some chemicals, notably organochlorine and organophosphorus insecticides, can affect the behaviour of mammals and birds at concentrations that do not affect their growth or survival, and it may be impossible to detect this on animals in cages. One example is provided by Grue *et al.* (1982), who carried out an experiment on starlings that were nesting outdoors. Some adults were given a single dose of an organophosphorus insecticide, others were undosed as controls. During the next 24 h the weight change of the dosed and control birds did not differ significantly. However, at the end of the 24 h the dosed birds had only half the concentration of cholinesterase in their brain tissue that the control birds had; this could have resulted in impaired functioning of the nervous system, which might reduce their ability to find food. During the 24 h after being treated the dosed birds visited their nestlings to feed them less frequently than before, and the nestlings lost weight. If such behaviour continued for some days it would probably result in death of nestlings. Thus this change in behaviour, which could not have been detected on birds in a cage, could influence abundance of the species in future years. Possibly a cholinesterase measurement in the laboratory could indicate when there is a risk of such behaviour change.

Effect of several chemicals together

Industrial waste often contains several harmful chemicals, so we have the additional problem of predicting how a plant or animal will respond to such a mixture. Some research has found effects of several pollutants to be approximately additive. Alabaster *et al.* (1972) studied the relationship between industrial pollution and occurrence of fish at 73 points on rivers in central England; the amount of pollution varied greatly between the points. They calculated a single toxicity value for each point, from the measured concentrations of ammonia, cyanide and heavy metals. The method of calculation took into account the individual toxicity of each chemical, as shown by its LC_{50} to trout, but otherwise the effects were assumed to be additive. The method was quite successful: fish were found to be absent from most of the rivers where the calculated toxicity was above a critical value. In this study most of the harmful effect was from inorganic chemicals; whether a single toxicity value can be devised to include the effects of complex organic pollutants has still not been adequately investigated.

How mycorrhizas influence plant response to heavy metals

In the real world outdoors species grow together and interact with each other. We need to predict how pollutants will affect such communities and ecosystems. I start with a two-species system, a mycorrhizal plant. If the effects of a heavy metal are tested on a plant growing in solution culture it will almost certainly be non-mycorrhizal, whereas plants growing outdoors in soil are usually infected by mycorrhizal fungus. Will this alter the plant's response to the heavy metal? In mine waste, where heavy metal pollution can make revegetation difficult, mycorrhizal inoculum is often lacking, so there is opportunity to decide whether to speed up mycorrhizal infection by inoculation or to discourage it. Mycorrhizal fungi increase plant uptake of various substances from soil, including Cu and Zn (Tinker & Gildon 1983). These two elements are essential micronutrients, so increased uptake is sometimes beneficial to the plant. But in higher concentrations they are toxic, so it is possible that in soils contaminated with large amounts of Zn or Cu (or perhaps other heavy metals) mycorrhizal infection could increase the plant's uptake and therefore the harm to the plant. This does sometimes happen. Killham & Firestone (1983) grew a bunchgrass, *Erharta calycina*, with some plants VA-mycorrhizal but others non-mycorrhizal. The nutrient solution added to the soil contained Cu, Ni, Pb and Zn in some pots but not in others. Mycorrhizas increased the concentration of the heavy metals in the shoots of the grass; Table 7.2 shows results for Cu. Shoot growth of mycorrhizal plants was significantly reduced by the heavy metal treatment, whereas non-mycorrhizal plants were little affected.

However, mycorrhizas reduce the sensitivity of other plant species to heavy metals. Bradley *et al.* (1981, 1982) showed that mycorrhizal infection caused the heather *Calluna vulgaris* and two other ericaceous shrubs to take up less Cu and Zn into the shoots, and to continue growth at high external concentrations of Cu and Zn which strongly inhibited growth of non-mycorrhizal plants (Table 7.2). Mycorrhizal

Table 7.2 Response of two plant species to high copper concentration in the nutrient medium, when grown with or without their normal mycorrhizal associate

			Dry weight of plant		
	Mycorrhizal status	Cu concentration in shoot ($mg\,g^{-1}$)	mg	% of low-Cu control	Source of data
Erharta calycina	VAM	0.28	125†	65	
(a grass)		*	*		1
	NM	0.18	159†	90	
Calluna vulgaris	ERM	6	9.7	20	
(heather)		*	*		2
	NM	44	0.7	2	

ERM, ericoid mycorrhizal; NM, non-mycorrhizal; VAM, vesicular–arbuscular mycorrhizal.
* Statistically significant difference ($P < 0.05$) between numbers immediately above and below.
† Shoot only.
Sources of data: 1, Killham & Firestone (1983); 2, Bradley *et al.* (1981).

infection of seedling birch trees can in a similar way reduce Zn uptake into the shoots and reduce growth inhibition (Brown & Wilkins 1985). Heather and other Ericaceae have ericoid mycorrhizas, birch has ectomycorrhizas. Both of these types of mycorrhiza have more fungal tissue in the infected roots than the VA-mycorrhiza that occurs in the grass *Erharta* (Table 7.2) and in most other herbaceous plants. There is some evidence that the fungi of ericoid and ectomycorrhizas can sequester heavy metals and hence reduce the amount reaching the shoots, and this could explain how the plant becomes more tolerant of the heavy metals (Bradley *et al.* 1982, Brown & Wilkins 1985). However, it is too early to generalize that ericoid and ectomycorrhizas always make plants more tolerant of heavy metals whereas VA-mycorrhizas make them less tolerant; we need information from more species.

Microcosms

We must now consider more complex systems where many species, perhaps at several trophic levels, interact with each other. It may be possible to simulate these in *microcosms*. I have already mentioned a simple example, a soil sample. Even 1 g of soil will contain many species, some competing with others, some eating others. To measure an 'ecosystem process' such as the overall respiration rate of the soil sample, without any information on individual species or even groups, is by no means useless but far from ideal. A more useful example of an ecosystem process to measure is rate of litter breakdown, since this will affect rates of nutrient cycling and development of soil organic matter. It can be measured by putting weighed litter samples in mesh bags, which can be placed at test sites outdoors, and weighing them again later. Berg *et al.* (1991) used this method to study the rate of decomposition of Scots pine needles at different distances from a mill and a smelter in Sweden that were sources of heavy metal pollution. At positions less than 1 km from the source decomposition was markedly retarded; beyond that distance the effect was small. Unfortunately the results do not show in detail how decomposition rate related to heavy metal content of the needles.

Lampert *et al.* (1989) compared the effect of the herbicide atrazine on the fresh-water crustacean *Daphnia* in a simple laboratory microcosm and in a 'mesocosm', an enclosed volume of water in a lake in northern Germany. Table 7.3 summarizes the treatments and the results. In the microcosm the alga *Scenedesmus* was cultured in flasks separate from the *Daphnia*, and a set volume of this algal suspension per unit time was supplied as food to the *Daphnia* flasks. When atrazine was added to the alga flasks the *Daphnia* dependent on them was much more sensitive to atrazine than if atrazine was used to treat *Daphnia* direct (Table 7.3). Since atrazine is an inhibitor of photosynthesis, this is perhaps not surprising. More surprising is the fact that when *Daphnia* was in the mesocosm, in the natural environmental conditions and among the full plankton species complement of a lake (though fish

Table 7.3 Response of *Daphnia* to the herbicide atrazine in its surrounding water. 'Effective concentration' is the lowest concentration at which an effect was observed

Where experiment was conducted	Other species present	What was measured*	Effective concentration of atrazine ($\mu g\,l^{-1}$)
Laboratory	None	Death (48 h LC_{50})	10 000
Laboratory	None	Growth and reproduction	2000
Laboratory	*Scenedesmus* (green alga)	Population biomass	50–100
Lake, enclosure	Phytoplankton + zooplankton (lake population)	Population numbers	0.1–1

* All measurements were on *Daphnia*.
From Lampert *et al.* (1989).

were excluded), it was more sensitive still: its population was reduced by atrazine two orders of magnitude more dilute than was effective in the laboratory microcosm. The reason for this marked difference between microcosm and lake is unknown.

A test in a stream

Figure 7.5 gives results from an experiment in Ohio which provided information on the response of the normal insect population of a stream to one pollutant. Over a period of 3 yrs copper salt was added at one point at a rate which maintained an approximately constant Cu concentration near that point. The concentration declined with distance downstream, but after 2.6 km was still about twice that in the unpolluted stream above the contamination point (Fig. 7.5(a)). The abundance of bottom-living insects decreased greatly where the Cu concentration was highest, and recovery in abundance downstream closely mirrored the decline in Cu. The insect species composition also changed (Fig. 7.5(b)). In particular, mayfly larvae almost disappeared from the most Cu-rich sites, whereas chironomids (midges) increased. The results show that a mean Cu concentration of $40\,\mu g\,l^{-1}$ was sufficient to have a marked effect on the insects. This concentration is lower than the 96 h LC_{50} values for Cu shown by any of the 10 fish species that have been tested (Table 7.1), in other words the stream insects were more sensitive. One possible reason for that is the difference in length of time, 4 days in the LC_{50} tests versus 3 yrs in the stream experiment. Hardness was not a reason for the difference, since the stream water, like the water in the fish tests, was high in $CaCO_3$.

The change in species composition in response to Cu shown in Fig. 7.5 raises the question whether the abundance of particular species or groups of organisms can be an effective way of monitoring for pollution. As a routine operation it is likely to require simpler equipment than some of the techniques previously described; but it requires skills in identification. To make the method reliable, we need to know that a particular species or group of species is consistently responsive to

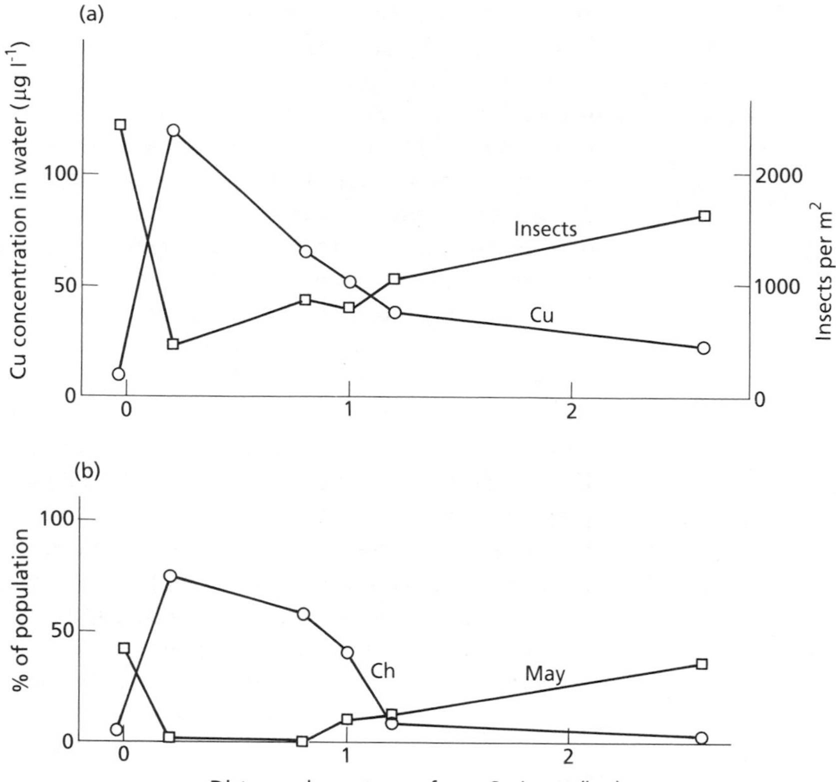

Fig. 7.5 Results from an experiment in a stream in Ohio, in which Cu was added at one point. (a) Mean Cu concentration in water and number of insect individuals in stream bottom. (b) Percentage of total insects that were chironomids (Ch) and mayflies (May). Data of Winner *et al.* (1980).

pollution; the response could be either an increase or decrease in abundance, as long as it is consistent. Chironomids are often abundant in polluted waters; so is the polychaete worm *Capitella capitata* (Gray, J.S. 1989).

Pollutant concentrations within living organisms

So far in this chapter there has been much about the responses of species to chemicals outside them or in their food. But it is the amount that gets into their tissues that is critical in determining the effects. The concentration in plant and animal tissues can, as we shall see, be either higher or lower than the concentration outside. As well as determining whether there is a harmful effect on that individual, the concentration will determine how much is passed on to another animal that eats it. There has been particular interest in whether chemicals become more concentrated along the food chain. Is it always true that the concentration of a chemical in the flesh of a herbivorous

Do pollutants become more concentrated as they pass up food chains?

animal is higher than in the plants it ate? If so, the carnivore that eats it will get a higher dose; but is the resulting concentration in the carnivore's flesh higher still, so the top carnivore that eats it gets a higher dose still? Such concentration along the food chain was at one time assumed by many ecologists to be a widespread phenomenon. An ecology textbook that is still on the shelf in my office (Collier *et al.* 1973) declares unequivocally: 'This process of biological concentration or magnification of materials is a general property of food chains.' Subsequent research has shown that things are not so simple.

My first examples are inorganic cations. Heavy metals such as lead, zinc and cadmium (e.g. from smelters) can be deposited from the air, or they can be present in soil (e.g. from application of sewage sludge). There are also persistent radioactive cations in soil which cause concern. The major accident at a nuclear power station at Chernobyl in Ukraine in 1986 resulted in ^{134}Cs and ^{137}Cs being deposited in various parts of Europe. The half-life of ^{137}Cs is 30 yrs, and since Cs (caesium) is leached only slowly from most soils, soil in these areas is likely to retain radioactive Cs well into the 21st century. Later I consider how organochlorine compounds enter animals, as examples of pollutants very different from inorganic cations in their physical and chemical properties. Box 7.1 gives more information about these organic and inorganic substances and their origins.

Inorganic cations

Availability of soil nutrients to plants

If metals or their salts are deposited on plants from the air, either dissolved in rain or as dry deposition, usually little will be taken up through the plant surfaces into the tissue, and the main interest is in the amount ingested by herbivorous animals. However, pollutant cations also get into soil, and we need to consider how these will be taken up by plants and the resulting concentrations in the plant tissues. Scientists have been interested in cation uptake from soil by plants for more than a century, but initially the interest was in major essential elements such as potassium and magnesium. One practical aim was to be able to predict whether growth rate of crops on a particular soil would be restricted by deficiency of a particular nutrient element, and if so how much fertilizer should be added. Agricultural scientists soon found that simply measuring the total amount of K (for example) in a soil sample was not a good indicator of the amount available to plants. In most soils K is present in a range of states which differ markedly in their availability to plants.

1 The ions dissolved in the soil water are the most readily available, though they still have to reach the surface of a root or its mycorrhizal fungi before they can be taken up.

2 Ions attached by their positive charge to negatively charged sites ('ion exchange sites') on clay or organic matter are less mobile, but can be taken up by plants.

3 Ions can also be adsorbed between clay micelles, which makes them essentially inaccessible to roots and only slowly exchangeable with more labile states.

4 Elements that are components of the mineral material itself are usually released extremely slowly.

Taking into account these states, it was found many decades ago that using fairly strong salt solution (e.g. 1 molar ammonium nitrate) to extract cations gave a useful indication of the amount available to plants. Such a solution will extract much of what is on ion exchange sites. This will not be exactly equal to what a plant will extract, since that will be influenced by features of the plant such as its root morphology, mycorrhizal associates and ability to excrete acids; but it is at least related to it. In more recent decades understanding of nutrient availability has greatly improved, taking account of such things as rates of nutrient transfer between different soil pools, concentration gradients and rates of movement around roots, and properties of the roots themselves. For more information see Nye & Tinker (1977), Wild (1988).

Uptake of ^{137}Cs by plants

Scientists working on uptake of pollutant cations have varied widely in their approach. When studying heavy metal input to soil from sewage sludge, the amounts available to plants have often been assessed using extractant solutions (e.g. Beckett *et al.* 1983). This has led to 'rules of thumb' used by agricultural advisers in Britain, quoted by MacNicol & Beckett (1985); e.g. if 0.5 molar acetic acid extracts 100 or more micrograms of Zn per gram of soil, Zn is likely to be toxic to plants. Yet in much of the research from 1986 onwards on the fate of ^{137}Cs from the Chernobyl explosion the *total* amount of ^{137}Cs in the soil has been taken as a measure of the amount available to the plant. Cs and K appear in the same column of the periodic table of elements, have similar chemical properties and are probably taken up by the same active carrier in plant membranes. Yet decades of research on K in soil and its uptake by plants seem to be largely ignored: uptake of ^{137}Cs is still being analysed in terms of a concentration ratio (*CR*):

$$CR = \frac{\text{concentration of substance in plant}}{\text{concentration of substance in soil}}$$

Among cations K is one that is particularly firmly held in clay micelles, and this is true for Cs also. It is not surprising, therefore, that marked variations in *CR* for Cs have been found. Sheppard & Evenden (1988) found leaf: soil *CR* values of 3–9 for blueberry plants growing on peat, but reported that other workers had found values ranging from 0.04 to 0.25 for other species on other soils. When the same species has been compared on different soils, ^{137}Cs uptake tends to be highest from soils that are high in organic matter (Barber 1964). But however carefully the soil is studied, that alone will not predict exactly the concentration in the plants, because features of the plants themselves can also have a marked influence. I have already pointed out that mycorrhizal infection can either increase or decrease heavy metal uptake by plants (Table 7.2). Concentrations can also change markedly through the growing season. For example, during 1987 the concentration of ^{137}Cs in leaves of the grass *Molinia caerulea* growing at a site in southern Germany declined from more than 3 Bq g^{-1} dry weight in early June to less than 0.3 in November (Bunzl & Kracke 1989). All the leaves had been newly grown in the spring of 1987, so they could not carry surface Cs from the fallout of 1986. K concentration declined approximately in parallel through the season, and the mechanisms causing change in K and Cs are probably the same. *M. caerulea* is known to lose much K by leaching from leaves from September onwards, as they senesce (Morton 1977); but reductions in concentration earlier in the season could be caused by dilution of Cs and K if plant growth is faster than uptake of Cs and K, and by redistribution within the plant.

Although there is some loss of cations such as K from leaves by leaching, most inorganic ions once taken up into a plant remain there while it is alive; most loss is in tissue that is eaten or dies. In contrast, a major determinant of the concentration of a metal element in a land animal is the balance between the rate of intake and the rate of loss in faeces and urine. The animal's growth rate can make some difference, too, since any extra tissue allows some opportunity for pollutant intake into the animal without increasing the overall concentration. The next example should help to make this clear.

^{137}Cs in animals

Alfalfa grown in northern Italy in 1986 became contaminated with ^{137}Cs from Chernobyl. Some of this alfalfa was made into meal and used to feed rabbits during a 6-week experiment. The ^{137}Cs content of whole animals was determined, so that average whole-body concentrations can be calculated. Table 7.4 shows that if rabbits were fed a diet high in ^{137}Cs, within 3 weeks the concentration had reached a plateau: it increased little more after 3 further weeks of feeding the high-^{137}Cs diet. If the animals were switched back to low-contamination food, within 3 weeks their ^{137}Cs concentration was back to normal. The low concentration in the rabbits' tissues and the rapid adjustment were possible because large proportions of the ingested ^{137}Cs were lost in faeces and urine (Table 7.4 (b)). However, 11% of the assimilated ^{137}Cs was retained in rabbit tissue during weeks 4–6. But because the animals were growing quite rapidly, there was about enough new tissue to accommodate the extra isotope with little increase in the whole-body concentration.

The rate at which cattle and sheep lose radioisotopes after they are removed from contaminated pasture is a subject of practical importance if they are being grazed on areas still containing substantial amounts

Table 7.4 ^{137}Cs concentration and balance of rabbits fed with alfalfa that had been contaminated during growth by fall-out from the Chernobyl explosion

(a) Whole-body concentration of ^{137}Cs in rabbits (Bq kg^{-1})

	Day		
Feeding regime	1	21	42
High-^{137}Cs food days 1–42	27	76	81
High-^{137}Cs food days 1–21, then low-^{137}Cs	27		20
Low-^{137}Cs food days 1–42	27		16

(b) ^{137}Cs balance, days 21–42, for rabbits fed high-^{137}Cs throughout

	Bq per animal	% of total	% of assimilated
Lost in faeces	2060	78.4	
Lost in urine	507	19.3	89
Retained in tissue	62	2.4	11

Data from Battiston *et al.* (1991).

of fall-out from Chernobyl. These animals are then moved to less contaminated areas, to reduce the concentration of isotope in their flesh. The resulting decline of ^{137}Cs in muscle of lambs in England was found to have a half-life of about 10 days (Coughtrey *et al.* 1989), in other words half of the remaining isotope was lost every 10 days. For example, the concentration in one set of lambs was reduced to about $\frac{1}{8}$ ($= \frac{1}{2} \times \frac{1}{2} \times \frac{1}{2}$) in a month (= 3 × 10 days). When cows were changed from contaminated to uncontaminated silage the ^{137}Cs concentration of their milk dropped to half in about 2–3 days, but the remainder was lost more slowly (Vreman *et al.* 1989).

Heavy metals in animals

Since the amount of a heavy metal in a land animal depends on the balance between gain in food and loss in urine and faeces, it follows that the concentration in the animal may be either higher or lower than in its food. Table 7.5 shows an example of this. Woodlice and dead leaves were collected from the ground-floor litter of a deciduous wood on the outskirts of Bristol, England; it borders on a busy motorway, and is 3 km downwind from a large smelting works. The concentrations of Pb, Zn, Cd and Cu on or in the leaves were much higher than for leaves from several woods in less polluted sites. Table 7.5 shows that after the woodlice had fed on the leaves for 20 weeks the four pollutant elements differed greatly in their concentration ratios, ranging from Cu far more concentrated in the woodlice than in their food to Pb far less concentrated.

Cu and Cd concentrations in a food chain

A detailed study of amounts of Cu and Cd in soil, plants and animals, involving many samples through a 12-month period, was made by Hunter *et al.* (1987a,b,c). Their samples were from three areas of rough grassland in Merseyside, England, one close to a copper refinery, another 1 km away and a third in a much less polluted area. Table 7.6(a) shows that there was a wide range of Cu and Cd concentrations among the invertebrates in each of the three diet groups, so that among plants and invertebrates there was more variation within each trophic level than there was between trophic levels. There was, nevertheless, a

Table 7.5 Concentrations ($\mu g\, g^{-1}$ dry weight) of heavy metals in leaves of a tree species (*Acer campestre*) collected from the litter layer of a polluted wood in England, and in whole bodies of woodlice (*Porcellio scaber*) that were fed on the leaves for 20 weeks in controlled conditions

	Cu	Cd	Zn	Pb
Leaves	52	26	1430	908
Woodlice	1130	73	1370	132
Concentration ratio	21.7	2.8	1.0	0.15

$$\text{Concentration ratio} = \frac{\text{concentration in woodlice}}{\text{concentration in their food}}$$

From Hopkin (1990a).

Table 7.6 Copper and cadmium concentrations ($\mu g\, g^{-1}$ dry weight) in plants and animals at three sites in Merseyside, England

(a) Range of concentrations in plants and animals at one heavily polluted site close to a refinery

	Cu	Cd
Plants, four species, above-ground parts	73–260	2.6–4.7
Detritivores		
Invertebrates	210–2390	19–231
Herbivores		
Invertebrates (all insects)	160–731	2.0–38
Small mammals, two species	12, 13	2, 3
Carnivores		
Invertebrates	300–1020	14–102
Small mammals, one species	29	71

(b) Concentrations in common shrew (*Sorex araneus*, a carnivore) and in its food, at three sites

	Cu			Cd		
Site	Least polluted	1 km from refinery	Near refinery	Least polluted	1 km from refinery	Near refinery
Concentration in food	52	104	652	2	16	55
Concentration in shrew	13	17	29	4	19	71

From Hunter *et al.* (1987a,b,c).

clear tendency for the invertebrate detritivores and carnivores to have higher concentrations than the plants. In contrast, the Cu concentrations in the mammals were lower than in any of their possible food items. Table 7.6(b) shows more detailed results for the carnivorous mammal, the common shrew, from the three sites which differed greatly in amount of pollution. At all three sites the concentration of Cu in the shrew's tissue was lower than in its food. The Cu concentration in its diet increased more than 10-fold between the least polluted and most polluted site, but its body concentration increased only twofold. In contrast, the Cd concentration in the shrew was similar to that in its food, rising as the food Cd rose. The two herbivorous mammals had diets lower in Cu and Cd, but like the shrew they were able to maintain their internal Cu concentration nearly constant as Cu in their food increased, whereas their body Cd concentration was always close to that of their food. These results show a strong ability of these mammals to control their body Cu even in heavily polluted areas, but not to control their Cd. Cu is an essential element for mammals, since it is a constituent of the enzyme cytochrome oxidase, whereas Cd is not beneficial even at low concentrations. The better ability of these mammals to adjust their body Cu concentrations than their Cd concentrations may relate to this.

Laskowski (1991) has drawn together from various sources data on the concentrations of Cu and Cd in animals (mammals and invertebrates; detritivores, herbivores, carnivores and top carnivores). For both elements some animals had higher body concentrations than their food and others had lower concentrations. That paper and the examples I have described show that it is certainly not consistently true that heavy metals are concentrated as they move up the food chain. The concentration of a particular element within the body of a particular animal species is strongly dependent on the ability of that species to eliminate the element in faeces and urine, an ability which varies greatly between elements and between species.

Concentrations of organic pesticides within living things

Organochlorine compounds differ from inorganic cations in several fundamental ways. One is that they are very insoluble in water, but are soluble in fats. A result of that is that before they can be excreted by an animal they have to be converted into more water-soluble compounds. So the ability to excrete a particular compound depends on the animal having a suitable enzyme. There is more information on the biochemistry involved later in this chapter.

Three possible ways that the concentration in an animal could respond to supply of one of these chemicals are:

1 If both intake and excretion are very slow the concentration within the animal might go on increasing slowly throughout its life. Some examples of this are known, e.g. DDT and dieldrin in lake trout in Lake Michigan in the early 1970s (Kogan 1986, Table 10.10).

2 The concentration may build up to a plateau level, where rates of uptake and loss are balanced. Figure 7.6 shows such a time-course for the concentration of dieldrin in an alga, an invertebrate and a fish when they were immersed in a very dilute solution of dieldrin. If the external concentration increases the plateau concentration in the organism is likely to rise too.

3 The excretory system may be able to maintain the body concentration at a level unaffected by the external concentration. This was approximately true for Cu in the shrew (Table 7.6).

Do aquatic animals get more pollutant from their food or from the surrounding water?

There has been much concern about organic pollutants, from pesticides or industrial waste, getting into fresh waters and oceans. Here the animals can acquire chemicals either from their food or by absorption through permeable membranes, e.g. in gills of fish. Experiments have been conducted to find out which of these routes is the more important. In one experiment guppy (small fish) were kept in water with dieldrin at 0.8–2.3 $\mu g\,l^{-1}$ and fed with uncontaminated *Daphnia*. Other *Daphnia* were kept in water with a similar concentration of dieldrin, and were then fed to guppy that were in uncontaminated water. After the concentration of dieldrin had reached a plateau in both sets of guppy, the concentration was about 10 times as high in those that received dieldrin from the surrounding water as in those that received it via

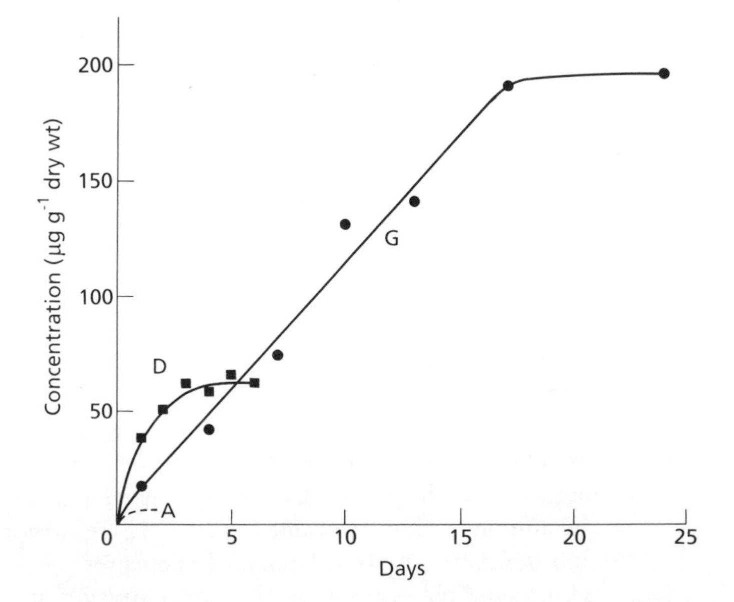

Fig. 7.6 Concentration of dieldrin in an alga *Scenedesmus obliquus* (A), a cladoceran *Daphnia magna* (D) and guppy (G) (a small fish). The species were immersed, separately, in dieldrin at 3–5 $\mu g\,l^{-1}$. The *Daphnia* were not fed during the test, the fish were fed on uncontaminated *Daphnia*. Data of Reinert (1972).

their food (Reinert 1972). Most experiments with other aquatic animals have also found that food is a minor source of pollutant intake compared with absorption from surrounding water (Moriarty 1988).

Concentrations of organic pollutants in aquatic species: relation to solubility in fat

Since animals (and also algae) are absorbing organic fat-soluble materials from the surrounding water, a very simplified 'model' for an aquatic animal would be to consider it as a volume of fat, living surrounded by water, with a solute capable of diffusing from one to the other. Given enough time for equilibration, the solute should reach a steady concentration in the animal which depends on the relative solubility of the solute in fat and water. This relative solubility is usually measured by the octanol:water partition coefficient. Octanol, water and the chemical are shaken up together, left for the octanol and water to separate into two layers, and the concentration of the chemical in each layer is then determined. Figure 7.7 shows that the octanol: water partition coefficient can sometimes be a good predictor of the concentration of a chemical that will build up in a fish. This in turn has led to the use of the octanol:water partition coefficient as a predictor of the toxicity of chemicals, on the grounds that, other things being equal, the more concentrated the chemical becomes in the animal's tissues the more toxic it will be. This may sound an excessive over-simplification, but Table 7.7 shows that it can work quite well. Toxicity of chemicals was measured on five quite unrelated species. In the first four tests shown the octanol:water partition coefficient correlated strongly with the measured toxicity. (The correlations are negative because chemicals that are more soluble in octanol give lower LD_{50} and LC_{50} values, i.e. they are more toxic.) However, the correlation coefficients were far enough from 1 to show that the prediction may not be precise or reliable. The method was very poor at predicting the toxicity of substances fed to rats.

However, the octanol:water partition coefficient does not always correlate so well with concentrations in animals as in the example of

Table 7.7 Correlation coefficients between log (octanol:water partition coefficient) of 54 or 59 organic pollutants and their toxicity to various species

Species	How chemical applied	What measured	Period	Correlation coefficient
Lettuce	In solution culture, roots immersed	Growth, EC_{50}	16–21 days	−0.76*
Fathead minnow	Animals immersed	Death, LC_{50}	96 h	−0.66*
Daphnia magna	Animals immersed	Death, LC_{50}	24 h	−0.59*
Photobacterium phosphoreum	Bacteria immersed	Luminescence, EC_{50}	30 min	−0.71*
Rat	Fed by mouth	Death, LD_{50}	96 h	−0.23 N.S.

Statistical significance: * $P < 0.001$; N.S. not significant ($P > 0.05$).
From Hulzebos *et al.* (1991) and Kaiser & Esterby (1991).

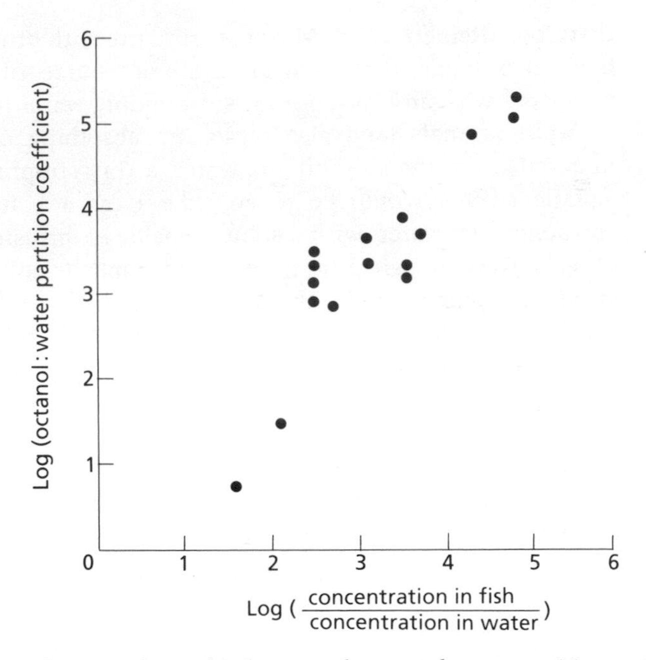

Fig. 7.7 Relationship between the octanol : water partition coefficient and the fish : water concentration factor for 16 insecticides. The fish were *Gambusia affinis*. From Kogan (1986).

Fig. 7.7 (Moriarty 1988, Walker 1990). If the animal has better ability to excrete some chemicals than others, clearly that could affect the concentrations in its body. Differences in concentration between species and between different parts of the body can sometimes, though not always, be explained by differences in fat content. Some of the results may be confusing because the experimenters did not wait until equilibrium concentrations had been reached. Figure 7.6 shows how the body concentrations of dieldrin in three species each rose to a plateau, but took different lengths of time to reach it. Conclusions about the three species' rates of uptake would be different if you harvested on day 2 or day 20.

Concentration up the food chain?

According to classical theory, each species builds up a higher concentration of the chemical in its tissues than was in its food, and so the chemical becomes more concentrated up the food chain. But if aquatic animals are not getting most of the pollutant intake from their food, will there be concentration up the food chain? Figure 7.6 shows that it can happen: in that experiment the animals ate no polluted food, the only possible source was the surrounding water, yet the final concentrations were in the order plant < herbivore < carnivore. However, other aquatic examples show the opposite trend. Moriarty (1988, Table 6.8) gives data from samples taken from the Mediterranean in which the concentration of PCBs in the phytoplankton and herbivorous zooplankton was about 10 times higher than in carnivorous shrimps. An important feature of Fig. 7.6 to emphasize is that the concentration of dieldrin in the water was 0.003–0.005 μg per gram of water, so all three species concentrated the insecticide greatly in their tissues. The concentration factors, from water to tissue (dry weight) were: alga 1300, *Daphnia* 14 000, guppy 49 000.

Among land animals, whose main source of pollutants must be their food, one might expect concentration up food chains to show up more clearly, but in fact the results are mixed. Freedman (1989) emphasizes in his text that concentrations increase up the food chain, yet the 'typical concentrations' that he gives in his Fig. 8.1 show no clear difference between land animals at three trophic levels. There are some field measurements that suggest concentrations increasing; for example Woodwell *et al.* (1967) found DDT residues in fish collected from an estuary on Long Island ranged from 0.2 to $2\,\mu g\,g^{-1}$, whereas in birds such as gulls that eat fish they were $1.5-75\,\mu g\,g^{-1}$. But some experiments have shown no increase in concentration. When hens were fed diets ranging 1000-fold in their DDT concentration, the concentration in their eggs was always very close to that in the food (Moriarty 1988, Fig. 5.15).

The conclusion, on both organic and inorganic pollutants, must be that they are not consistently more concentrated as they pass up food chains. Some animals contain higher concentrations in their bodies than was in their food, but other animals do not. Perhaps the most important conclusion to come out of this research is that aquatic species can have in their bodies concentrations of organic pollutants thousands of times more concentrated than in the water that surrounds them. These can be passed on to fish-eating birds, which sometimes receive harmful doses, as the following case history shows. And from these fish-eaters pollutants can pass, through their droppings and predators, back to land. So pollutants washed into fresh water and oceans cannot be safely ignored, even if they are very much diluted.

A case study: death of birds of prey

Up to now this chapter has always worked from a named pollutant to its effect. But in real life the problem may present itself the other way round: animals or plants become unhealthy or die, and we want to know the cause. Is it pollution or something else? The example I take here is well known, and it may seem in retrospect obvious that pollution was to blame, but at the time it was not. The problem started in the late 1950s and early 1960s, when bird-watchers in North America and western Europe noticed declines in abundance of some bird species, especially birds of prey, and also failure of adult pairs to produce young. Marked declines in peregrine falcon were reported from various European countries, including Britain, Sweden, Germany and Poland,

Declines in numbers of birds

and from U.S.A. and Canada (Ratcliffe 1980). In eastern U.S.A. the peregrine had disappeared altogether by the early 1960s. There were quantitative data for some areas, e.g. Fig. 7.8. Among other species declining in abundance was the bald eagle, a symbolic bird for Americans.

So what might be the cause of these declines? Among possibilities were unusual weather conditions, reduced food supply (e.g. because of changes in vegetation cover), increased predation on birds or their eggs, disease, direct effects of people by increased shooting or removing eggs. Some of these were made unlikely because the declines were occurring over a wide area on both sides of the Atlantic more or less in synchrony. One thing that happened during the 1950s, in western Europe and North America, was the first widespread used of the new synthetic organochlorine insecticides: DDT was introduced in the late 1940s, the cyclodienes including aldrin and dieldrin in the mid-1950s. A clue that these might be affecting bird populations was first provided when people noticed unusually large numbers of dead birds, especially of seed-eating species, near fields where these insecticides had been used. This suggested that cyclodienes might be involved, since they were commonly used as seed-dressing insecticides and were shown experimentally to kill many bird species (Newton & Wyllie 1992). However, DDT and its breakdown products are much less toxic to adult birds than the cyclodienes, and seemed rarely to be present in sufficient concentrations to cause death. Evidence that DDT was involved in population decline of peregrines started from observations that there were often broken eggs in nests of the birds of prey. The number of broken eggs became so large that it caused a substantial percentage reduction in the number of young successfully reared per

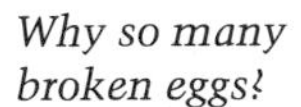
Why so many broken eggs?

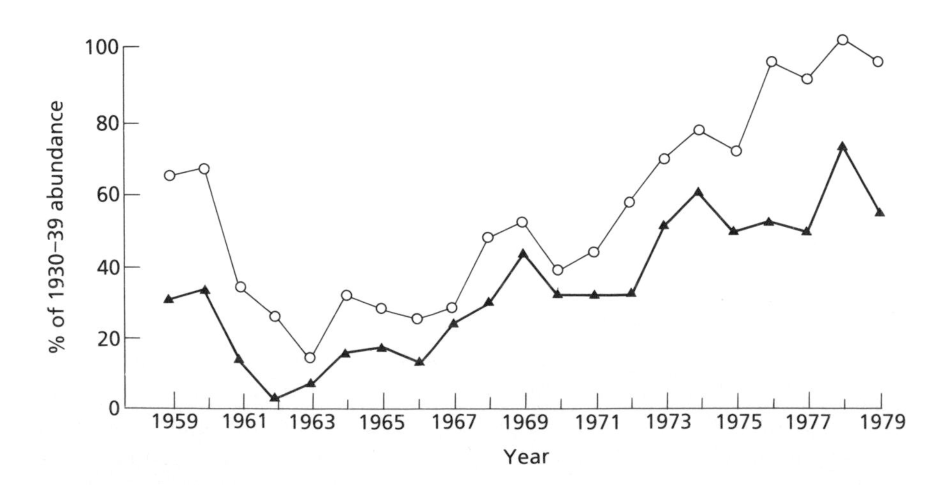

Fig. 7.8 Abundance of peregrine falcon (*Falco peregrinus*) in north-west England. ○, number of pairs producing eggs; ▲, number of pairs rearing young, both expressed as a percentage of the number in the same area in 1930–39. From Ratcliffe (1980).

adult pair per year. It was at first not known whether the breakages were due to change in behaviour of the adult birds, making them attack their own eggs, to increased predation, or to thinner and more fragile shells. The question 'Have egg-shells got thinner?' could be answered, because there were many eggs, collected at various dates in the 20th century, in museums and private collections. Figure 7.9 shows that the eggs of peregrines in California had maintained a nearly constant mean shell thickness from 1891 to 1946, but the mean for 1947–52 was significantly lower. Similar results were found for peregrine in Britain (Ratcliffe 1970). Other species whose egg-shells became significantly thinner at about the same time were bald eagle and osprey in U.S.A. and golden eagle, merlin, sparrowhawk and shag in Britain (Hickey & Anderson 1968, Ratcliffe 1970).

There could be several possible reasons for the egg-shells being thinner: disease or change in the birds' diet seemed possible, even ^{90}Sr (strontium) from atmospheric bomb tests was suggested. It was now possible to investigate whether organochlorine insecticides were a cause. Insecticides were fed to captive birds, though not, as far as I know, to peregrines. Porter & Wiemeyer (1969) fed a mixture of DDT and dieldrin to sparrowhawks. Their dose rate, which was intended to be about equal to the amount the birds would take in with normal food items during the 1960s, resulted in egg-shells 8–17% thinner than for control birds. Eggs of wild sparrowhawks collected between 1947 and 1967 did in fact have shells 17% thinner than eggs collected earlier (Ratcliffe 1970). Experimental confirmation that DDT on its own can cause egg-shell thinning was obtained for other bird species. The other approach was to find out whether the amount of shell thinning was correlated with the concentration of organochlorine pesticides and their breakdown products in the eggs. Some of the

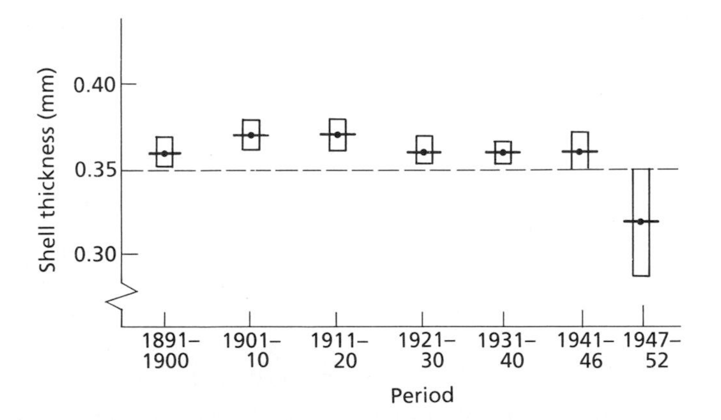

Fig. 7.9 Mean shell thickness, with 95% confidence limits, of eggs of peregrine falcon collected in California in periods from 1891–1952. From Hickey & Anderson (1968).

earlier studies showed a very wide scatter in the results, but later studies have shown very convincing correlations (Newton *et al.* 1989, Newton & Wyllie 1992). Figure 7.10 shows that when mean values are taken there was a clear tendency for species with higher organochlorine concentrations to have suffered greater egg-shell thinning. The three black circles at lower right are for peregrine, sparrowhawk and merlin. Birds make up a substantial part of the diet of all three; they could have received their pesticides by eating other birds which had eaten seed or crops dressed with insecticide. Another possible source is if they ate tame pigeons that had been treated with insecticide for control of external body insects. The lowest black triangle is shag, whose sole food is ocean fish. Shag's eggs contained significant amounts of the DDT breakdown product DDE and of HEOD which is the active ingredient of dieldrin. This illustrated the point that these pesticides can be recycled from the ocean to birds in dangerous amounts. However, three other fish-eaters (kittiwake, guillemot and razorbill), shown in Fig. 7.10, also had substantial concentrations of organochlorines in

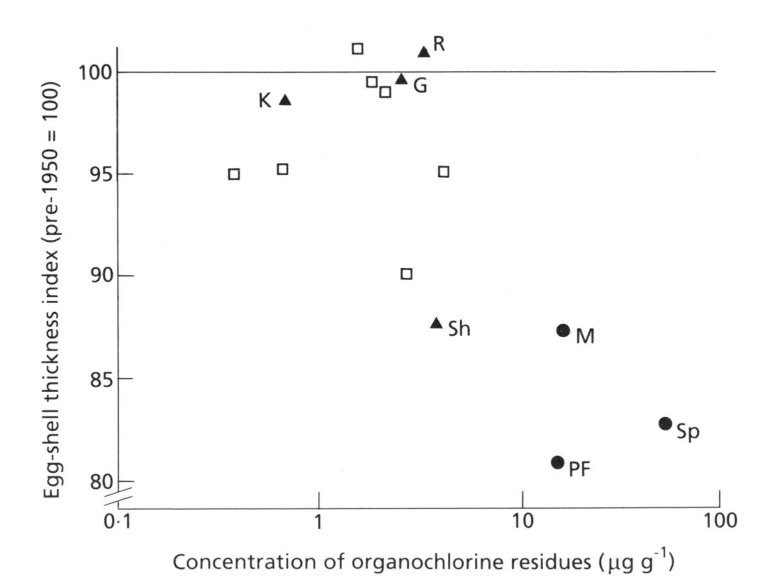

Fig. 7.10 Data for 14 carnivorous or omnivorous bird species in Britain, showing the relationship between the egg-shell thickness and the concentration of organochlorine pesticide residues in eggs. Each point is the mean for a species. The residue concentrations, plotted on a log scale, are the means of measurements made in the 1960s.

$$\text{Egg-shell thickness index} = \frac{\text{Weight of shell}}{\text{Egg length} \times \text{breadth}}$$

The vertical axis shows the change in this index from before about 1950 to later. G, guillemot; K, kittiwake; M, merlin; PF, peregrine falcon; R, razorbill; Sh, Shag; Sp, sparrowhawk. Principal food of the species: ▲, fish; ●, birds, □, other (small mammals, invertebrates, plant matter, carrion). From Ratcliffe (1970).

their eggs but showed no significant change of shell thickness. So the response of egg-shell thickness and breakage to these pesticides is still not fully understood. The pesticides may act by altering oestrogen levels in the birds, but they could have other effects as well.

As I have described it, this may sound like a routine piece of research, but I am sure it did not seem like that at the time. Ratcliffe (1980) recounts the events in a semi-popular style that conveys more what it must have actually felt like. The research involved widely differing skills, from watching birds that nest on remote cliffs to measuring minute amounts of complex organic chemicals. There were also strong feelings involved. These new insecticides were extremely useful, among other things for control of farm pests and vectors of human diseases. Firms had invested large sums of money in developing and manufacturing them. It was not surprising that some people were very reluctant to see restrictions on their use if the evidence that they were harmful was not fully convincing. In fact reductions in use of some of the insecticides started in the early 1960s and were extended later, though DDT, aldrin and dieldrin were not banned in Britain until 1986. Through the 1960s and 1970s concentrations of pesticides in eggs decreased, shells got thicker (Newton *et al.* 1989) and numbers of birds recovered (Fig. 7.8). The peregrine, which had completely died out in northeastern U.S.A., was reintroduced into New Jersey in the late 1970s and has maintained itself.

You may think it odd that I have devoted so much space to the effects of DDT, an insecticide now banned in most developed countries. There are several reasons why I thought it was worthwhile.

1 DDT and other organochlorine insecticides are still widely used in many tropical countries. Peregrine eggs measured in New Jersey in 1985–88 had shells on average 16% below pre-1947 thickness (Steidl *et al.* 1991), which is close to the thinning associated with population decline in the 1950s and 1960s (Fig. 7.10). These birds lived near the Atlantic coast, and may have ingested pesticide residues by eating birds migrating northwards from other countries.

2 The use of other toxic organochlorine compounds, e.g. PCBs, still continues in most countries, and is increasing in some. In Britain PCB concentrations in peregrine eggs and in sparrowhawk livers were still as high in the 1980s as they had been in the 1960s (Newton *et al.* 1989, Newton & Wyllie 1992). However, there is no evidence that they were concentrated enough to harm the birds.

3 We can learn a lot from looking back at an investigation that allows us to view the whole episode from start to finish.

Concentrations in animals and plants as indicators of pollution

The previous section showed that we now have some ability to predict the concentrations of toxic chemicals that will build up in plants and animals, on land and in water; but it also pointed out the difficulties and limitations. An alternative approach would be to measure concentrations within plants or animals in the field and to use these as indicators of whether amounts of particular pollutants have reached harmful levels at the site. One problem is that any chemical is likely to vary in concentration between different parts of an individual plant or different organs of an animal, and different parts are likely to have different sensitivities. Beckett & Davis (1977) minimized such variation by growing plants of a single species, barley, to the five-leaf stage and then analysing the shoot material, which would be mainly young leaf tissue. They grew the plants with various amounts of Cu, Ni, Zn or Cd in the nutrient solution, and found that for each element the relationship between growth rate (as measured by final dry weight) and heavy metal concentration in the shoot was like that shown in Fig. 7.11. There was a range of concentrations that gave a plateau weight, but above a critical concentration the weight decreased. If they plotted the concentration on a log scale the right-hand, declining region formed a straight line, and it was possible to decide the position of the corner with fair precision.

Critical heavy metal concentrations in plants

The question then is, can this method be applied to a range of species growing under a range of conditions outdoors? MacNicol and Beckett (1985) collected together data from a variety of crop species grown in soil, peat or solution culture. The results for Cu were the most encouraging. There were 28 sets of results that allowed determination of the critical concentration, in other words the corner in Fig. 7.11 where Cu begins to cause a reduction in growth. These 28 values of critical concentration ranged from 5 to >64 $\mu g\,g^{-1}$ dry weight, but 21 (75%) of them were in the range 10–30 $\mu g\,g^{-1}$. So if a crop plant is found to have less than 5 $\mu g\,g^{-1}$ of Cu in its shoot tissue it is unlikely

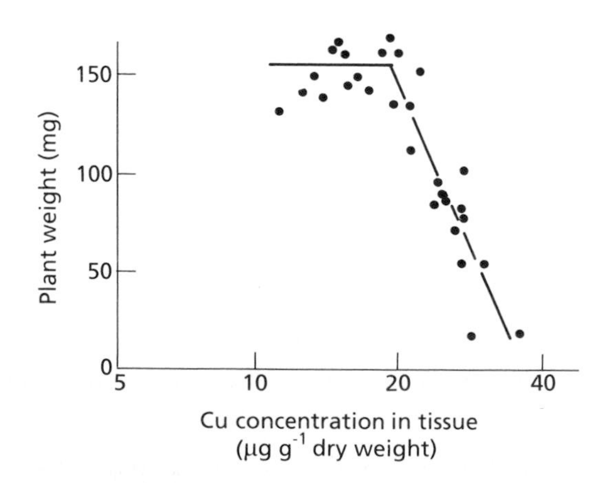

Fig. 7.11 Relationship between shoot dry weight and Cu concentration in shoots of young barley plants grown with a range of Cu supplies. From Beckett & Davis (1977).

that Cu is harmful at that site, whereas above 20–30 $\mu g\,g^{-1}$ it probably is. The results for other elements were more variable. Among 46 data sets, critical concentrations for Cd ranged from 4 to 200 $\mu g\,g^{-1}$, though 27 of them (59%) lay between 10 and 40 $\mu g\,g^{-1}$. One can at least say that a Cd concentration below about 5 $\mu g\,g^{-1}$ suggests that Cd is not harmful at that site. So this method shows promise. All the data came from experiments where the heavy metals were supplied to the roots. In sites where heavy metals are deposited on the outer surface of the foliage the method would work only if all the external heavy metal could be washed off.

A similar method could be very useful for animals, to identify which pollutants are near to danger levels. The concentration in whole invertebrates could be measured, but they would need to be selected species or groups, not just any invertebrates collected at the site, since concentrations can vary greatly between species (see Table 7.6). Concentrations can vary greatly within the body of a single invertebrate. For example, in the woodlice of Table 7.5 most of the heavy metals in each animal were in the hepatopancreas, dead-end tubes off the main digestive system. Accumulation in the hepatopancreas is a detoxification mechanism, and the concentrations in some other, more sensitive parts of the animal are presumably what determines its response to the heavy metals; so measurement of mean concentration in the whole animal may in this case not be a good indicator of whether it is near to a critical point for toxicity. One example of an informative measurement on animals is organochlorine compounds in eggs (Fig. 7.10). Enough is now known about the relationship between their concentration, egg-shell thickness and number of chicks reared so that measurements of organochlorine concentrations in eggs can be a useful indicator of whether these chemicals are reaching danger levels in the diet of the birds.

Methods of reducing the harmful effects of pollutants

Reducing production of the pollutant

If techniques described in the earlier parts of the chapter have led to the conclusion that a particular chemical is harming living things, there are various courses of action that could improve the situation. These are summarized in Box 7.3. The most obvious action is alternative 1(a), to stop producing and using the harmful substance, and this has already happened for some of the pesticides listed in Box 7.2. But it will be difficult to give up producing some pollutants soon. There is, for example, strong reluctance to stop completely the burning of fossil fuels, although this produces greenhouse gases (Chapter 2), SO_2 and polynuclear aromatic hydrocarbons. It is difficult to imagine life without metals, but most metals can be toxic. Human excreta can also be a pollutant, as discussed later, and we might find it awkward if we were told to stop producing that. Therefore we need to consider also the other possible courses of action listed in the box. As in the earlier parts of the chapter, this final section concentrates on biological aspects. It will not, therefore, consider such topics as how to clean SO_2 from power-station flues, how to store radioactive waste or what are the best substitutes for PCBs. Biologists can, however, help to reduce the need for some polluting chemicals. Possible alternative energy sources to replace fossil fuels were considered in Chapter 2, and ways of controlling pests that use less (or no) chemical pesticides in Chapter 6. Here I consider one other pollutant whose origin is largely under biological control: nitrate.

Nitrate as a pollutant

Leaching of nitrate from farmland has caused concern for two reasons. One is the potential risk to human health from nitrate in drinking water: nitrite formed from nitrate can interfere with haemoglobin function in young children ('blue baby disease'), and there is also evidence linking nitrate to stomach cancer. The other problem is eutrophication, undesirable effects of increased nutrients on rivers, lakes and some enclosed seas. Phosphorus as well as nitrogen is a major contributor to this, but phosphorus loss from farmland is usually slow (see Chapter 3), and the main source of it as a pollutant is sewage. For example, in a part of Norfolk, England, 80% of the P reaching the rivers came from sewage and 20% from farmland; whereas only 14% of the N reaching the rivers came from sewage and 86% from farmland (Moss 1988). So here I concentrate on losses of inorganic N by leaching from farmland. The question is, can these losses be reduced by altering farming practices, without serious reduction in farm yields?

Box 7.3 Methods of reducing harm by pollutants.

1 Reduce the amount of the pollutant reaching sensitive organisms.
(a) Reduce or cease production of the chemical. This may require the development of alternative, less damaging chemicals to perform the same functions, or alternative techniques not requiring chemicals, e.g. biological control of pests.
(b) Ensure the chemical remains too dilute to be harmful.
(c) Store the chemical away from organisms it can harm. This method is used, for example, for highly radioactive waste from nuclear power stations.
2 Remove the toxic chemical from sites where it is doing harm.
3 Convert the chemical to a less toxic form. For organic toxins this often involves microorganisms.
4 Manage the living things to minimize the harmful effects. Examples considered in this chapter: sowing plants that are more tolerant of heavy metals; managing lakes to reduce effects of eutrophication.

How can we reduce nitrate leaching from farmland?

I have looked for data on nitrate losses from arable cropland, comparing fields that received inorganic N fertilizer, organic manure or no N input. The best data I can find come from measurements made from 1878 to 1881 in the Rothamsted long-term winter wheat plots (Table 7.8). The farmyard manure added each year contained about 225 kg of N, substantially more than the inorganic fertilizer, but it maintained a similar grain yield. From 1850 to 1950 the grain yields from the farmyard manure plot and from a plot receiving 144 kg $ha^{-1} yr^{-1}$ inorganic N (+P, K and Mg) were closely similar, usually between 2 and 3 tons $ha^{-1} yr^{-1}$, whereas the unfertilized plot yielded only about 1 ton $ha^{-1} yr^{-1}$ (Jenkinson 1991). When studying these results we can bear in mind that the World Health Organization's recommended maximum concentration for nitrate in drinking water is 50 mg l^{-1}, though U.S.A. has adopted 45 mg l^{-1}. Expressed as milligrams of N (rather than NO_3^-) per litre, these are 11 and 10 mg l^{-1}. So any figure in Table 7.8 that is about 10 or higher is cause for concern. The results show several points of interest.

1 All plots showed marked variation through the year. These relate to the stage of development of the wheat crop: nitrate losses tend to be low when the root system is well developed and active, in the spring and summer.

2 Plots that had received no N lost the least nitrate in leaching. However, it is not realistic to recommend farmers to make no addition of fertilizer or manure to their fields (see Chapter 3).

3 N applied as ammonium resulted in less nitrate leaching than did nitrate addition at the same time. There was probably some loss of ammonium by leaching (unfortunately that was not measured), but NH_4^+ is less readily leached than NO_3^- (e.g. Table 5.8).

4 Farmyard manure resulted in nitrate losses higher than those from

Table 7.8 Concentration of nitrate in water from drains under winter wheat plots at Rothamsted, England. The figures are mean concentrations of NO_3^- (expressed as mg N l^{-1}) for four periods and for the whole year

Fertilizer or manure added each year	From time of spring fertilizer addition to 31 May	From 1 June to harvest	From harvest to autumn sowing	From autumn sowing to spring fertilizing	Whole year
None	3	0	5	5	4
Inorganic P, K, Mg (but no N)	3	0	5	6	4
Farmyard manure, 35 tons $ha^{-1} yr^{-1}$	4	1	6	10	8
Inorganic N, 96 kg $ha^{-1} yr^{-1}$ (+PKMg)					
As NO_3^- in spring	50	9	15	8	12
As NH_4^+ in spring	27	1	7	5	7
As NH_4^+ in autumn	7	3	8	28	19

Data from Cooke (1976).

unfertilized plots, especially in winter. Ammonium added in spring gave about the same average losses as farmyard manure; however, this was less evenly distributed through the year, with a burst of loss soon after fertilizer application, reaching nitrate concentrations well above the 10 mg l^{-1} danger level. The other two inorganic fertilizer treatments also gave high nitrate losses soon after they were added.

These results, and others from more recent experiments, show that N input, whether as inorganic fertilizer or organic manure, increases nitrate losses in leaching. Organic manure was in this experiment able to maintain as high a yield as inorganic fertilizer, but gave less high seasonal nitrate loss. However, organic manure does still increase nitrate loss, so it cannot be guaranteed as an automatic solution to the problem at all sites. One basic message from Table 7.8 and other research is that most of the nitrate loss occurs at times when there is no crop present or when the crop is poorly developed. Control of nitrate loss can therefore be helped by: (1) reducing the length of time between harvesting one crop and sowing the next; (2) promoting early development of a dense root system; and (3) adding the nutrients at several times through the season, in amounts related to plant demand and uptake ability. Method 3 can be achieved with inorganic fertilizers but less easily with most organic manures, which cannot be applied when the crop is well grown. These suggestions are not new; they have already been incorporated into farming practice in the developed world. Earlier sowing of autumn cereals has become widespread. An alternative way to reduce the time the soil is bare is to sow a *cover crop*, a species which develops quickly, takes up any available nutrients, and can then be ploughed in when the real crop is sown.

Table 7.9 shows N losses by leaching from cattle farms, comparing fields of grass or grass–clover unfertilized, and grass receiving high N fertilizer application. Although the grass is perennial and has active

Table 7.9 N lost by leaching, calculated for cattle farms in Netherlands and England under contrasting management

	N loss by leaching (kg ha^{-1} yr^{-1})
Dairy farming, Netherlands	
1 Grass–clover; no fertilizer (1937)	11
2 Grass; inorganic fertilizer applied*	44
Beef farming, grass, England	
1 No fertilizer, poor drainage	0
2 No fertilizer, good drainage	5
3 Inorganic fertilizer*, poor drainage	43
4 Inorganic fertilizer*, good drainage	166

* 400 kg N ha^{-1} yr^{-1}.
From Frissel (1978) and Scholefield *et al.* (1991).

roots throughout the year in England and Netherlands, N fertilizer does increase nitrate losses substantially, especially in well-drained sites. Table 3.3 shows the milk and beef yields from these two Dutch farms; yields were much higher from the one receiving fertilizer. Nitrate leached from grass pasture does not necessarily come directly from the fertilizer. Ryden *et al.* (1984) measured nitrate leaching under grassland in southern England that received 420 kg N $ha^{-1} yr^{-1}$ from fertilizer. The average loss of N by leaching from ungrazed plots cut for hay was 29 kg $ha^{-1} yr^{-1}$, but it was 162 kg $ha^{-1} yr^{-1}$ from plots grazed by cattle. Much of the extra loss is likely to come from urine, since uric acid is quickly converted in soil to ammonium and nitrate. If the N concentration in what the cattle eat increases, a large proportion of this extra N intake is passed to the urine (Scholefield *et al.* 1991), and this will happen whether the high-N feed is clover or fertilized grass. This suggests that converting from fertilized grass to grass–clover will not automatically solve the problem of nitrate leaching from pasture. Other aspects of management can help, including drainage (see Table 7.9).

Removing or degrading the pollutant

Using plants or microbes to clean heavy metals out of water

If there are dangerous amounts of an inorganic pollutant—e.g. a heavy metal or radioisotope—in water or soil, they need to be removed; living things can sometimes help with this. In sewage works metals that are in particulate form are mostly removed by letting them settle out. However, removal of dissolved substances involves a 'biological treatment system', in which the contaminated water passes over a mixture of microorganisms—usually mainly bacteria and protozoa but sometimes with algae too. These can remove a substantial proportion of some heavy metals, e.g. Cu and Pb, but are less effective for others, e.g. Ni and Mn (manganese) (Beveridge & Doyle 1989). Some of the heavy metal is taken up actively into the cells, but much of it remains outside the living cells, adsorbed in the cell walls or on the outside. Beveridge & Doyle also describe some artificial lakes constructed to help clean up very polluted effluent from mines, in which algae remove heavy metals from the water. None of these systems detoxifies the heavy metals; the microbes or plants that have accumulated high concentrations of the metals have to be collected and disposed of.

Degrading organic pollutants

Organic pollutants can, in contrast, be converted into less toxic substances, and this is usually carried out by living things. Often this happens naturally in fresh water, oceans and soil. However, some synthetic organic chemicals break down very slowly. After application of some pesticides, a proportion can still be found in the soil several years later; Freedman (1989, Table 8.3) gives the persistence time in soil of some organochlorine insecticides. In an experiment in England, sewage sludge containing polynuclear aromatic hydrocarbons was mixed into soil in 1968. Twenty years later substantial proportions of some of them were still there (Wild *et al.* 1991). It is not only some synthetic chemicals that degrade slowly, some natural substances do, too. One example is humus, a complex mixture of organic substances which can remain in soil for several thousand years (see Chapter 3). One might perhaps assume that if any chemical stays around long enough, some species will arrive or evolve that can break it down and will then multiply until the chemical is broken down rapidly; but clearly this is not true of all chemicals, natural or synthetic: rapid breakdown of some of them is blocked. Here I consider ways of promoting decomposition, especially of substances that normally decompose very slowly.

Which organic chemicals are more easily degraded?

The ability of microorganisms to break down organic chemicals under laboratory conditions has been much investigated, and some generalizations can be made about which sorts of chemicals are more easily degraded. Many organic pollutants are aromatic, and the splitting of the benzene ring is often the limiting step in their breakdown. Whether a particular enzyme or a particular bacterial species can accomplish this depends on what side groups are attached to the ring and in what positions (Betts 1991). For example, more species can split a ring if there are two OH groups *ortho* to each other than if they are *para* (Fig. 7.12). However, some side chains elsewhere on the ring can block cleavage of the *ortho* compound and have to be hydrolysed off first. The presence of one or more chlorines attached to the ring often prevents an enzyme from splitting the ring; the Cl atom is large relative to C or O, and so perhaps prevents the molecule fitting into the right gap in the enzyme. Chapter 2 of Betts (1991) considers in detail the degradation of Cl-containing aromatic compounds. Some enzymes can remove a Cl from a ring, replacing it with OH, but others can split the ring before the Cl is removed. PCBs have the framework shown in Box 7.2 with various numbers of Cl attached in various positions. What is used in industry is always a mixture of several different PCB compounds. In general the more Cl in the molecule

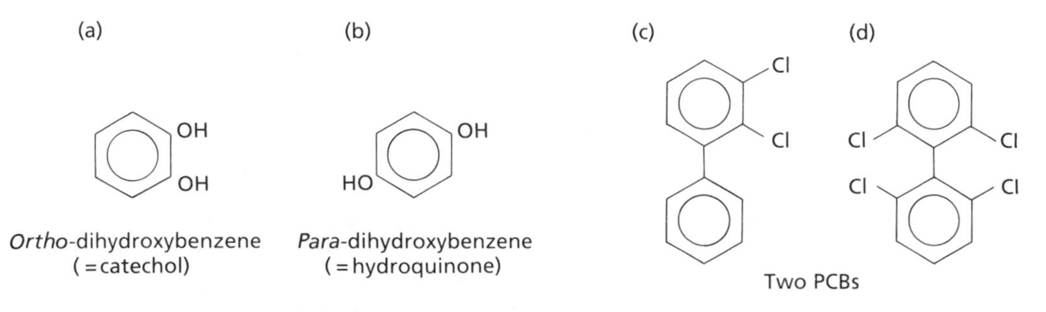

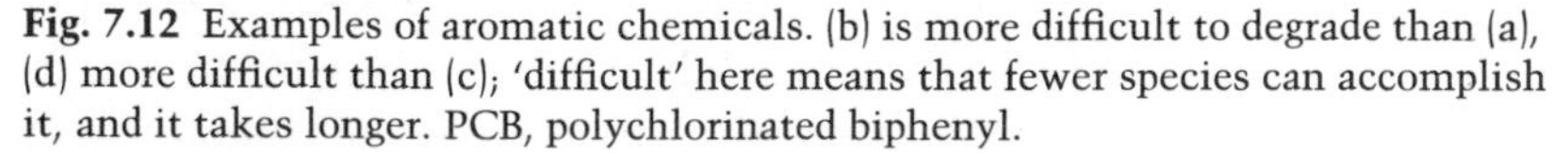

Fig. 7.12 Examples of aromatic chemicals. (b) is more difficult to degrade than (a), (d) more difficult than (c); 'difficult' here means that fewer species can accomplish it, and it takes longer. PCB, polychlorinated biphenyl.

the slower the breakdown, and molecules with more than 5 Cl are extremely resistant to breakdown. Degradation is faster if one of the two rings has no Cl, or failing that at least the *ortho* and *meta* positions of one of the rings are free. Figure 7.12 shows an example of a more-easily and a less-easily degraded PCB. Samples of grass and clover harvested from a field at Rothamsted, England, since 1965 and stored have been analysed for PCBs (Jones *et al.* 1992), to indicate how the abundance of different PCBs in rural air has changed between 1965 and 1989, a period when their production has been declining. Although all PCBs showed a substantial decline, those with 5 or more Cl per molecule declined less fast than those with 4 or fewer Cl.

To promote microbial degradation of a pollutant one can try two approaches.

1 If suitable microbes are already present but are limited by physical conditions or lack of some requirement, then we should try to improve the conditions or supply the requirement.

2 If there is no microbial species capable of breaking down the pollutant then we need to find such a species and inoculate with it.

Both approaches have been tried successfully.

Increasing microbial breakdown of oil

Public concern about oil pollution is greatly increased after a major spill from a tanker accident; but much more oil is released each year from many small individual sources such as leaks, industrial waste and natural seeps. Crude oil is a mixture of various aliphatic and aromatic hydrocarbons, some of which can be rapidly lost by evaporation. The less volatile components are attacked by a range of microbial species, including bacteria, streptomycetes, yeasts and fungi. These can be found sparsely in water of the open oceans and in coastal mud, so if there is an oil spill an inoculum of suitable microbes is normally present, which can be then multiply. Ward *et al.* (1980) reported on the abundance of bacteria that could use hydrocarbons as substrate, in coastal muds in Brittany a year after the wreck of the tanker *Amoco Cadiz* caused the world's largest oil spill there. At unoiled control sites there were a few hundred of these bacteria per gram of mud, but at heavily oil-polluted sites numbers ranged from tens of thousands to tens of millions per gram. Inoculation with oil-degrader species has sometimes been carried out at oil-polluted sites, but there is little evidence that it does speed up oil breakdown (Prince 1992). More useful is to try to promote the multiplication and metabolic activity of microbial species already present. Two common limitations to oil breakdown are availability of water and oxygen, since microbial breakdown of the hydrocarbons under anaerobic conditions is very slow (Ward *et al.* 1980). Often the most useful procedure to speed up oil breakdown is to promote fairly stable emulsions of oil and aerated water (Hughes & McKenzie 1975). Mineral nutrient supply is another possible limiting factor. Following the large *Exxon Valdez* spill on the southern coast of Alaska in 1989, much effort and expense was invested in applying slow-release N+P fertilizer to contaminated beaches; this markedly speeded up the disappearance of oil (Prince 1992).

Adding another compound as an energy source for the microbes

Sometimes a bacterium can degrade a synthetic chemical but does not obtain metabolic energy in the process. Therefore it may be necessary to add another organic chemical which it can use, perhaps a chemical related to the pollutant. An example is that the addition of biphenyl to soil can greatly enhance the breakdown of PCBs. (Biphenyl is the two-ring framework without any Cl attached, see Box 7.2.) Brunner *et al.* (1985) found that if a PCB was added to a soil only 2% was broken down in 10 weeks, but if biphenyl was added as well about 17% of the PCB was broken down. There was a lag of about 3 weeks before PCB breakdown became rapid, presumably while the degrader bacterial population multiplied. Another example concerns simple chlorinated aliphatic hydrocarbons, which are found as contaminants in water in industrial areas and are often only slowly degraded by microorganisms. In a laboratory experiment Wilson & Wilson (1985) found that when water containing trichloroethylene (TCE) passed through a column of soil the TCE was not detectably degraded; but if natural gas (mainly methane, ethane and propane) was added to the air passing over the top surface of the soil then most of the TCE was degraded. Presumably microorganisms in the soil were using the natural gas hydrocarbons as metabolites.

Inoculation with a special strain of bacterium

In all these examples microorganisms with the necessary ability to break down the pollutant were evidently already present; but in other cases inoculation with a particular species or strain of bacterium is necessary. Kilbane *et al.* (1983) isolated a strain of *Pseudomonas cepacia* (named AC1100) which can use the herbicide 245T (2,4,5-trichlorophenoxyacetic acid) as a sole carbon source. In a laboratory experiment, when 245T was added to a soil there was no detectable degradation over 6 weeks, but after strain AC1100 was added most of the 245T had disappeared within a few weeks. If there was no 245T in the soil the population of AC1100 declined rapidly; so it would need to be cultured and added to soil whenever required.

These examples show that adding a metabolite or in other cases inoculating with a particular bacterial strain can lead to degradation of organic pollutants. However, this has so far been mostly shown only in laboratory experiments, and the techniques need to be developed for practical use on a field scale.

Management of plants and animals to reduce harmful effects

Revegetating sites that are polluted with heavy metals

An alternative approach, instead of reducing the amount of pollutant, is to manipulate the living things so that although the pollutants remain the effects are reduced. I describe here two examples. The first concerns land that has become strongly polluted with heavy metals, for example mine waste and sites used for sewage sludge disposal. Assume that it is now desired to revegetate the area. Many plant species cannot tolerate the pollutant levels. But there are some plants that can grow where heavy metal concentrations are high: this is clear because there are sites, around old mines or metal mills, that have become contaminated over a century or more, and some plants do grow on them. Usually there are few species on such sites, and the same species recur at different polluted sites. In Britain a few grasses are characteristic of old mine waste, e.g. *Agrostis capillaris*, *Festuca ovina* and *Anthoxanthum odoratum*, and also a few forbs, e.g. *Minuartia verna*. The absence of other species could be partly because of other limitations besides heavy metal toxicity, for example low levels of available nitrogen and phosphorus. But it raises the question of whether the species found at these sites are always tolerant of heavy metals, or whether they have races that are specially tolerant.

Genetic variation in heavy metal tolerance

The genetics of heavy metal tolerance has been extensively investigated (see reviews by McNeilly 1987, Baker 1987). It has been consistently found that plants growing on heavily polluted sites are genetically different, even from members of the same species growing close by on unpolluted soil: they are ecotypes with genetically based heavy metal tolerance characters. Figure 7.13 shows results of tests for copper tolerance on three grass species. Tolerance was assessed by the length of roots that grew in a solution containing Cu at 0.25 mg l^{-1} as a percentage of the length in Cu-free solution. Two populations of *Agrostis capillaris* were compared, one from an old copper mine, the other from uncontaminated soil. When numerous plants were grown

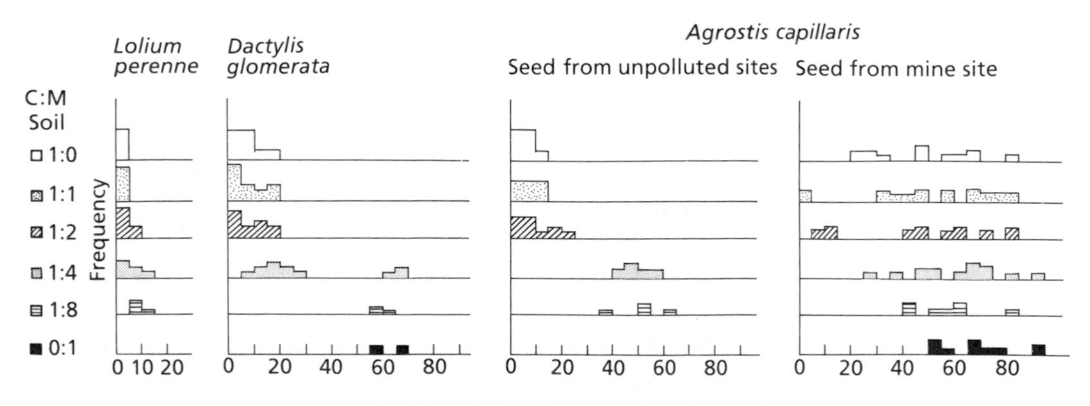

Fig. 7.13 Frequency of plants of different copper tolerances in three British grass species. All plants were grown from seed collected at unpolluted sites, except for the mine population of *Agrostis capillaris*. The seed was sown on mixtures of potting compost and Cu-polluted mine soil; the figures under C:M show the ratio of compost : mine soil. The surviving plants were assessed for Cu tolerance, by their root extension in Cu solution. From Gartside & McNeilly (1974).

on uncontaminated soil and then tested (see top line of Fig. 7.13), the mine population showed a wide range of Cu tolerance, some plants being extremely tolerant; in contrast, none of the 'normal' population showed more than slight tolerance. If ability to grow on strongly polluted soil depends on such genetic adaptation, why do not more species occur on these sites? Are only a few species able to undergo the necessary genetic change? To investigate this, Gartside & McNeilly (1974) looked for Cu-tolerant individuals within nine British herbaceous species—seven grasses and two forbs. They made up mixtures of potting compost and Cu-polluted mine soil in various proportions, and sowed 10 000 seeds of each species on each mixture. The seeds all came from populations on unpolluted sites. Some of the seeds from each of the species germinated, but three of them had no survivors on the mixtures with a high proportion of mine soil. Of the remaining species, some (e.g. *Lolium perenne*, Fig. 7.13) had no individuals that showed high Cu tolerance in the root extension test. In contrast two grass species, *Agrostis capillaris* (unpolluted site ecotype) and *Dactylis glomerata*, had a few individuals, among the survivors on the most polluted soils, that had high Cu tolerance (see Fig. 7.13). It seems then, that in these populations from unpolluted sites, two out of the nine species had the ability to produce mutants that were Cu-tolerant, but the other seven did not. Of these two, *A. capillaris* is suitable for mine waste, because of its tolerance of low mineral nutrient supply; *D. glomerata* requires richer soils but could be suitable for areas used for sewage sludge disposal.

The fact that only a few species can provide heavy metal-tolerant races is a disadvantage in reclamation of polluted areas. For example, clovers would often be useful species, to improve the nitrogen status of mine waste; but white clover was one of the species that had no survivors at all on the more Cu-rich soils in the tests of Gartside & McNeilly described above. (However, other legumes can show heavy metal tolerance.) If we can discover the mechanism of heavy metal tolerance, and especially if we can find a one-gene mechanism, there might be a possibility of transferring it to other species. Populations from polluted sites often show a wide range of tolerance, for example the mine population of *A. capillaris* in Fig. 7.13, and this suggests that several or many genes are involved. There is, however, evidence for a few species in a few cases that tolerant and non-tolerant races differ in only a single gene, e.g. for Cu tolerance in the forb *Mimulus guttatus* (Macnair 1983). Some heavy metal-tolerant bacteria have carriers in the membrane that actively pump the ion out; these are likely to be a single protein, and sometimes the structure of the protein and the base sequence of its gene are known (Beveridge & Doyle 1989). Meharg & Macnair (1990, 1991) compared two races of the grass *Holcus lanatus*, one more tolerant than the other of arsenic (As). The As-tolerant race took up As more slowly than the other. It also took up P more slowly.

Mechanisms of heavy metal tolerance

P and As are chemically similar and it is likely that they are taken up by the same carrier or carriers. This indicates how genetic change towards greater tolerance of a heavy metal might involve a 'trade-off', a loss in some genetically or functionally related beneficial character; in other words, tolerance has a 'cost' (Baker 1987). In this example As tolerance reduced the ability to survive on soils low in available P.

In plants tolerance of a toxic inorganic element seems only rarely to involve ability to maintain a low content of the element within the plant. The As tolerance in *Holcus* is one of the few exceptions to this rule. Most heavy metal tolerance in plants involves the heavy metal being concentrated in parts where it is less damaging, or being sequestered in less damaging forms, or involves changes in the metabolism that make the plant less sensitive. In some animals and microorganisms there are low-molecular-weight proteins called metallothioneins which bind certain metals and are probably important in accumulating them within the organism in a non-toxic form (Hopkin 1990b). Some plants produce peptides which may act in a similar way (Robinson & Jackson 1986). They bind to metals, and the amount of them in some plants is increased by Cd, Cu and Zn. In a few species it has been found that heavy metal-tolerant races contain more of these peptides than do non-tolerant races. These findings suggest that these peptides could be involved in heavy metal tolerance, but the evidence for this is not yet conclusive.

Thus our understanding of the mechanisms of heavy metal tolerance in plants is still at an early stage, and it has not yet reached the point where it can help much with finding or producing more tolerant races.

Reducing effects of eutrophication

Waters that contain high concentrations of mineral nutrients are called eutrophic waters. The mineral nutrients concerned, which are especially nitrogen and phosphorus, are essential for life, but high concentrations can have undesirable effects (Moss 1988). They can lead to very dense populations of phytoplankton in the surface waters which physically interfere with people swimming and fishing. If the lake is supplying piped water the algae can block the filtering system. Some cyanobacteria produce chemicals which give an unpleasant taste to drinking water. Although fish production may increase, there are likely to be more of the coarser, less desirable species. In temperate regions the plankton bloom dies in autumn and sinks. Its decomposition can lead to anaerobic conditions in the lower water layers and the mud which, as well as nasty smells, can lead to death of fish, especially in lakes that are frozen in winter. In this section I am mainly concerned with eutrophication of lakes, but it has also had serious effects on some partly enclosed seas, for example the northern Adriatic near to Venice (Winteringham 1985).

Effects of N and P on phytoplankton abundance

A major algal bloom usually occurs only if both nitrate and phosphate are in ample supply, though N-fixing cyanobacteria can respond to P alone. It follows that omitting P from the input should be sufficient to prevent a bloom. This was convincingly demonstrated by experiments in an uninhabited part of northern Ontario, where combinations of nitrate, phosphate and sugar were added to whole lakes for several years (Schindler & Fee 1974); the sugar was intended to simulate soluble organic matter input from sewage. Phytoplankton abundance in the epilimnion (upper water layer) was measured about once a month, by chlorophyll-*a* concentration, and Table 7.10 gives the highest value for the year. N and P together (with or without sucrose) resulted in the phytoplankton increasing greatly, but if P was not added the phytoplankton remained at control lake levels; and in Lake 304 after P addition was stopped the plankton abundance dropped to control values within the first year. The clear message from this and other research was that the unpleasant effects of eutrophication could be eliminated by reducing P input.

Curing algal blooms by reducing P input

The principal source of P to most eutrophicated lakes is sewage. One way to reduce P in sewage, which has been brought into force in many developed countries since the early 1970s, is to stop using polyphosphates in detergents. They were not, however, the only contributor of P, so sewage is normally treated chemically to precipitate out much of the remaining P. The first classic success story of curing

Table 7.10 Peak phytoplankton abundance in epilimnion (upper layer) of lakes in northern Ontario, measured by chlorophyll-*a* concentration. Some of the lakes received additions of inorganic N, P and sucrose

Treatment	Year of measurements	Chlorophyll-*a* ($\mu g\,l^{-1}$)
Several unfertilized lakes	1970–72	Maximum 7–11
Lake 227:		
N + P added 1969 onwards	1973	185
Lake 304:		
Before treatment	1969	11
N + P + sucrose added 1971 and 1972	1972	115
N + sucrose (but no P) added 1973	1973	11*
Lake 226:		
Before treatment	1971	11
Then divided into two parts:		
(a) N + P + sucrose added 1972 and 1973	1973	32
(b) N + sucrose added 1972 and 1973	1973	9

* Values for early 1973 ignored.
Data of Schindler & Fee (1974).

eutrophication, however, used the device of pumping the sewage elsewhere. Lake Washington, which borders Seattle, received its sewage and became increasingly eutrophic. Between 1963 and 1967 arrangements were made to divert all the sewage to Puget Sound. The effect on the algal blooms was rapid: there had already been some reduction in phytoplankton in 1965, and by the mid-1970s it was down to about one-tenth of its peak abundance (Mason 1991). However, when sewage P inputs to lakes in other parts of the world were reduced, this sometimes failed to produce any marked reduction in plankton. An example was Lake Trummen in southern Sweden. There the lake sediment had accumulated large amounts of P during the years of high P input, and after the polluted input was diverted release of P from the sediment maintained high P concentrations in the water. To cure the problem all the sediment was removed with a suction dredger (Mason 1991). A key difference between these two lakes is their depth: Lake Washington has a maximum depth of 76 m; Lake Trummen's is only 2 m, so there is much more opportunity for the sediment to influence the chemical composition of the surface waters.

The influence of fish on plankton abundance

Another example where reducing P input did not work is the Norfolk Broads. These are shallow artificial lakes in eastern England, formed centuries ago by peat cutting. The water used to be clear, with low abundance of plankton, but there were many macrophytes, i.e. large angiosperms growing submerged or with floating leaves (e.g. water-lilies, *Nymphaea alba*). In recent decades the amount of plankton has increased greatly, the water has become much less clear and most of the macrophytes have gone (Moss 1989). In order to reverse this, the sewage P input to some of the Broads was greatly reduced in the early 1980s, but phytoplankton abundance declined only a little, and it still remained a problem. Moss (1989) discussed carefully the possible reasons for this, and concluded that it is inadequate to consider only the phytoplankton and its nutrient supplies, we must also consider the other species. Figure 7.14 sets out a very simple hypothesis on the abundance of species in different trophic levels in a lake where the nutrient supply to the phytoplankton is adequate (i.e. non-limiting) at all times. The prediction is that if no fish are present, phytoplankton will be sparse even when nutrients are abundant, because the zooplankton are food-limited and so multiply until they eat most of the phytoplankton. The ample nutrient supply allows the phytoplankton to have high productivity, but most of this is quickly eaten. However, in a three-trophic-level lake the fish are now food-limited and eat most of the zooplankton, so the phytoplankton can now attain high standing biomass. The presence of top carnivores can greatly reduce the abundance of planktivorous fish, so zooplankton abundance becomes high and phytoplankton low, similar to the two-level lake. This hypothesis is not meant to suggest that nutrient inputs are irrelevant to the control of algal blooms: if inputs are low enough (oligo-

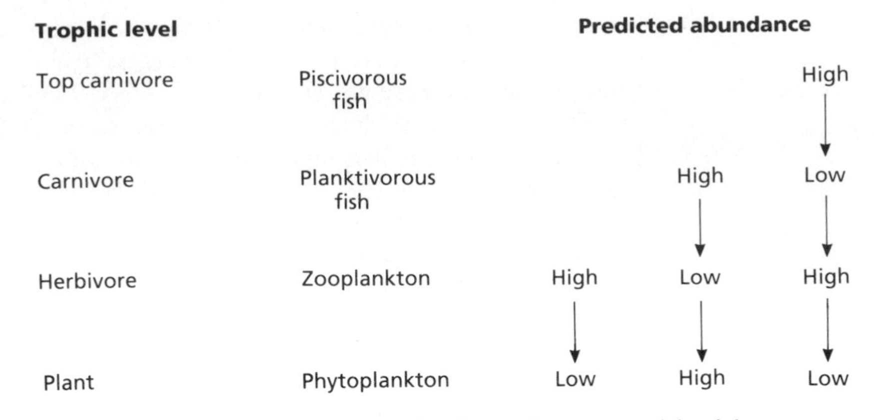

Fig. 7.14 Diagram to show predicted abundance of species in a lake if there are two, three or four trophic levels present, and if nutrient supplies to the phytoplankton are non-limiting.

trophic lakes) then the biomass of all trophic levels will be low. But the suggestion is that if nutrients can be reduced only to an intermediate level, then it is necessary for planktivorous fish to be sparse or absent, if the phytoplankton is to be greatly reduced.

Controlling phytoplankton by altering the fish population

This model predicts that in order to control excessive phytoplankton at intermediate nutrient levels we should kill off most of the planktivorous fish; or alternatively we should increase the abundance of top carnivore (piscivorous) fish, so that they will keep down the planktivores. Both these control methods have been tested experimentally. Removal of planktivorous fish, rather than resulting in a simple increase in zooplankton abundance, often results in the zooplankton becoming dominated by large species such as *Daphnia*, which are efficient grazers; but in other respects the results often agree with expectation. Andersson *et al.* (1978) inserted large enclosures in two eutrophic lakes in Sweden. In each lake all the fish were removed from one enclosure, the other was stocked with native fish that eat zooplankton and benthic invertebrates. Table 7.11(a) shows that the presence of fish greatly increased the abundance of phytoplankton in both lakes. Fish often affect the nutrient concentration in lake water, probably by their faeces or by stirring up mud, and it has sometimes been questioned whether this is how they promote algal growth, rather than by eating the zooplankton. However, in the experiment of Table 7.11 fish reduced the P concentration in one lake but increased it in the other, so P cannot be the main explanation for the changes in phytoplankton.

Shapiro & Wright (1984) described an attempt to improve the quality of a small lake near Minneapolis by altering the fish population. During the 1960s and 1970s the density of phytoplankton had increased; there had also been an increase in P concentration in the water and in the abundance of planktivorous fish, but the main piscivore species had

Table 7.11 Phytoplankton density, measured by chlorophyll-*a* ($\mu g\,l^{-1}$)

(a) In top 1.5 m of two Swedish lakes during July–August

	Planktivorous fish present	No fish
Lake Trummen	440	20
Lake Bysjon	286	<5

From Andersson *et al.* (1978).

(b) In tanks in California, May–June

Fish had access to:	Whole tank	Part of tank
	1036	20

From Smith (1985).

almost disappeared. In September 1980 all the fish were killed with a chemical. The lake was then restocked using a ratio of piscivores to planktivores aimed at dominance by the piscivores. During the next two summers the hoped-for results were seen: phytoplankton was less abundant and the water was clearer. However, there was some suggestion that this effect was disappearing towards the end of 1982. Such changes in fish species composition may need to be actively maintained.

Interacting effects of P and grazing

Sarnelle (1992) drew together results from these and other experiments, in which the abundance of zooplankton (*Daphnia* especially) in lakes or enclosures had been altered experimentally, usually by altering fish stocks but in some experiments by adding *Daphnia*. The phosphorus concentration in the water ranged from 9 to 463 $\mu g\,l^{-1}$; 9 $\mu g\,l^{-1}$ is classed as mesotrophic; there were no truly oligotrophic lakes involved. Figure 7.15 shows that if the P concentration was high, allowing the zooplankton to increase (e.g. by removing planktivorous fish) caused a marked reduction in phytoplankton (note the log scale). But if P was only about 10 $\mu g\,l^{-1}$ there was no such effect. From this P concentration downwards evidently algal biomass is strongly controlled by P supply. Increasing P above 10 $\mu g\,l^{-1}$ results in a great increase in phytoplankton (about 100-fold within the range of the graph) if zooplankton are sparse, but a much more modest increase if zooplankton are abundant. This indicates that if phytoplankton abundance below about 10 $\mu g\,l^{-1}$ on the scale of Fig. 7.15 is acceptable, there is a wide range of P concentrations within which manipulation of fish could provide a useful contribution to controlling phytoplankton. But if clearer water than that is required, then severe control on P input will be necessary.

These examples have ignored various complicating factors which may be important. One is macrophytic plants. Returning to the Norfolk Broads eutrophication problems, Moss (1989) suggested that

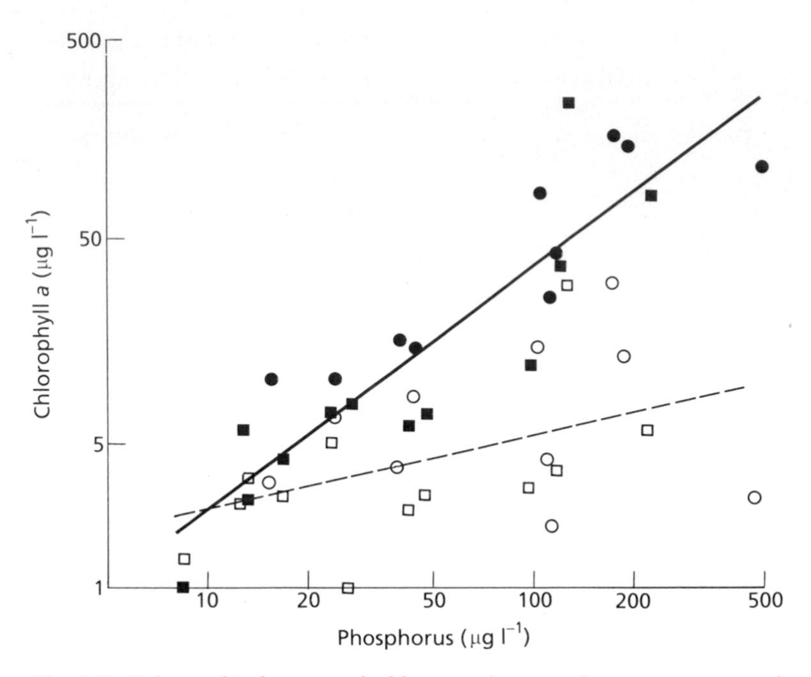

Fig. 7.15 Relationship between algal biomass (expressed as concentration of chlorophyll *a*) and phosphorus concentration in lakes in which the abundance of zooplankton was altered experimentally. Both axes are log scales. Zooplankton abundance: ■ ●, low; □ ○, high. From Sarnelle (1992).

the loss of macrophytes was a key factor hindering restoration to low-phytoplankton conditions. He suggested that they could provide refuges for the large zooplankton (mainly cladocerans), where they could escape being caught by fish, and that this had previously allowed them to exert a controlling influence on the phytoplankton. Although Moss's attempts to confirm this experimentally in the Broads were not entirely convincing, there is supporting evidence from elsewhere. In California tanks were filled with well-fertilized lake water and stocked with silver carp, which feed on zooplankton and larger phytoplankton (Smith 1985). In some of the tanks fish were excluded from half of the water volume by partitions of 1.5 × 1.5 cm plastic mesh. Table 7.11(b) shows that the presence of this refuge greatly reduced phytoplankton density, even though the plankton could move freely through both parts of the tank. Smith's explanation was that without the refuge the carp ate most of the zooplankton and the larger phytoplankton, but the smaller phytoplankton became abundant; with the refuge present enough zooplankton survived to eat most of the small phytoplankton. These results suggest that phytoplankton control could be aided either by establishing macrophytes or by placing artificial refuges in the water.

Conclusions

- Tests lasting only a few hours or a few days can predict, with useful accuracy, the longer-term toxicity of a chemical. The solubility of a chemical in fat can provide some prediction of its toxicity to aquatic animals.
- These predictions need to be assigned a considerable margin of error. One reason is that the toxicity of a chemical can vary widely between species.
- The structure of an organic chemical gives some indication of whether it will be readily broken down by microorganisms.
- Aquatic animals usually take in more pollutant from the surrounding water than from their food.
- The concentration of an inorganic cation in an animal depends substantially on the animal's ability to excrete it, which can differ markedly between elements. Because of rapid excretion, the concentration of ^{137}Cs in herbivorous mammals adjusts within a few days to changed ^{137}Cs supply in food.
- Pollutants, organic or inorganic, are not consistently more concentrated at higher levels of food chains.
- The concentration of a pollutant in an animal or plant (or a particular part of it) is sometimes a useful indicator of whether the pollutant is near toxic levels in the environment.
- Nitrate concentrations in drainage water from heavily fertilized fields can be above official safety levels for drinking water. Various changes in crop management are known to help reduce this.
- Sometimes breakdown of an organic pollutant will occur once a suitable microorganism has been inoculated. But often a suitable species is already present, and its activity can be promoted by supplying requirements, e.g. extra mineral nutrients or carbon source.
- Only a few plant species seem capable of evolving heavy metal tolerance. Understanding the mechanisms of such tolerance may allow us to produce other heavy metal tolerant races in future.

Further reading

General:
  Moriarty (1988)
  Freedman (1989)

Gaseous pollutants:
  Wellburn (1988)

Freshwater pollution:
  Moss (1988)
  Munawar *et al.* (1989)
  Mason (1991)

Managing eutrophication:
Welch & Cooke in Jordan *et al.* (1987)

Marine pollution:
Salomons *et al.* (1988)

Plants and heavy metals:
Baker (1987)

Terrestrial invertebrates and pollution:
Hopkin (1989)

Bacteria and heavy metals:
Beveridge & Doyle (1989)

Soil microorganisms:
Somerville & Greaves (1987)

Breakdown of pollutants by microorganisms:
Morgan & Watkinson (1989)
Betts (1991)

Chapter 8: Conservation and Management of Wild Species

Questions

- If a large area of natural vegetation is reduced to small, separated fragments, how will this affect the plant and animal species living in it? Can fragments be too small to maintain some species?
- How can we predict the minimum viable population for a species, i.e. the smallest population that has a high chance of survival?
- Can corridors promote migration between habitat fragments, and survival of species?
- How do we decide which species and communities deserve highest priority for conservation?
- Are there 'keystone' species, whose extinction would have a major effect on the whole ecosystem?
- How can management promote high diversity of plant species? Of animal species?
- If we want to create new 'natural' areas on abandoned farmland, should we leave succession to take its course or is active intervention necessary?

Background science

- Mechanisms that allow many species to coexist. What prevents a few, fittest species ousting all the others?
- Species–area relationships. Why do there tend to be more species the larger the area?
- Reasons why some species fail to survive in small habitat areas. Genetic drift in small populations. Modelling fluctuations in size of small populations.
- Why there are more species in some areas than others. Influence on plant diversity of heterogeneity in the habitat and of interactions between species.
- Influence of vegetation structure on animal diversity.
- Evidence that people were responsible for extinction of large mammals at the end of the Ice Age.
- How interactions between carnivores, herbivores and vegetation affect species abundance.

The title of this chapter may seem to contain a contradiction: if we manage a species, how can it be wild? In this chapter I adopt a very broad meaning for 'wild', to cover any species that we are not purposely growing or maintaining—for example to provide food or timber, to decorate our gardens or as a pet. This means that birds in suburban gardens are considered as wild, and so are plants and animals in a roadside verge or hedge. In this chapter 'natural' also has a wide meaning, to include areas of vegetation that have been much influenced by people. Examples are rough, unsown pastureland, even if grazed by domestic animals; and mixed-species forests, even if cut down in the past and then left to regrow. Arable crops, sown single-species grassland, single-species tree plantations clearly are not natural, though they may well have wild species within them.

The aims of conservation

'Conservation' implies trying to keep things as they are. A recurring theme of this book is that things cannot remain unchanged everywhere, because there is an increasing human population in a world of fixed size (Chapter 1). This will generate strong pressures to convert some areas of natural vegetation to other uses such as farmland, forestry plantations, roads or towns. There will be pressures to change the management of cropland, pasture and forest in ways that can increase production but may harm wild species. And other by-products of human activity—such as nutrient enrichment of lakes and climate change—will also influence wild species. Some people may see it as their duty to try to prevent all such changes. This chapter, on the contrary, assumes that they will sometimes happen, and that ecologists should be prepared to help with difficult decisions on which natural areas should be saved and which sacrificed, and to give advice on how harm to wild species can be minimized. However, this chapter is not all about saving existing natural areas: towards the end it also considers creation of new 'natural' communities.

In Chapters 3 and 5 I considered farmland and forests purely as places to produce food and timber. For example, hedges were ignored, except for their possible effect on crop pests (Chapter 6); and regeneration of forests after felling was considered as if regrowth of timber trees was all that mattered, there were no other plants or animals. This chapter now redresses the balance: it acknowledges that farmland and forest contain many wild species, whose conservation deserves careful attention.

Who is conservation for?

When making choices of which areas to conserve and how to manage them, one basic question is: Who is conservation *for*? Is it for the benefit of the species conserved or for the benefit of people? Many people gain great pleasure from the countryside. But the countryside that gives most pleasure is not necessarily ideal for conservation of all wild species. To preserve some species may require that access by people to their habitat be restricted. If we are concerned about preventing extinction of species, this leads to special concern about rare species, because they are often particularly at risk. Yet most rare species are, by definition, rarely seen by people. There has been much concern about the grizzly bears of Yellowstone National Park (see later in this chapter), yet most people who camp in the Park for a holiday never see a grizzly. In some countries a legal requirement is placed on mining companies to restore the vegetation, after open-cast mining, to as near as possible its previous state. This seems a good idea in principle; but if the mine is in an uninhabited area and far from public roads (e.g. some mines in Wyoming and Western Australia), would it be better to spend the money on management of other countryside where people could enjoy it? There has been surprisingly little research on what people actually want or most enjoy in the countryside; some of the research that has been done is described and discussed by Van Doren *et al.* (1979) and Patmore (1983). Part of the problem in deciding how to manage the countryside is that time-scales are long—decisions taken now may affect what is there in a century's time—and that opinions change. Gardens provide visible evidence of changing attitudes. Famous formal gardens of Europe—such as Versailles (near Paris), the Generalife at Granada and the Boboli Garden in Florence—with their straight lines, enclosed spaces and vistas towards the palace, indicate owners who saw the garden as a safe haven from the dangers beyond, as an extension of the palace. This contrasts strongly with 18th-century English landscape gardens and parks, whose designers saw the garden as an extension of the countryside.

Priorities in conservation

As in the rest of this book, I try not to take sides in these value judgements, but rather to show how ecologists can help in the decision-making process. I assume that conservation is at least partly for the benefit of species other than humans. There has been much concern about the species that have become extinct within the last few centuries as a result of human activities, and a topic of this chapter is how to prevent some of the remaining threatened species from the same fate. Table 1.2 gives figures for the number of species, worldwide, in major groups. Presumably more species remain to be discovered, but these figures, even if too low, show clearly that we cannot take active steps to conserve every individual species in the world—there are far too many of them. A widespread solution to that problem has been to place species in a *de facto* ranking of value, with birds and large mammals at the top, trees also ranking high up, followed by herbaceous plants if they have pretty flowers, and a few butterflies. Within each of these major groups systems have been proposed for identifying endangered species in need of special attention. Below those groups we get into a grey area where species have less certain value—reptiles, amphibians, small mammals, grasses, large mushrooms, and so on down into groups such as nematodes which have no conservation value at all, apparently. The U.S. Congress set up a special committee to advise on the conservation of the northern spotted owl. So far they have not set up a committee devoted to a single nematode species. In this chapter, while accepting that we cannot have a separate nature reserve for each species, I try to avoid getting involved in emotionally based ranking. Several other approaches are possible, and feature in this chapter.

1 We can ask whether there are 'keystone' species, which are specially deserving of preservation because their extinction would lead to a major change in the whole community and hence affect many other species.

2 We may aim to conserve ecosystems rather than individual species.

3 If we aim to maintain or promote diversity this will by definition preserve many species. Another reason for promoting diversity is that it often contributes to people's enjoyment of countryside; it is a key difference from cropland, forestry plantations and parking lots.

4 Species that have been reduced to a few small populations pose special problems and special urgency, whatever the intrinsic virtues of the species, and some of these special problems are considered in this chapter.

Maintaining and promoting diversity

In this chapter diversity is usually taken to mean species-richness, the number of species present. Another component of diversity is evenness: if an area has 10 species all of equal abundance it is in a sense more diverse than another area that has one very abundant species and nine others that are rare. There are diversity indices that take this into account, and this chapter occasionally makes use of one of them.

Diversity can be considered at different scales, often referred to as α-, β- and γ-diversity. For example, in the Swiss uplands the hay meadows have high α-diversity (e.g. many plant species in a square metre), while the contrast between meadows and forests provides β-diversity (i.e. change between sites, or diversity in the landscape). The number of plant species in the whole of Switzerland is its γ-diversity, the pool available.

Why are there so many species?

If we are to promote diversity we first need to understand why it occurs at all. Darwin (1859) argued persuasively that species compete and only the fittest survive. Thus natural selection provides a strong pressure towards loss of species, as the less fit are eliminated. From this has arisen the *competitive exclusion principle*, which is based on the assumption that no two species can be exactly equally fit. It states that if two or more species exist in the same habitat, ultimately all but one of them will be excluded. We thus have the *paradox of diversity*:

we expect few species but we see many. Why are there so many species?

That question has been discussed at great length. The answers proposed by scientists can be grouped into three categories, which are summarized in Box 8.1. It can be argued that there are really only two categories, because number 3 is not a separate cause of diversity, but may contribute to causes 1 and 2. The four 'competition preventers'—disturbance, stress, predation and disease—can slow down the exclusion of a less fit species, and so contribute to cause 2, but they can alternatively provide niche separation. For example, if two plant species are eaten by different insect species, or if one bird species is better able than another to survive low temperatures, that can provide niche separation and could allow them to coexist. It is likely that mechanisms 1 and 2 both operate, though to different extents in different ecosystems. Which of them is the more important could influence methods of conservation. For example, how should we promote continued species-richness in tropical rainforest? According to one school of thought (e.g. Hubbell & Foster 1986b), this diversity occurs because there are many species that are extremely close in their fitness, and the very stable environment results in very slow loss of species; in other words mechanism 2 predominates. According to this view, we should aim to leave the forest as undisturbed as possible. On the other hand, much tropical rainforest has been used for shifting cultivation in the past, and has been subject to other disturbances such as fire (see Chapter 5). The diversity could then be in part a response to disturbance in the past, the forest of today representing a mosaic of small patches at different stages of succession from disturbance in the past. Connell (1979) argued this point of view strongly. Under this view (mechanisms 3 and 1), disturbance should not be reduced, as it provides niches and contributes to diversity.

I shall return later to discuss in more detail possible ways of managing for diversity. But first we must consider how the number of

Box 8.1 What allows species to coexist?

Mechanisms that may be responsible for preventing loss of species by competitive exclusion, and hence for allowing diversity to be maintained.

1 Each species has a different *ecological niche*, a set of conditions where it is fitter than its competitors.

2 The species are very *evenly balanced in fitness*. The less fit species are in the process of being eliminated, but so slowly that there will be time for other species to arise by evolution or to invade from other regions.

3 *Competition is reduced or prevented*, because the main controls on abundance are physical disturbance, stresses (e.g. low temperature, toxic substances), predation and/or disease. Hence competitive exclusion does not occur.

species relates to the area available, and what may be the causes of the relationship.

Species–area relationships

There tend to be more species the larger the area

It was noticed several decades ago that there tend to be more species on larger islands than on smaller ones. This was brought to the attention of many ecologists by MacArthur & Wilson (1967), though they were not the first to notice it. Figure 8.1(a) shows an example of the relationship, for birds. If species number and area are both plotted on a log scale, as here, the points usually fall approximately on a straight line. As in this example, there is often a considerable scatter of points, but the relationship is significant (here $P = 0.005$). The relationship has been shown for other groups of animals, vertebrate and invertebrate, and for plants (Connor & McCoy 1979). It also applies to other sorts of 'island', in other words any region suitable for some species surrounded by an area inhospitable to them. Thus a patch of forest surrounded by farmland acts as an island for plants and animals that can live in forest but not in farmland. Figure 8.2(a) shows an example, the number of

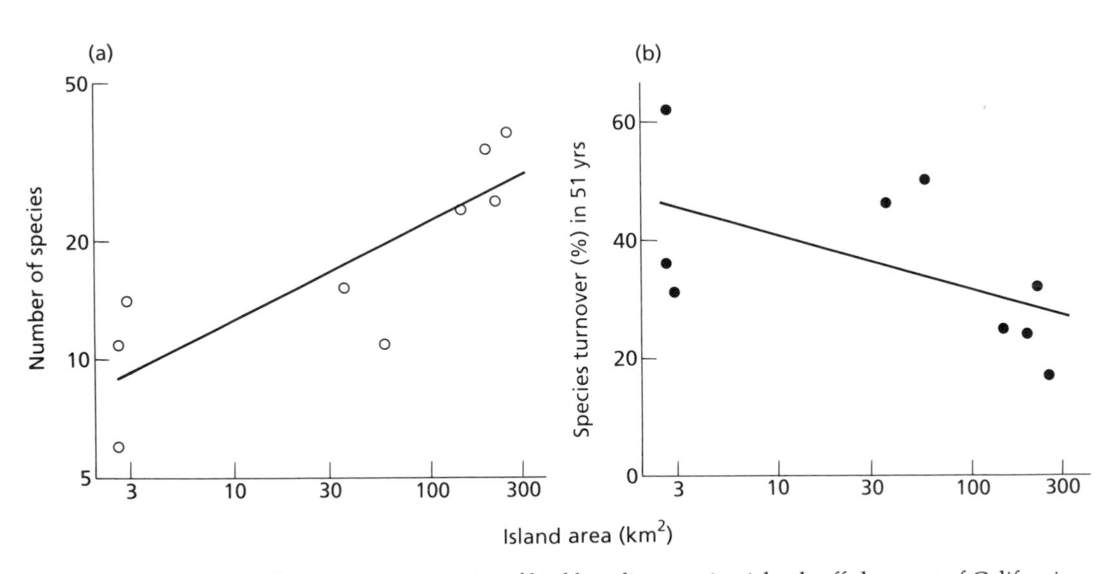

Fig. 8.1 Data on species of bird breeding on nine islands off the coast of California.
(a) Relationship between number of species in 1968 and area of island. Both axes are on a log scale.
(b) Species turnover 1917 to 1968, in relation to island area.

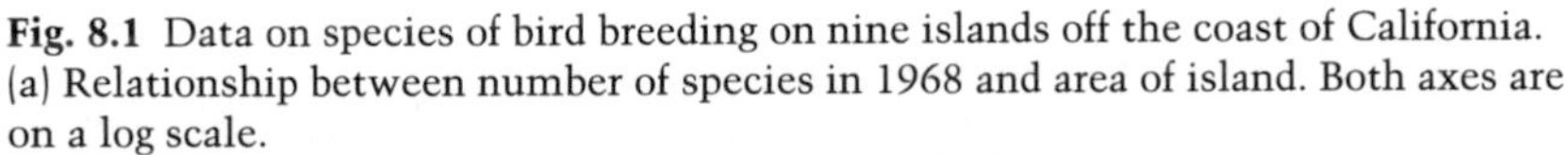

$$\text{Turnover} = \frac{(I + E)}{(N_1 + N_2)} \times 100$$

where I = immigrations, i.e. number of species present in 1968 but not in 1917;
E = extinctions, i.e. number of species present in 1917 but not in 1968;
N_1, N_2 = number of species present in 1917 and 1968 respectively.

One or two bird species that were intentionally introduced to some islands are omitted from the calculations. From data of Diamond (1969).

species of herbaceous plant in patches of ancient forest. This graph comes from research by Game & Peterken (1984) which I shall make use of several times in this chapter. They studied 362 forest fragments ('woods') separated by farmland in Lincolnshire, eastern England. They were able to classify the woods as 'ancient', i.e. there has been forest on the site continuously since before AD 1600, or 'recent', i.e. on land that has been clear of forest for some period since 1600. This was done by reference to old maps, written records and signs of former cultivation

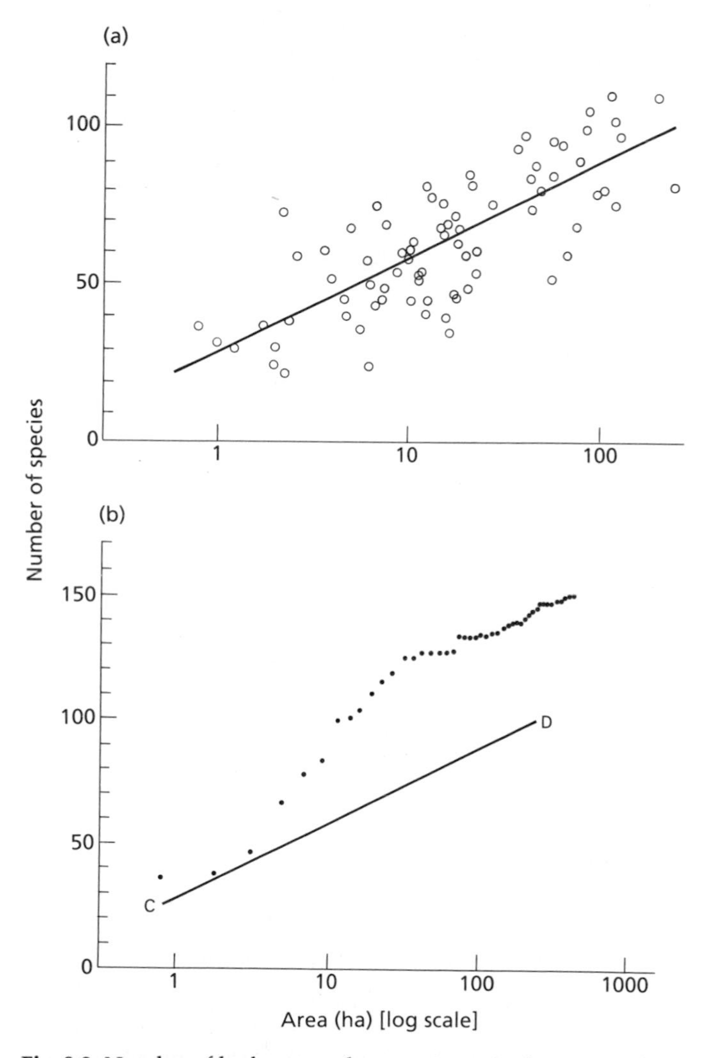

Fig. 8.2 Number of herbaceous plant species in 78 fragments of ancient forest in central Lincolnshire, England. Between the forests is mostly farmland. (a) Number of species in each fragment in relation to its area. (b) Points show total number of species in one or more fragments, in relation to their total area, combining the smallest first: i.e. the left-hand point refers to the smallest fragment alone, the next point to the two smallest combined, the next to the three smallest, and so on. Line CD is the regression line from part (a). From Game & Peterken (1984).

within the woods. Many of the ancient woods have probably been continuously forested since trees first colonized after the Ice Age.

Most nature reserves are effectively islands, and this species–area relationship sparked off much discussion of how we should design nature reserves. Clearly the larger each reserve is, the better. But supposing only a certain total area can be spared for nature conservation, is it better to save a few large or many small? The answer to that question hinges on what is the cause of there being more species in larger areas. One possible cause is that the larger the area the greater, on average, will be the variation in environment within it, so the more opportunity for niche separation among species. If that is the only reason for there being more species, then survival of the maximum number of species would probably be promoted by having a large number of small reserves well spread out, chosen to provide examples of as wide a range as possible of environmental conditions. Figure 8.2 (part b) illustrates this by imagining that in the area of England where the fragments of ancient forest shown in part (a) occurred, it was possible to save only a certain total area of them. The straight line in part (b) is the regression line from part (a) and shows the expected number of species in a single wood. So if, for example, only 10 ha could be preserved, and the decision was to use it to save a single wood, the number of herbaceous plant species would be about 60. The black points show a different strategy, aiming for as many separate woods as possible by saving the smallest preferentially. Then 10 ha would allow the seven smallest to be saved, and they contain about 100 species. Whatever total area is allowed, the strategy of saving many small woods would include more species than saving one large one. One likely explanation for this is that the area of this study included several major soil types, whose woodland flora could be conserved only by saving at least one wood on each soil type.

Are there more species because there is more habitat diversity?

MacArthur & Wilson (1967) suggested that the number of species on an island is not determined solely by the diversity of habitats, but also by the rate at which species become extinct and at which other species arrive. Figure 8.3 summarizes their hypothesis. (They were much concerned about whether the lines are curved or straight, but that does not affect the principle.) It is assumed that each year a certain percentage of species become extinct on the island, hence the E lines rise in proportion to the number of species present. The I lines fall towards the right because the more species are present the more chance that an individual arriving will belong to a species already present. A large island may have more species arriving (the I line is higher) because the larger the area the greater the chance of a passing animal or seed landing. The chance of extinction is hypothesized to be greater on smaller islands (the E line is steeper) because the number of individuals of a species will be smaller and there is therefore more chance that year-to-year fluctuations will one year take it to zero (see

Is diversity controlled by rates of invasion and extinction?

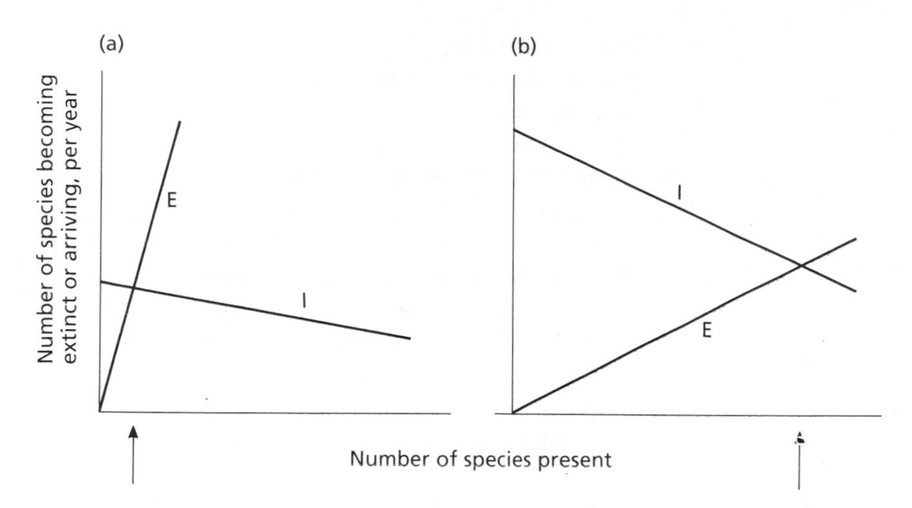

Fig. 8.3 Diagrams to explain MacArthur & Wilson's hypothesis for control of the number of species on islands. (a) Small island, (b) large island. I, number of immigrants, i.e. new species arriving per year; E, extinctions, i.e. number of species ceasing to exist on island per year. Arrows show equilibrium number of species. Simplified from MacArthur & Wilson (1967).

later in this chapter). The number of species will stabilize where the two lines cross, and so the number of species will be greater the larger the island.

These two explanations for there being more species on larger islands—greater habitat variation or balance of invasion and extinction—are analogous to the two main mechanisms for maintaining species diversity in the whole world, alternatives 1 and 2 in Box 8.1. So in this sense each nature reserve or patch of natural vegetation is a microcosm of the world.

Testing the immigration–extinction hypothesis

Two predictions from the immigration–extinction hypothesis are: (1) on an island species become extinct from time to time; but this is approximately balanced by arrival of new species, so the total number of species changes little; (2) this turnover (when expressed per species present) is slower the larger the island. To test whether these predictions are correct requires counts of immigrations and extinctions on islands of different sizes. This is not easy to do. To be sure that the species have really left or arrived we need to be sure that every species present on the island was recorded, on at least two occasions. Even for conspicuous species, such as birds or higher plants, it is difficult to be sure that some have not been overlooked. Some of the most careful studies have been on recently formed islands, where successional changes were taking place, whereas the MacArthur–Wilson hypothesis is most interesting when applied to steady-state turnover. The results available for various species-groups in various parts of the world show consistently that species turnover does occur: some species disappear and others arrive, but the number of species on each island often stays approximately constant. To the question, is there more turnover per species per year on smaller islands, the answer is more equivocal. Figure 8.1 part (b) shows one example. Over 51 yrs there was gain and loss of species on all the islands. This turnover, expressed as a percentage of the number of species present, does show some tendency to fall towards the right; but this is not statistically significant ($P = 0.12$). Quinn & Hastings (1987) give five other graphs of species turnover against area: three for birds, one for vascular plants and one for plankton in lakes. Some give a more convincing correlation of turnover to island size than in Fig. 8.1(b), some a weaker correlation. So the evidence in favour of the immigration–extinction hypothesis is mixed, a conclusion supported by more extensive surveys of the relevant research (Burgman *et al.* 1988).

Habitat fragments

In many parts of the world natural and semi-natural vegetation remains only as fragments. This is conspicuous in many parts of western Europe and the eastern two-thirds of the United States: the landscape is a mosaic, with patches of forest, rough grassland, heathland, wetland, rivers and their banks, set among the farmland, roads, towns. In those regions people have contributed much to the fragmentation, in Europe over many centuries. In England the forest had already been reduced to 15% of the land area by AD 1086, according to the Domesday survey (Rackham 1986). In many parishes the forest was even then fragments among farmland. Fragmentation is not always of human origin: there are many natural causes of small patches of vegetation including hills and valleys, rivers and lakes, windthrow, fire, and many more, at various scales. An individual shrub may be a fragment to an insect that lives in it. Fragmentation is in fact an essential basis for β-diversity. A mosaic landscape supports more species than a uniform one not only because it provides different habitats in different patches, but also because some species inhabit the edges. To mention one example, in an area of Michigan where there are patches of abandoned farmland interspersed with patches of deciduous forest Gates & Gysel (1978) found that there were more birds' nests, belonging to more species, within 10 m of forest edges than further into the forests or in the farmland. Some species, e.g. cardinal, nested only in these edge regions. Another example is provided by one of the smallest nature reserves in the world, 400 m^2 containing a small pool and marsh in Gloucestershire, England, where a rare buttercup (*Ranunculus ophioglossifolius*) grows. The plants survived only because trampling and grazing by cattle from the surrounding area kept the vegetation open (Frost 1981), so the buttercups require a site that is a fragment within another, larger vegetation type. Deer provide an example of species that are favoured by a mosaic, because they need forest for cover but can rarely obtain all the food they need throughout the year from old forest. Management that favours deer in North America and in Europe involves intermingling patches of old and younger forest, or maintaining clearings where there are herbaceous plants as food source (König & Gossow 1979, Alverson *et al.* 1988).

So a mosaic landscape of fragments is not necessarily unnatural, nor in all respects a bad thing. Nevertheless, the effects of increasing habitat fragmentation on wild species need careful consideration. We have seen that larger areas consistently support more species. If this is due only to their having more habitat variation, then we should plan for many small fragments to remain, provided they contain examples of the whole range of habitats. But if the invasion–extinction theory is correct, then if (for example) most of a large forest is cut down, leaving a few small fragments, the species that were present in the fragments at the time they were isolated will not all survive. The natural local extinctions will continue, but normal reinvasions will be prevented by the open land around.

How will fragmentation affect individual species?

The Amazonian forest fragment experiment

A bold plan to investigate experimentally the effects of isolating patches of tropical rainforest was started in 1980 near Manaus, in central Amazonia. Much forest was being cut down in the area to create grazing land, and the plan was to arrange for patches of forest to be left, providing replicate patches 1, 10, 100, 1000 and 10 000 ha in area. Control areas within a large extent of undisturbed forest would also be studied for comparison. In the event several patches of 1, 10 and 100 ha were isolated between 1980 and 1990, but no larger patches. Most of the research was on animals. The most extensive published summary of results is by Lovejoy *et al.* (1986). It is unfortunate that few longer continued studies have been published. Most of the scientists who initiated the project have now returned to U.S.A. Nevertheless, very interesting results have come out of the project, and I mention some of them in this chapter. Figure 8.4 shows the abundance of several species of bee in forests of different sizes, assessed by the number visiting chemical baits on sampling days. Two of the species were much affected by the size of the forest fragment, and were extremely sparse in 1 ha fragments. In contrast, two other species were equally abundant in all fragment sizes.

Why do some species fail to survive in small fragments?

Some need to migrate

This Amazonian experiment and other research has shown that isolating a fragment of an ecosystem can lead to extinction of some species. There are three sorts of reason why some species fail to survive in small habitat patches (Box 8.2). Large herbivorous mammals in East Africa provide an example of species that need to migrate between different areas each year (Pratt & Gwynne 1977, McNaughton 1985). Serengeti National Park, in Tanzania, and the adjoining Masai Mara reserve in Kenya form an area about 200 × 250 km, of open savanna woodland and grassland. Some of the large herbivores, notably wildebeest, zebra and Thomson's gazelle, migrate long distances each

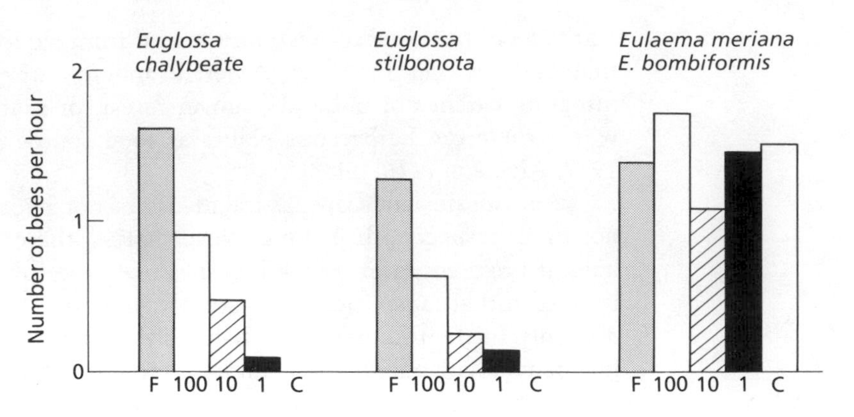

Fig. 8.4 Abundance of bees of named species in tropical rainforest area near Manaus, Brazil, assessed by number visiting baits. F, in large area of continuous forest; 100, 10, 1, areas (ha) of forest fragments; C, cleared area, recently deforested. From Lovejoy *et al.* (1986).

year between the rainy and dry season. During the rainy season they are in the eastern part, where they can drink from temporary water pools. But in the dry season these dry up, and the animals move to the western and northern parts where there are permanent rivers. During the dry season the areas within reach of the rivers are very heavily grazed, and this presumably provides one advantage to the animals of moving back to the grasslands of the east when the rains come again. If Serengeti were to be broken up into small fragments, with no opportunity for migration, it is unlikely these species could survive. Some other large African wildlife parks may already be too small, in the sense that the migration routes of some of their large herbivores formerly extended outside the park boundaries. Examples are wildebeest in Kruger National Park and elephants in Tsavo (Diamond, in Jordan *et al.* 1987). The survival of some species may depend on migration

Box 8.2 Some reasons why a species may disappear from an area if the area becomes an isolated fragment surrounded by different land cover.

1 The species *needs more than one site.*

(a) The species needs to migrate between different areas during each year, e.g. because food or water supply are not available in one area throughout the year.

(b) The species needs to inhabit different sites in different years, e.g. to survive different weather conditions.

2 *Edge effects.* The species cannot survive near the edge of the fragment, e.g. because of different microclimate or because of species invading from other habitats.

3 The number of individuals the area can support falls below the *minimum viable population* size. Risks to the species then are:

(a) *genetic*, from inbreeding and genetic drift; and/or

(b) *demographic*, from fluctuations in population size which could in due course take it to extinction.

routes through farmland, around reservoirs and across new roads. Johnsingh *et al.* (1990) give examples of such migration routes for elephants in India. The importance of corridors for migration between fragments will be discussed later.

Another example, on a different scale, is the Bay checkerspot butterfly, which occurs only in patches of grassland on outcrops of serpentine southwest of San Francisco Bay in California (Harrison *et al.* 1988, Murphy *et al.* 1990). In this Mediterranean climate of mild, wet winters and hot, dry summers, the checkerspot's food plants senesce in late spring. The larvae of the checkerspot feed actively during spring, then enter diapause. The larvae develop faster on south-facing slopes, and in years when the spring weather is cool and wet larvae from these slopes are the major contributors to the future population. However, in drier years the host plants on these south-facing slopes senesce before the larvae have reached diapause, so they die, and continuation of the butterfly population depends on there being north-facing slopes where the soil remains moist longer and the host plants senesce more slowly. Thus long-term survival of the species requires these contrasting topographical habitats, and this may be the reason why during three consecutive dry years, 1975–77, the species became extinct in some small patches of serpentine grassland but survived in larger ones.

Edge effects

A second reason why small fragments may be unfavourable for some species is edge effects. The smaller a fragment the nearer even its centre is to an edge. An example of how this can affect birds is provided by the Amazon rainforest fragment experiment. Table 8.1 shows that a year after a 10-ha patch was isolated by cutting down the forest around it, the number of bird species caught in nets was fewer than in a 10-ha area inside undisturbed forest. (Unfortunately the number of species in the fragment was not determined before it was isolated, so there is no proof that the difference in species-richness was solely caused by the isolation.) Traps 10 m from the edge of the large

Table 8.1 Numbers of species of bird caught in mist nets in tropical rainforest in Amazonia. Each figure is from 24 nets deployed for a total of about 100 days

10-ha region deep in large forest	
In tree-fall gaps	47
Patches with no gaps	50
10-ha region isolated 1 yr earlier	
In tree-fall gaps	37
Patches with no gaps	39
In large forest but near edge	
50 m from edge	47
10 m from edge	28

From Lovejoy *et al.* (1986).

forest area yielded fewer species than the average for the 10-ha fragment, but traps 50 m in yielded as many species as those deep in the forest. So there was an edge effect, but it extended less than 50 m in. Since the 10-ha fragment was square, about half of its area was within 50 m of an edge. Thus edge effect could have contributed to its lower species-richness, but probably was not the full explanation. A study of three forest fragments in Michigan by Gates & Gysel (1978), mentioned earlier, found, in contrast, more species of bird nesting near the margin than further from it. The reasons for these differences in bird diversity near edges are not fully understood. The edge is likely to be markedly different in microclimate, e.g. light, air humidity, wind, and later in vegetation as shrubs and vines grow up and light-demanding seedlings establish. Unfortunately little has been published about these aspects of the Amazon fragments. Another possible influence on birds near edges is predators from the open land outside. In the Michigan study, although there were more nests near the edge, a smaller percentage of the eggs gave rise to fledgelings that survived to fly; this was primarily because of greater predation (Gates & Gysel 1978). Similar conclusions were reached by Andren & Angelstamm (1988) in a region of central Sweden where there are patches of forest surrounded by farmland. They placed eggs from domestic hens on the ground, simulating nests, in farmland and in forest. The percentage of eggs lost to predators was much lower in forest than in farmland, provided the eggs were more than 50 m from an edge. Within 50 m of an edge the predation rate rose sharply.

The area needed to support an animal or a family group

A third type of reason why a species may become extinct in a small habitat patch is that it requires a large area per individual to obtain adequate food. Many pairs or family groups of bird and animal species have territories or home ranges. These are likely to be largest for top carnivores, but an example of a species that eats fruit and insects is the golden-handed tamarin, a small primate that lives in the region of the Amazon forest fragment experiment (Lovejoy *et al.* 1986). They live in groups of four to ten individuals with a home range of 10–30 ha. The groups are often separated by as much as 1 km. We can then ask: could a 10-ha forest fragment support a group? Could a 100-ha fragment support more than one group? The longest record in Lovejoy *et al.* (1986) is for a 10-ha fragment isolated in 1980. At isolation it contained one group of tamarins, but by 1985 they had disappeared. Of the 10-ha fragments isolated in 1983, two had tamarins and both still retained them 2 yrs later. One 100-ha fragment contained part of the home range of four groups when it was isolated in 1983. Two years later three of them remained. This last may seem surprising, given that family groups in the forest are often less than one per 100 ha. Tamarins have been observed to favour habitats with some disturbance and secondary plant growth, so a fragment with disturbed edges may actually be more favourable for them than the previous forest.

Minimum viable populations

Examples of isolated populations

The absolute minimum population which could possibly allow a species to continue is one individual of each sex, but nowadays people involved in conservation of rare species get worried long before the number falls that low. An example is the giant panda. Fossil remains show that the species was formerly widespread in China, but today it is confined to six isolated mountainous areas. A census in 1974–77 indicated that the total number of pandas was 1000–1100, but there has probably been a decrease since then (O'Brien & Knight 1987). The six habitat areas are further divided, e.g. by rivers and deforested areas, so some populations have fewer than 20 adults, and the largest is probably less than 200. Are populations of pandas of this size viable? In other words, are they in danger of dying out just because they are too small?

Another example is the northern spotted owl, which lives in old conifer forest in the Pacific northwest region of U.S.A. Each pair requires 800–2000 ha of forest. This old-growth forest is commercially attractive for logging, so the question has been asked: how large a fragment of forest needs to be left to support a viable population of owls? Clearly a patch that could support only one pair would leave the long-term survival of the species much at risk. A third example: there are about 100 adult grizzly bears in Yellowstone National Park, isolated from all other grizzlies. Is this population large enough to persist indefinitely, or is it at risk?

Plants can also have small, isolated populations. In tropical rain-forest, because of the great species-richness, many of the tree species have only a few individuals per hectare, or are even sparser. A census was carried out in a 50-ha plot in tropical forest on Barro Colorado Island in Panama, in which every woody plant with stem diameter 1 cm or more was identified and listed. Of the 303 species, 21 were present as only one individual within the 50 ha, and a further 18 species had only two individuals (Hubbell & Foster 1986a). So the effective size of the population of interbreeding individuals is for some species only one or a few. Does that put the species at risk?

If we wish to take active steps to prevent the extinction of a species, we need to know what is the minimum size of population that is safe, the *minimum viable population*. For the moment I define this as the smallest population that gives the species a high chance of surviving for a long time. To get a more precise definition we need to start including the probability of survival; this will be considered later. A book edited by Soulé (1987) discusses in some detail how to predict the minimum viable population size, in general and for particular species. There are two types of risk that become greater in small populations, genetic and demographic (Box 8.2). Before considering each in more detail, I first ask whether we can get direct evidence, by observation or experiment, that a population of a particular species is likely to go extinct if it is below a particular minimum size. This would need to be done under field conditions, and would require long-term counts on numerous populations of different initial sizes. Some problems are that (1) often rare species do not have many remaining populations; (2) the individuals are often difficult to count; and (3) we may need decisions soon about conservation of habitats, so we cannot afford to wait for a long study. In fact there is one species for which enough long-term records are available to give a clear indication of the minimum viable population: bighorn sheep in southwestern U.S.A. This species lives as isolated populations in open vegetation on mountains, so numbers can be counted, at least approximately, from a distance. Berger (1990) collated data extending up to 70 yrs for 129 populations. Figure 8.5 shows that all populations which had 50 animals or fewer when first recorded had gone extinct within 50 yrs, whereas most populations larger than 100 remained. So the minimum viable population for this species appears to be about 50–100.

Genetic risks to small populations

A genetically uniform population is at greater risk of being wiped out, for example because it lacks individuals resistant to a disease or tolerant of unusual weather conditions. The genetic risks to small populations are explained in some detail by Lande & Barrowclough in Soulé (1987). One problem is genetic drift, which means that a particular gene, even if beneficial, can be lost from a population by chance: the eggs and sperms that happened to meet and fuse did not include it.

Genetic drift

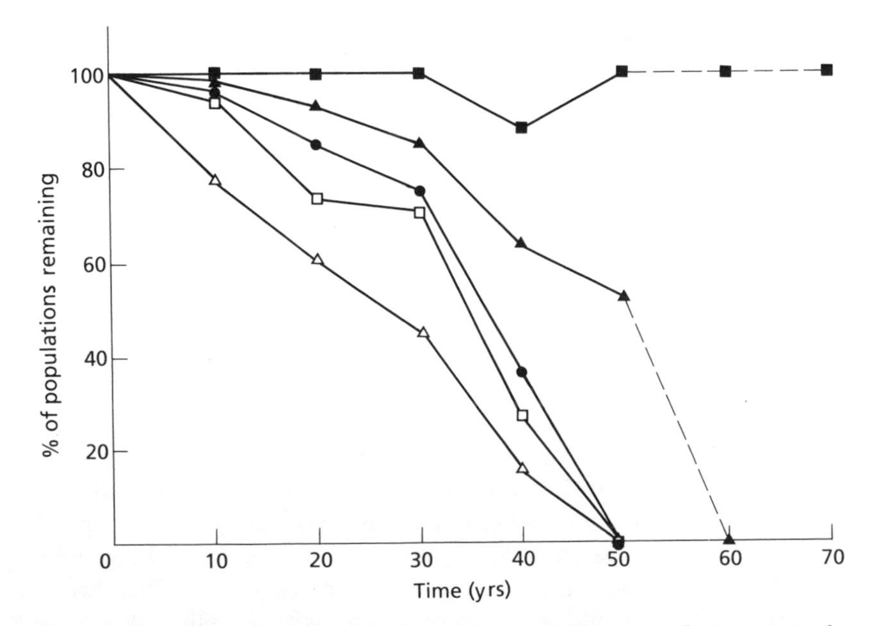

Fig. 8.5 Percentage of bighorn sheep populations remaining, in relation to initial size of populations: △, 1–15 sheep originally; □, 16–30; ●, 31–50; ▲, 51–100; ■, >100. ---, data from fewer than four populations. From Berger (1990).

When studying how the rate of loss of genes is related to population size, we are interested in whether each locus on each chromosome carries the same gene throughout the population or whether there are alternative genes, i.e. in the amount of heterozygosity. A simple classic model predicted that the amount of heterozygosity in a particular generation, H_{t+1}, is related to that in the previous generation, H_t, by:

$$H_{t+1} = H_t(1 - 1/2N)$$

where N is the number of individuals in the population. This shows, for example, that after $2N$ generations the loss of heterozygosity will be about two-thirds. In real species the situation is more complex, because the rate of genetic drift is affected by the sex ratio, fluctuations in population size and other things. Lande & Barrowclough in Soulé (1987) discuss how to deal with such complexities.

Small populations have less genetic variation

Populations also gain new genes, by mutation. The amount of genetic variation within the population will depend on a balance between losses and gains, but a small population will contain less genetic variation because genetic drift is faster in small populations. For example, among 17 populations of a coniferous shrub, ranging from 20 individuals to about 400 000, in the South Island of New Zealand, there was a clear positive correlation between the size of the population and the amount of genetic variation between individuals (Billington 1991). These montane populations were probably isolated by climate warming after the Ice Age, about 100 generations ago. Another example is a population of about 12 wolves on an island in Lake Superior, which was found to be less genetically variable than populations on the mainland (Wayne *et al.* 1991). The island was first colonized by wolves about 40 years earlier, perhaps by a single pair, and this 'founder effect', rather than genetic drift, could be the main reason for the limited genetic variability.

Predicting how large a population needs to be to maintain a desired amount of genetic variation is difficult, not least because it involves knowing the rates of mutation within the population. Lande & Barrowclough suggest that for quantitative polygenic characters a population of the order of 500 would be adequate, but for single-locus genes it could be 100 000 to a million.

Another problem with small populations is inbreeding depression: in a small population mating between closely related individuals is more likely, and this increases the probability that they both carry the same recessive, harmful gene, which can then be expressed in their progeny. According to Lande & Barrowclough, most inbreeding depression can be avoided if the population size is more than a few dozen.

We have, then, three different recommendations for the population size, from three different genetic risks: the recommendations are a few dozen, a few hundred and a few hundred thousand. The last one raises the question of how large an interbreeding population can be: even if there are no physical barriers how far do animals spread to find mates, how far does pollen travel? There may, therefore, be a *metapopulation*, a large number of individuals comprising several sub-populations, with only limited interchange of individuals or genes between the sub-populations. It is clearly not realistic to plan for reserves containing hundreds of thousands of individuals of a mammal or bird species, though it might be possible for some plants and invertebrates.

Risks from population fluctuation

The other type of risk to small populations, demographic risk, occurs because the numbers fluctuate from year to year. Each year some animals die and some young are born. On average births and deaths may be equal, but that is unlikely to be true each year: chance will dictate that in some years deaths will exceed births, and the question is, what is the chance that one year all the remaining animals will die (or all those of one sex). To predict this involves models fundamentally different from conventional statistics. The familiar statistics of *t*-tests, correlation coefficients and so on assume that we have a small sample drawn from a very large population, and we are trying, from measurements on the sample, to reach conclusions about the population. But now the population itself is small, its size is a crucial feature, and we need *stochastic* models, that is models that take into account chance events. Some of these models are presented and discussed by Goodman in Soulé (1987) and by Dennis *et al.* (1991). Here I shall just describe an example in a non-mathematical way.

The grizzly bear population in Yellowstone

Grizzly bears occur in Yellowstone National Park and surrounding countryside, a total area of about 20 000 km^2. Background information about grizzlies in Yellowstone is provided by Knight & Eberhardt (1985) and Mattson & Reid (1991). Grizzly bears occur widely in Canada and Alaska, and the species as a whole is not at risk. But the Yellowstone population is separated by more than 200 km from other grizzly populations further north, and there has been concern whether it is large enough to survive. The grizzly is an omnivore; its natural food in Yellowstone includes ungulate mammals, moths and pine seeds. Until the late 1960s garbage from camp sites and visitor centres was left uncovered at dumps, and grizzlies came there to look for food. This provided an opportunity to count numbers of bears and to mark them for studies of age-related mortality. The dumps were closed in 1970–71, and since then estimating population numbers has been more difficult. Observations from aeroplanes have been supplemented by trapping animals and fitting them with radio transmitters, so their future position can be determined (Knight & Eberhardt 1985). Figure 8.6 shows the estimated number of adult females each year from 1959 to 1987. If these numbers are accurate there have been some large changes from one year to another, which may seem surprising for a long-lived animal with low reproductive rate. One of the largest falls was soon after the garbage dumps were closed. Up to 1981 there seemed to be a downward trend, and people were worried that this might continue. However, during the 1980s the numbers rose, and the response of many people to Fig. 8.6 now might be that there is nothing to worry about. Dennis *et al.* (1991) propose a more sophisticated way of deciding whether the grizzlies are at risk, involving stochastic modelling. The manager is required to state a 'threshold' population size, a minimum acceptable number of animals, and the model will then predict the chance of that threshold being reached and when that is likely to occur. The input required is population numbers over a period of years (e.g. as in Fig. 8.6), sufficient to give an indication of the amount of variation from year to year. Table 8.2 shows the application of this model to grizzlies in Yellowstone. Even if the manager decides

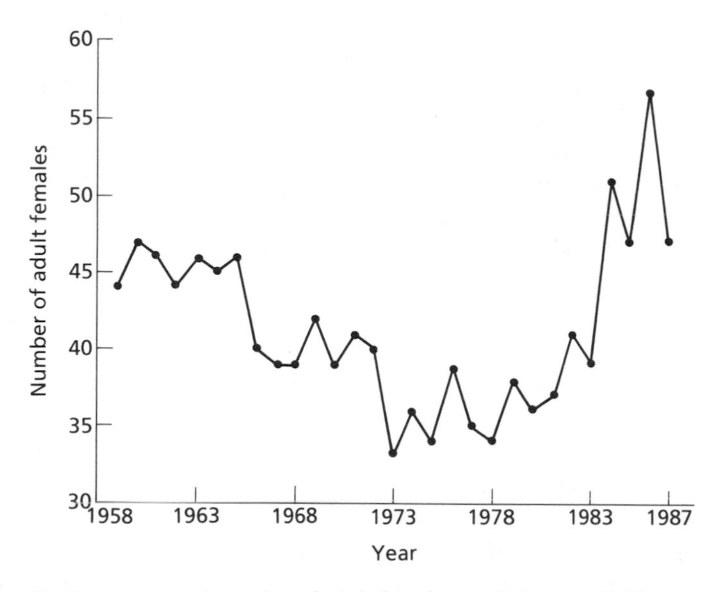

Fig. 8.6 Estimated number of adult female grizzly bears in Yellowstone National Park, 1959–87. From Dennis *et al.* (1991).

Table 8.2 Prediction, by stochastic model of Dennis *et al.* (1991), of probability that numbers of grizzly bears in Yellowstone and of Puerto Rican parrot will fall to a specified threshold number, and if so when that may happen. For each species calculations have been performed using two alternative chosen thresholds

	Grizzly bear, Yellowstone*		Puerto Rican parrot	
Present population size (at last count)	47		38	
Chosen threshold population size	10	1	10	1
Probability (%) of reaching threshold	100	100	0.1	$<10^{-6}$
Median time (years) to reach threshold (95% confidence limits)	152 (0–679)	448 (0–2280)		

* Number of adult females.

that one female grizzly bear is an acceptable threshold population (i.e. a 'crisis minimum') for Yellowstone, the model predicts that this is certain to occur, though probably not for several centuries. This suggests that there is ample time to plan changes to increase the bears' chance of survival. However, the confidence limits on the median time to reach the threshold are very wide, so it could happen next year. If the manager sets a higher threshold the probability is that it will be reached sooner.

Dennis *et al.* (1991) also applied their model to six bird species. The model made more optimistic predictions about some of them than it did about grizzly bears. Although there were only 38 Puerto Rican parrots remaining in 1989, living in one forest, the species was predicted to have more than 99.9% chance of remaining above a threshold of 10 birds. The limitations in this and any stochastic model lie in the data as well as the calculations. The key input is the amount of variation in the real population. If it is difficult to determine the number of animals, then there must be uncertainty how much of the apparent change from year to year is real and how much is due to errors of counting. The paper by Dennis *et al.* gives details of how a single year-to-year change in numbers of Fig. 8.6 can have a large influence on the predicted chance of extinction.

This research on minimum viable populations strongly suggests that some large mammals and birds are at risk because their populations are too small. Where habitats are in the process of being destroyed, predictions of minimum viable population size are likely to be very important in deciding how much of the habitat should be left and in what size of patches. Where species are already confined to fragments that are calculated to be too small, the ideal solution for conserving them would be to enlarge the area. Occasionally this may be possible, but usually it is not practicable, (1) because of other demands on the land, and (2) because of the difficulty of regenerating the ecosystem (see later). An alternative is to promote migration between habitat fragments; this provides genetic exchange and allows recolonization of fragments where the species has become extinct.

Migration between fragments

Examples of limited migration

If the number of species in an island is determined, at least in part, by a balance between rate of invasion and rate of extinction, then the ease with which species can migrate from one habitat island to another will influence diversity. MacArthur & Wilson (1967) devoted much attention to this aspect. There are examples which show clearly that migration from one habitat fragment to another on land is important for individual species. Figure 8.7 is from the study of forest fragments ('woods') in Lincolnshire that was described earlier. The ground-floor perennial dog's mercury occurred in many of the ancient woods (i.e. those known

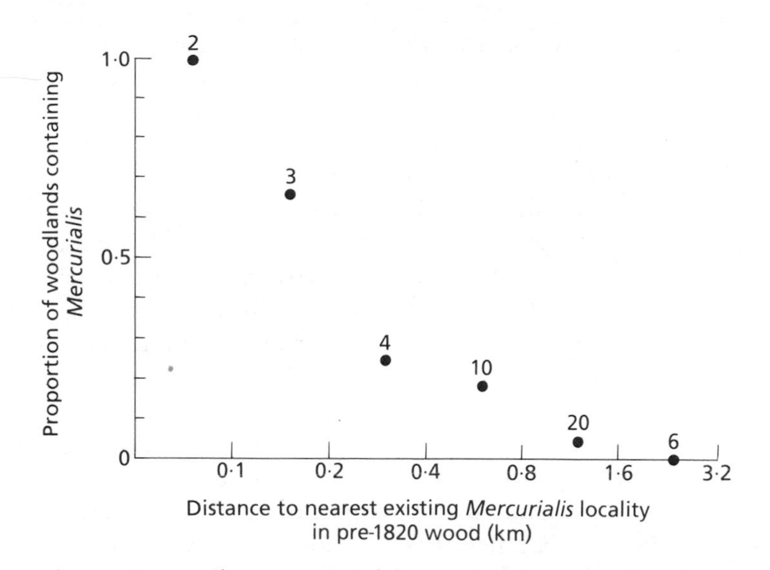

Fig. 8.7 Data on the occurrence of the perennial forb dog's mercury (*Mercurialis perennis*) in small woods in Lincolnshire, England that originated between 1820 and 1887. The woods are grouped according to how far each is from another wood already present in 1820 which contains dog's mercury. The figure above each point shows the number of woods in that class. From Peterken & Game (1981).

to have been present since before 1600). Its seeds are neither fleshy nor wind-borne, so it would not be expected to spread rapidly. Figure 8.7 shows that whether it occurred in more recently established woods depended on how far they were from a wood already containing the species. Evidently dog's mercury quite often spreads a few hundred metres, but rarely spreads more than 1 km, even during 100–150 yrs.

Another example is provided by the Bay checkerspot butterfly which, as I described earlier, inhabits patches of serpentine grassland near San Francisco Bay. Following a drought period in 1975–77 the butterfly was still present in Morgan Hill, an area of suitable grassland about 10 km long × 1–2 km wide, but absent from many smaller patches nearby. Harrison *et al.* (1988) identified 59 serpentine outcrops within 21 km of Morgan Hill all of which carried the chief food plant of the butterfly and which seemed suitable for it in other respects. In 1987 nine of these sites contained populations of the butterfly. These nine varied in size and other characteristics, but all were 4.4 km or less from Morgan Hill. This strongly suggests that reinvasion from the large 'reservoir' area was the main thing determining how many sites it occupied. If it takes about a decade for the species to migrate that far, there is serious risk of another major drought before it has reached all the sites. Thus the overall abundance of the species may depend on its ability to migrate across unfavourable countryside. Thomas (1991) describes how some British butterfly species fail to colonize suitable habitats because of a gap of a few hundred metres between habitat fragments. The uncolonized fragments were shown to be suitable, because butterflies that were later introduced there artificially survived and multiplied.

Corridors for migration across farmland

There has been much discussion of corridors that could allow migration of species from one fragment to another across otherwise unfavourable landscape. Hedges and rough grassland (e.g. road verges) could provide such corridors through farmland. If we are going to create or maintain such corridors for the benefit of wild species we need to know whether they really are effective in promoting migration of the species and maintaining their abundance; and if so what features of the corridor are needed to make them effective. For example, hedges vary greatly. Some consist of a narrow strip of a single shrub species. Others are wide, rich in plant species and include trees; in Europe some of these are the remnants of ancient forest. In some parts of North America there are hedges that are rows of trees several trees wide. In other parts there are 'fencerows', mixtures of trees, shrubs and herbs that have established along old fence lines. Hedges can be valuable in their own right, as habitats for wild species and as contributions to the appearance of the landscape; but here I discuss their possible importance as corridors. Unfortunately, information about species movement along corridors is limited. Movement of large animals, e.g. elephants, can be observed directly, but not movement of smaller animals and plants. Trapping has shown that in farmland some small mammals are found only in or near hedges. Pollard & Relton (1970) found that field mice were captured about equally often in any part of a 130 × 130 m field in eastern England, whereas bank voles were captured only near the surrounding hedges. Two small mammal species that occurred in a wood in Ontario, chipmunks and white-footed mice, were also both captured in nearby fencerows; the mice were also captured in open fields but chipmunks never were (Wegner & Merriam 1979). These results might suggest that bank voles and chipmunks require hedges to allow them to migrate from one wood to another. However, further work on chipmunks by the same research group in a nearby area of Ontario did not support that (Henderson *et al.* 1985). All chipmunks living in two woods were removed. Although the woods were separated from all other woods by a road on one side and a river on the other, they were recolonized by chipmunk within a few weeks. When chipmunks were marked and later recaptured, their movements must have involved crossing open gaps tens of metres wide, and in a few cases hundreds of metres. This shows the difficulty of deciding whether a hedge corridor is important for a particular species.

Some woodland plants migrate only very slowly along hedges

Figure 8.7 shows the limited ability of the woodland herb dog's mercury to spread between woodland fragments during more than a century. One question is whether hedges connecting woods could help it to spread. In a nearby area Pollard (1973) recorded the distribution of dog's mercury and two other woodland herb species along a hedge that had been planted between 1791 and 1835. The hedge was about 600 m long, joining at each end to much older hedges in which the three species were abundant. About half of the new hedge ran beside a road, and here there were five separate colonies of dog's mercury. In the other section, which ran through fields, dog's mercury occurred only in 25 m at one end nearest to the old hedge. This indicates very slow spread along hedges but perhaps faster, irregular spread along roads. The other two species recorded, bluebell and wood anemone, occurred in the old hedge to within a few metres of the new hedge but had not spread into the new hedge at all. This is of course only one example, but it shows that a hedge does not automatically provide a corridor for woodland plants.

Rapid migration of a vole along road verges

A convincing example of migration of a small mammal along corridors through farmland involves a vole in rough grassland in road verges. Up to 1970 the southern limit of *Microtus pennsylvanicus* in central Illinois was about 40 km north of Champaign-Urbana. In the late 1960s and early 1970s interstate highways (motorways) were built in the area. Trapping in 1976 (Getz *et al.* 1978) showed that the vole had extended its range about 90 km further south. It was found not only in the verges of the interstate highways, but up to several kilometres away, in other rough grassland, e.g. beside old roads and railways, provided it was connected to an interstate. However, it was not found in similar grassland if it was unconnected to the interstate system. Interstate highways do not pass through towns, whereas other roads and railways do, which could explain why interstates are more effective corridors.

Thus there is still very limited information about whether corridors really help to maintain species, and about what features of corridors make them more effective. Until we have better information, there seems inadequate justification for creating or maintaining corridors through farmland and urban land solely as migration routes for plants and animals. Spacing fragments close enough together to promote migration seems more promising. Examples of migration in relation to distance between fragments were given above, for a butterfly and a plant species.

Possible ways to aid survival of small, isolated populations

There are other, more artificial procedures that could promote the survival of species that have been reduced to a small population. It may be possible to increase the population size by alterations to the habitat, e.g. creating a particular type of disturbance, encouraging a food plant or reducing the abundance of a predator. Genetic variability can be increased by: (1) moving animals, eggs or seeds from one population to another; or (2) interbreeding individuals from two populations in captivity and releasing the offspring. The risk of the population falling below a threshold of numbers may be reduced by feeding animals at critical times, e.g. unusually cold winters or dry periods; or a captive reserve population can be maintained in case the wild population dies out. Any of these is likely to involve considerable time and effort, but people are prepared to dedicate such effort to the preservation of certain specially valued species. There has been much care devoted, for example, to the breeding of giant pandas in captivity.

A conservation plan for the northern spotted owl

The northern spotted owl provides an example of a species where conservation plans needed to take into account minimum viable population size and also ability to disperse between fragments. This rarely seen bird has become a test case for conservation, because it inhabits old-growth conifer forest in the U.S. Pacific Northwest which is particularly suitable for commercial logging. Congress ordered the setting up of a Spotted Owl Scientific Committee to 'develop a scientifically credible conservation strategy for the Northern Spotted Owl'. Murphy & Noon (1992) summarized the scientific reasoning that led to the committee's detailed plan. The committee aimed to predict the minimum area of forest, and its arrangement in patches, that would give a high probability of the species persisting for at least a century. Because some forest has already been cut down, and because it was decided that all the forest conserved as owl habitat should be on publicly owned land, the plan was for many fragments, with owls expected to disperse between them, rather than a single large forest area to maintain a permanently viable population. To produce a model to answer the questions 'How large should the forest fragments be?' and 'How far apart should they be?', the committee needed information on population fluctuations and on ability to disperse, which are available for some other species (see earlier examples), but not nearly adequately for the northern spotted owl (Doak 1989, Murphy & Noon 1992). The committee did, nevertheless, with the help of models, produce maps showing proposed areas of forest that should be preserved from felling as habitat for the owl. These may well be the soundest proposals possible given the limited information available; but they are unlikely to be considered by loggers and logging companies as adequate argument for restricting logging so much.

Keystone species

If the keystone is removed the building falls down. Are there analogous species in communities? In other words, if one particular species becomes extinct, will there be major resulting changes in the structure or species composition of the community? There are many herbivorous insect species which eat only one or a few plant species; these include some rare butterflies whose conservation receives priority (Thomas 1991). So if the plant species disappears the insect will die out too. There are also pairs of species that require each other. Fig trees and their specific pollinators are an example, mentioned in Chapter 6. The bees of Fig. 8.4 are the only pollinators of some plants in the tropical rainforest. So if the bees fail to survive in small forest fragments, their plants will go extinct there as well. So extinction of one species can have knock-on effects on another. But here I am concerned with whether extinction of one species could have an effect on not just a few other species but many, causing a major change in the whole ecosystem. If we can identify such species in advance, there is a special reason for trying to maintain them.

Ecologists have in the past suggested that the more species-rich an ecosystem is the more stable it will be; in other words, diversity causes stability. If this were a general 'law', it would mean that the disappearance of any species would increase the instability or fragility of an ecosystem and so might put it at risk. Early, distinguished proponents of this view were MacArthur (1955), Elton (1958) and Hutchinson (1959). Many models have been devised to test this hypothesis. Some support it (e.g. Law & Blackford 1992), but many predict the opposite, that increasing number of species and of interactions between species tend to make an ecosystem *less* stable, and there is some observational evidence to support that (May 1974, 1981; chapter 23 of Begon *et al.* 1990). The message for us is that there is no established law to say that disappearance of *any* species will necessarily put others at risk. There may be *particular* species that have a keystone role in the community, and we need to consider how we can identify which they are.

Are there keystone plant species?

Study of interaction between plant species indicates that in general these interactions are not specific, that an entire complement of species does not need to be present for the ecosystem to function. One sort of evidence is that along many transect lines, if the physical environment changes gradually, some plant species decline gradually in abundance and then disappear, others appear and increase. This was particularly clearly shown by Whittaker's description of forest vegetation in the Great Smoky Mountains (Whittaker 1956), but has been recorded in many other vegetation types elsewhere. There *can* be abrupt changes in vegetation, but they are not necessary. Another source of evidence is that, as the world warmed after the Ice Age, plant species did not migrate at the same rate, or even always in the same direction; so plant communities did not migrate as units but changed (see Chapter 6). A well-recorded large-scale extinction of a plant species is American chestnut, which almost entirely disappeared during the first half of the 20th century, due to chestnut blight (see Chapter 6). This did not result in the disappearance of any other tree species, or of any large vertebrates as far as is known. In contrast to plant deletions, invasion by a single species can have a marked effect on other plant species. Examples are Klamath-weed in Californian rangeland (see Chapter 6) and *Rhododendron ponticum* in British woodlands; there are many more. Here an invader becomes abundant and reduces or eliminates less competitive native species. The invader is unlikely to be at risk of extinction or in need of protection; if it did disappear from its new range this would more likely lead to increased species-richness rather than other extinctions.

Thus as long as we consider plants alone there seems no ecological reason for pinpointing particular species as specially worthy of preservation.

Grazing mammals as keystone species

Large grazing mammals can have a conspicuous effect on vegetation. Figures 3.10 and 3.11 show marked effects of cattle and sheep on the abundance of plant species. At the upland site of Fig. 3.10(d) the most abundant species initially was a wiry, unpalatable rush, *Juncus squarrosus*, which after grazing was excluded virtually disappeared. Table 8.3 shows results from a nearby grassland site which was better drained and richer in species. Excluding grazing led to increases in some species, particularly some grasses, but to decreases in many smaller forbs, bryophytes and lichens. Some of these were not recorded at all by the end; this does not prove that they were completely absent, but they were certainly much rarer. Unfortunately there was no proper control area in which similar records were made while grazing was continued, but there can be no serious doubt that these major changes were a response to cessation of grazing. Data from another year (Table 8.3(b)) confirm this by showing fewer species in the exclosure than in a grazed area. This example and Figs 3.10 and 3.11 show that removing a single domestic grazer species can result in drastic decline and sometimes local extinction of plant species, presumably because other, more competitive plant species increase.

An example of near-extinction of a naturalized grazer is the sudden killing of most rabbits in Britain by myxomatosis in the mid-1950s (see Chapter 6). Rabbits, which were introduced to Britain in the 11th

Table 8.3 Effect of excluding sheep from an area of grassland in the Pennine Hills, northern England. Based on records from 1000 point quadrats

(a) Plant species that changed significantly in cover percentage during 22 yrs inside a fenced area from which sheep were excluded

	Grasses	Other angiosperms	Bryophytes	Lichens	Total
Significant increase in cover	5	3	0	0	8
Significant decrease, but still present at end	2	6	12	2	22
Decreased to zero	0	4	7	3	14

(b) Number of plant species present (total)

Where sheep excluded for 24 yrs	24
Nearby area, grazed throughout	31

Data of Rawes (1981).

or 12th century, had been abundant in many lowland grasslands. After the near disappearance of the rabbits many of these became colonized by shrubs and trees, and some of them are now forests. The changes in one part of eastern England, Breckland, are described later.

Sudden disappearance of large mammals 11 000 yrs ago

It would have been useful at this point to describe the effects of the extinction of a *native* herbivore. Large herbivore species have been made extinct by people, but not in controlled experiments. In North and South America about three-quarters of the genera of large mammal that had been present through the last glacial period became extinct as it ended; in Australia the percentage becoming extinct was even higher. In Europe, Asia and Africa there were some extinctions but far fewer. Stuart (1991) has described the extinctions that occurred in North America and Eurasia, and discussed their possible causes. There is good reason to think that in North America the primary cause of species disappearing was not climate change but people hunting and killing them. One piece of evidence is that no similarly massive extinction event occurred during equivalent warming at the start of earlier interglacials. More direct evidence comes from precise dating of extinctions. Extinction of the Shasta ground sloth can be dated precisely from its dung. It lived in caves in the U.S. southwest, where remains of its dung can still be found; they can be dated by ^{14}C, so the age of the last dung deposited is known. At most of the known sites the species became extinct within a 600-yr period, between 11 300 and 10 700 yrs BP (Spaulding *et al.* in Porter 1983). Other species' extinctions can be dated somewhat less accurately, from remains of their carcasses; they all disappeared between 11 500 and 10 500 BP, and probably many of them during 300 yrs, 11 300–11 000 BP (Stuart 1991). This coincided with the rapid spread southwards from Alaska, 11 500–11 000 BP, of the Clovis people, hunters who made particularly fine stone spearheads, and who are known to have been able to kill large animals, even mammoths. There seems little doubt that these hunters were the major cause of this 'mega-extinction event'. The key similarity between North America, South America and Australia was probably rapid invasion by people with the ability to kill large animals. In Africa and Eurasia people were present much earlier and their hunting skills developed gradually. In these areas extinctions of large mammals were fewer and less concentrated in time. This probably explains why Africa's grazing lands are much richer in large mammals than are the Americas or Australia.

These past major extinctions by people raise important questions for conservation. In North America, how much had previous herbivore extinctions affected the apparently natural lands that the first European explorers saw—for example the sparsely grazed bunchgrass prairies? This could influence decisions about what conservationists today should be aiming for, which I discuss later. In Africa, how strongly are the large herbivores controlling the nature of the vegetation and the abundance or even existence of other animals? If some large herbivores become extinct, through hunting or loss of habitat, will it have major knock-on effects on the remaining grazing land? Owen-Smith (1989) discusses messages for conservation from this extinction event, with particular reference to Africa and elephants. There is no doubt that elephants do have a major effect in some savanna areas of Africa, opening up wooded areas and increasing grazing for some smaller mammals. Whether elephants should be actively managed by culling in some areas, and if so how much, is a question which Owen-Smith discusses, and which remains unresolved.

Herbivorous insects as keystone species

We should not assume that only *large* herbivores can have major effects; small ones can too. This is shown by the example of Klamath-weed and its control by the beetle *Chrysolina* (see Chapter 6). Klamath-weed and *Chrysolina* are both sparse in California at present. If the beetle becomes extinct Klamath-weed will presumably return to its previous status, extremely abundant in rangeland, reducing other plant species and preventing grazing by mammals. Specific insect herbivores may in a similar way be preventing dominance of some plant species in natural vegetation also. Janzen (1972) found that when seeds of a tree, *Sterculia apetala*, in tropical forest in Costa Rica fell to the ground below the parent tree they were quickly attacked by bugs, *Dysdercus*, which sucked out the contents and left them inviable. Janzen showed that seed had to be transported more than 30 m from a *S. apetala* tree to escape attack by the bugs. This could be the reason why trees of this species are widely spaced in the forest. Janzen suggested that much of the plant species diversity of tropical forest could be explained if each plant species has a specific animal or pathogen that limits its reproduction close to existing plants of the species. Although this hypothesis attracted a lot of interest, few other examples from tropical forest have been reported since. Chapter 6 gives a more thorough explanation of how a pathogen or herbivorous insect can control the abundance of its host. The message for conservationists is that these inconspicuous species, whose conservation is rarely discussed, may have major effects on the abundance of more visible species. When a host population falls to low numbers this increases the chance of a pathogen becoming extinct (see Chapter 6), so this could be an additional risk for populations confined to small fragments.

Carnivores as keystone species

This book has already given examples, from aquatic systems, of how a single carnivorous species can influence the species composition of a whole community. The preceding chapter described how carnivorous fish in lakes can markedly affect the abundance and species composition of not only the zooplankton on which they feed but through them the phytoplankton also (Fig. 7.14). Peruvian anchoveta is a marine fish whose near disappearance altered a whole ecosystem (Chapter 4). Here we are concerned with large terrestrial carnivores. In large parts of temperate Europe and North America most of the native large carnivores are much reduced in numbers and range compared with a few thousand years ago. The main causes of their decline are deliberate killing by people and reduction of habitat area. A carnivore species usually needs a larger area per individual to provide enough food than does a herbivore of the same size, and carnivores are thus likely to be particularly sensitive to fragmentation. The northern spotted owl is an example already mentioned. The question to be considered here is whether loss of carnivores in Europe and North America has had an effect on remaining species. This is difficult to prove conclusively, in the absence of closely similar areas with and without carnivores.

Deer and their effects on forest regeneration

Decline in carnivores may be expected to lead to increase in the herbivores that they ate. Animals whose abundance has probably greatly increased through decline of their predators are the deer of North America and Europe. White-tailed deer have become much more abundant and widespread in North America during this century. We have no basis for accurate estimates of deer population numbers several thousand years ago, but it does seem likely that their present abundance in many forested areas of northeastern U.S.A. is higher than it was then (Alverson *et al.* 1988, Whitney 1990). These elegant creatures give pleasure to many people, which is one reason why I chose it as the animal to be on the cover of this book. But if the deer become too abundant their browsing on tree seedlings and saplings could be intense enough to prevent regeneration of important tree species.

Experiments have confirmed that deer can have a marked effect on tree seedling survival. Tilghman (1989) set up large fenced areas (13 or 26 ha) in four hardwood forests in northern Pennsylvania, and placed white-tailed deer in them to give a range of densities. Figure 8.8 shows that after 5 yrs deer had reduced the number of tree seedlings, but had not eliminated them. However, most of these seedlings were very small: if only seedlings more than 30 cm tall are counted, deer were reducing the numbers to a point where regeneration of trees could be affected. More important than the total numbers may be the species composition: in the highest deer density some major forest tree species, including two of the dominants in the canopy, sugar maple and red maple, had no seedlings at all more than 30 cm tall. In that treatment many sample plots had abundant seedlings of only one species, black cherry. So this experiment indicates that although deer over this range of density would not prevent tree regeneration altogether, they would probably prevent some species from regenerating and so alter the composition of the forest. Exclosures set up in other forests to keep out deer and other large grazers have confirmed this (Alverson *et al.* 1988). The highest deer density used by Tilghman is probably rare, but densities of 10–15 per km^2 are common.

Wolves, deer and trees in a Polish forest

Large carnivores have disappeared from most of Europe, but one place where wolves still survive is Bialowieza Forest, an ancient, little-disturbed mixed deciduous forest on the Belorussian–Polish border.

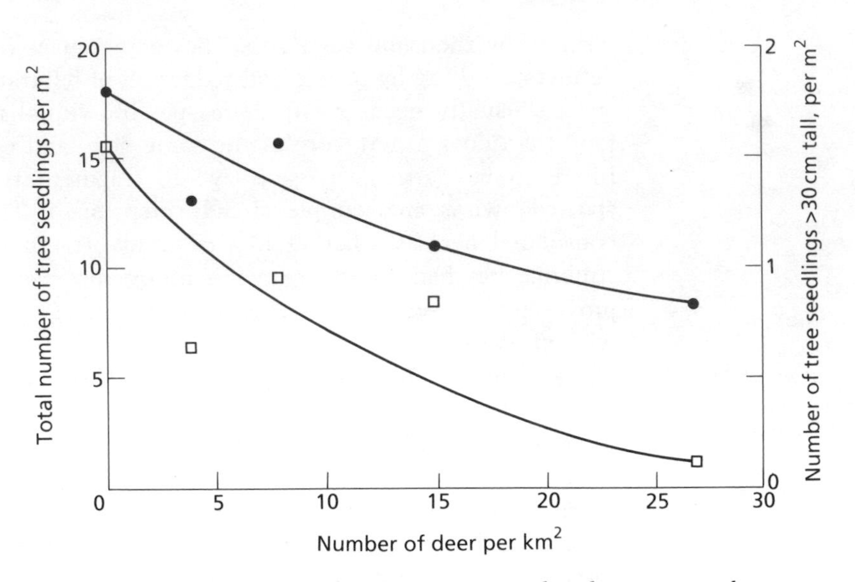

Fig. 8.8 Abundance of tree seedlings in experimental enclosures in northern Pennsylvania where white-tailed deer were maintained at different densities for 5 yrs. ●, all seedlings (left-hand scale), □, seedlings more than 30 cm tall (right-hand scale). Data of Tilghman (1989).

In a study conducted from 1985 to 1990 Jedrzejewski *et al.* (1992) found that the diet of the resident wolves was mainly red deer and wild boar, and that wolves were the main cause of death of red deer. Even during severe winters shortage of food did not become the major cause of death of the deer. Since red deer were the major browsing species, this suggests that the presence of wolves could, by controlling deer numbers, have an important influence on tree seedling survival. In 1973 Pigott (1975) measured the girth sizes of lime (*Tilia cordata*) trees and saplings in the central part of this forest, and compared them with measurements made by a Polish scientist in 1928. These results strongly suggest that lime had not been able to regenerate for some decades before 1928, but regenerated abundantly after then. This central part of the forest was made a National Park in 1923; before then it had been used for hunting. Pigott speculated that the survival of lime seedlings and saplings became possible from 1923 onwards because shooting of wolves was banned within the National Park and they then became more abundant and reduced the deer population. The deer may previously have been food-limited, but became predator-limited. This makes a nice story of how a carnivore can, through effects on grazers, alter the composition of the vegetation; but it is speculative: there is no direct evidence that the abundance of wolves or deer changed after 1923.

Should deer be culled?

The message from this and other research is that the abundance of deer, and their effects on tree seedlings, need monitoring; and active management to reduce populations or their effects may be necessary in some areas. A question for conservation managers is: if natural control by carnivores no longer operates, should we replace it by culling? There is a long tradition of hunting deer in many parts of North America and Europe, but managers of nature reserves and national parks are usually reluctant to kill large animals. In Britain forests were for centuries used for several purposes, including growing deer for meat and growing timber. It was recognized that these two uses are in conflict, and there were rules for management of forests which aimed to keep them in balance (Rackham 1976). Where grazing intensity became too great forests sometimes changed to heathland through lack of regeneration. Rackham (1976) describes examples of this, based on written records, though the grazing was partly by domestic animals, not just deer.

There is no doubt that disappearance of a single species can sometimes have a major effect on the balance between the remaining species and even the continued existence of some of them. All the examples described in detail were large mammals, some of them herbivores, some carnivores. However, there may well be invertebrates and even pathogens that play an important role, and detection of these inconspicuous keystone species is a task for future research.

Managing for high diversity

Diversity in vegetation

Is plant diversity much dependent on niches in the physical habitat?

If an aim of management is to maintain or increase species diversity in an area, we need to take into account how diversity is maintained naturally, and what causes there to be more species in one area than another. I want to consider first what maintains diversity of plant species, moving on later to animals. Using as a basis the alternative mechanisms for maintenance of diversity listed in Box 8.1, the first question to consider is: how far is plant species-richness in the vegetation of an area dependent on heterogeneity in the physical habitat, providing niches? The short answer is that for maintaining β-diversity such heterogeneity is very important, for α-diversity it is not. In other words, on a landscape scale patches and mosaics of varying vegetation can often be related to differences in exposure, steepness, soil depth, drainage and wetness, rock type affecting soil properties and other factors of microclimate and soil. Each species responds differently to the environmental factors and so the proportions of species change. A clear example of this was provided by Whittaker (1956), who showed how each woody species in the Great Smoky Mountains (on the borders of Tennessee and North Carolina) had a different distribution in relation to altitude and exposure (ridge-top to valley-bottom). There are many other published examples.

β-diversity

α-diversity

If we wish to explain the coexistence of 20 herbaceous species in

1 m^2 of grassland, or 20 woody species in 0.1 ha of forest, physical niches can play only a limited part. A legume and a non-legume can coexist because they draw nitrogen from different sources, but otherwise all plants require the same basic resources from their environment. There are a few opportunities for physical separation on a small scale. Sheikh & Rutter (1969) showed that roots of a heath shrub tended to occupy wider soil pores than roots of a coexisting grass. Species whose shoots grow side by side may have root systems that exploit different depths; this has been shown in U.S. desert and in British grassland (Yeaton *et al.* 1977, Fitter 1986). In theory such physical separation between soil pores or by rooting depth could allow a very large number of species to coexist, if each had its very precisely defined and limited range of pore sizes or depths. The few known examples suggest that in practice each of these factors allows only two species to coexist within an area, so much of the observed diversity must have other causes.

Grubb (1977) proposed that the 'regeneration niche' is very important in the maintenance of plant species-richness; i.e. that species differ in the way their seed production, dispersal, germination or seedling establishment responds to environmental factors and this provides niche separation and allows coexistence. Grubb gave many examples of relevant differences between species. A few species have been shown to differ in the size or shape of microsites in soil in which their seeds germinate and seedlings establish best (Harper *et al.* 1965). But apart from this almost all of Grubb's examples relate either to variation of weather between seasons or years, or to biotic interactions. Varying weather could result, for example, in some species setting much seed in one year but different species doing so the next year. Biotic interactions include differences between species in seed or seedling loss to herbivores, and differences in response to different gap sizes. Figure 5.2 shows, for a temperate forest, how some species regenerate better in larger gaps and others in smaller gaps. This has also been found in temperate heathland (Miles 1974).

Pattern and process in the community

So it seems that coexistence of plant species is little dependent on heterogeneity of the physical environment, and is much dependent on how the species interact with each other. This view was put forward by Watt (1947) in a classic paper where he coined the phrase 'pattern and process', meaning that patterns in plant communities are caused by processes going on within the community, by the community as a working mechanism. If each plant in a community were occupying its own particular site where soil or microclimate is most favourable to it, then we should expect each species to remain in its site. But Watt (1947) showed that, on the contrary, the patterns of plants within communities often move about and replace each other. His long-term studies, e.g. in grassland (Watt 1960), provided further examples of this.

One message for vegetation managers is that strategies for promoting diversity depend on whether it is α- or β-diversity they are interested in. For β-diversity we must pay attention to heterogeneity in the physical environment. Perhaps there are waterlogged patches which can be left undrained. Perhaps there is a quarry which, instead of being filled with municipal garbage and then covered with top-soil, can be left to be colonized by plants tolerant of shallow, infertile soil. Disturbance also often contributes to vegetation mosaics. So managers need to consider carefully their attitude to fire in forests—how often should it be allowed? (see Chapter 5). Maybe we should not always wait for natural death of trees, but sometimes create gaps of different sizes.

Effects of environmental conditions on plant species diversity

If our aim is to promote α-diversity, then niches are less significant, and we need to pay attention to alternative 2 in Box 8.1, to maintaining near-equality of fitness and competitive ability. We want to slow down competitive exclusion, and for this paragraph 3 of Box 8.1 is relevant: stresses, grazing, disease or disturbance may play a part in promoting diversity. Grime (1973) proposed a generalization that diversity is highest at sites of intermediate stress or favourableness. He called this his 'hump-back model'. Figure 8.9 shows one of his examples, in which the deeper, more nutrient-rich soil at the left end of the transect

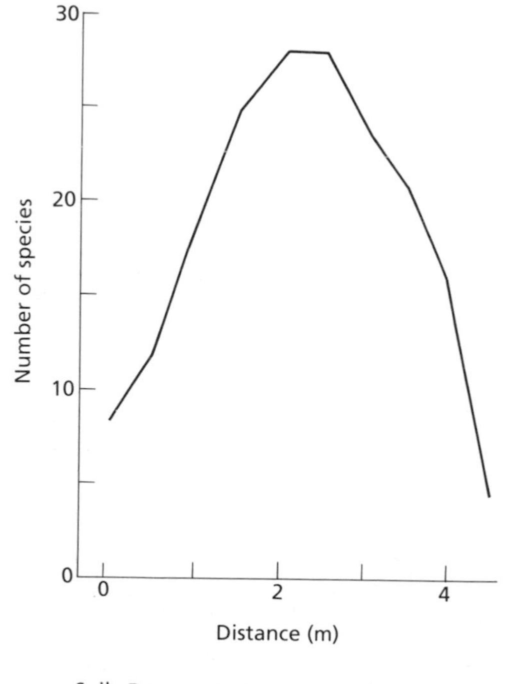

Soil: Deep  Shallow

Fig. 8.9 Number of plant species in 0.5 × 0.5 m quadrats along a transect from rough pasture to a limestone outcrop in Derbyshire, England. Simplified from Grime (1973).

provided the most favourable conditions, becoming progressively less favourable along the transect. There are few species in the least favourable sites because few species can survive the conditions; and there are few species in the most favourable sites because a few strongly competitive species out-compete and exclude all the others. Grime proposed this model for temperate grassland, and we should not expect it to apply to all vegetation types, or to all environmental factors. In the moist tropics the temperature, solar radiation and moisture regime are often very favourable, but this does not prevent forests being extremely species-rich. (However, the soils are often distinctly unfavourable, e.g. nutrient-poor.) Grime's model has important messages for conservation of temperate grassland, one being that soils rich in available nutrients often support grassland poor in species. An experimental demonstration of this is the Park Grass Experiment at Rothamsted, England, where different fertilizer treatments have been applied to plots in a hay meadow since 1856. No species have been sown, but the species composition has changed in response to the treatments. The treatments have altered the soil pH, and the plots of lower pH tend to have fewer species (Silvertown 1980); but in Fig. 8.10 I have illustrated the effects of nutrient supply by picking treatments where, by various additions

Temperate grasslands on poor soil are often species-rich

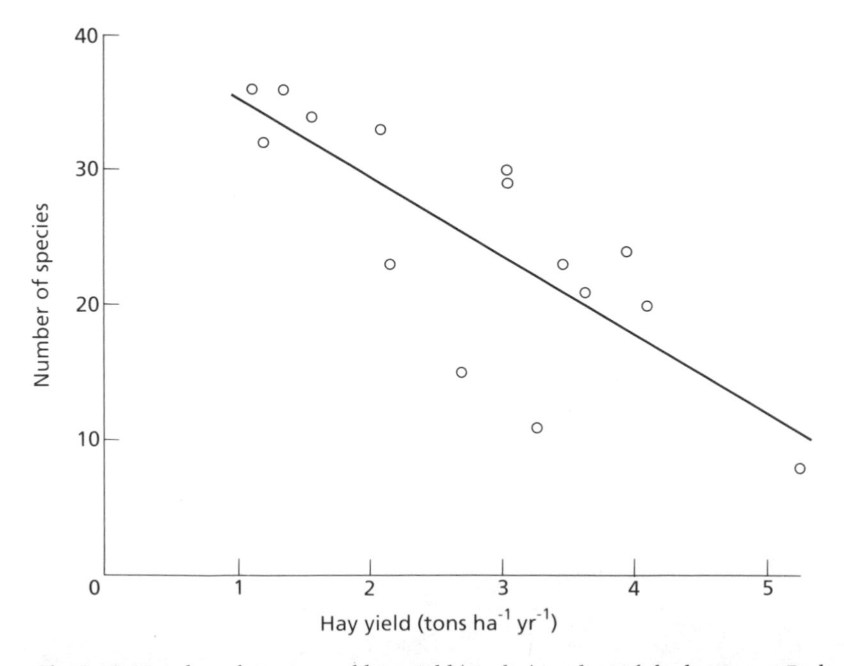

Fig. 8.10 Number of species and hay yield (air dry) in plots of the long-term Park Grass Experiment, Rothamsted, England. Each plot had received yearly farmyard manure, inorganic fertilizer or no addition. Soil pH was within the range 4.5–5.7. Number of species recorded 1948 or 1949; hay yield is mean for 1936–49. The correlation coefficient is significant at $P < 0.001$. Data from Brenchley & Warington (1958).

of lime (or none), the pH was kept within the range 4.5–5.7. Figure 8.10 shows that whether organic or various inorganic manures were added, there was a strong negative correlation between the hay yield and the number of species: the higher the yield, i.e. the more favourable the soil nutrient regime for growth, the lower the diversity. There were many other treatments in the experiment (see Brenchley & Warington 1958, Williams, E.D. 1978), and the relationship between treatments, yield and diversity has been analysed further by Silvertown (1980).

The loss of species in very fertile grassland is mainly due to competition: a few species, mainly grasses, grow strongly and out-compete others. Shading effects are important in this competitive exclusion, and grazing can often help to maintain a more species-rich sward. The long-term grazing exclusion experiment of Table 8.3 showed that sheep were necessary for the survival of some low-growing species in this grassland and increased the abundance of other species, so that the overall diversity was increased. However, grazing does not always increase diversity. For example, at some other sites near the experiment of Table 8.3, but on less fertile soils, the number of species inside and outside exclosures was about the same (Rawes 1981). The Park Grass Experiment (Fig. 8.10) shows that hay meadows in which the vegetation is allowed to grow tall can nevertheless be species-rich, provided the soil is not very fertile. So there can be more than one method of preventing dominance by a few species: grazing and low fertility can be alternatives.

Diversity of animals

Diversity of animal species can much more often be explained by niche separation than can diversity of plants. Niche separation among insects is often provided by their limited diet: the larvae may eat only a single plant species, probably only some parts of it; or they may be parasitoids on one other insect species. Adults may obtain nectar from flowers of one species. A single plant provides more than one type of food, and may provide physical niches as well. An example is the bracken plant (see Begon *et al.* 1990, p. 714). The leaf (frond) has a different set of herbivorous insects living off the pinna (thin lamina) from those that live off the stem-like main axis and lateral branches; and each of these habitats can support four groups of insect which differ in their way of feeding: chewers, suckers, miners and gall-formers.

Some insects have conservation priority, e.g. some rare butterflies (Thomas 1991). If each plant is the food for a different animal species, or several species, then maintaining high plant diversity should promote high animal diversity also. The bracken example shows that the structural complexity of individual plants is also important.

Among vertebrates there are much broader feeding categories. There are carnivores, omnivores and herbivores; among herbivores there are seed-eaters and leaf-eaters; grazers show some selectivity (see Chapter 3), but usually will eat many species. So the diversity of vertebrates cannot be attributed primarily to different food species, and we need to look for other sorts of niche separation. There have been many studies of this, and I summarize just one classic paper, by MacArthur (1958). He studied five bird species, warblers, all living in conifer forest in Maine and Vermont. All eat insects, and MacArthur chose to study them because they seemed ecologically very similar and it was not obvious how they managed to coexist. He was able to show differences between species in the height of their feeding zones within the canopy, whether they fed mainly near branch tips or nearer the tree trunks, the proportion of time they spent searching tree surfaces versus catching flying insects, the height of their nests. These differences taken together showed how each of the five species had a different physical niche from which it obtained its food, though the niches did overlap. The structural complexity of the forest, and the resulting variety of niches for insects, were crucial.

Animal diversity is related to structural complexity of the vegetation

MacArthur & MacArthur (1961) investigated how bird species diversity was related to vegetation structure at 13 sites ranging from Maine to Panama. 'Foliage height diversity' was calculated from the way the foliage was distributed vertically. In Fig. 8.11(a) lowest foliage height diversity means that all the foliage was in one layer, high diversity means a forest with about equal amounts of leaf area in low vegetation, shrubs and in the canopy. Bird species diversity was calculated by the Shannon Index, which takes into account evenness of abundance as well as number of species. The graph shows that there was a strong correlation between bird diversity and structural diversity of the vegetation. Taking into account plant species diversity did not help to account for the remaining scatter on the graph.

Morris (1971) determined the number of species of three insect groups at more than 100 grassland sites in southern and eastern England, in which the height of the vegetation ranged from 2 cm to about 40 cm. Figure 8.11(b) shows the results for leaf-hoppers. Although there is a wide scatter of points, there is a clear and statistically significant tendency to more species in taller vegetation. This was true also for the other two groups, Heteroptera (plant bugs) and weevils. Morris did not report how many species of plant were at each site, but as explained earlier taller grassland is likely, if anything, to have fewer species. Therefore there could be a conflict between managing for maximum plant diversity and maximum animal diversity.

Much has been written about animal species diversity and what controls it. The main point here is that promoting maximum plant species diversity can sometimes help to maintain animal diversity but is not the only thing to consider. Structural diversity is also important, and managers can promote this at various scales: it could involve

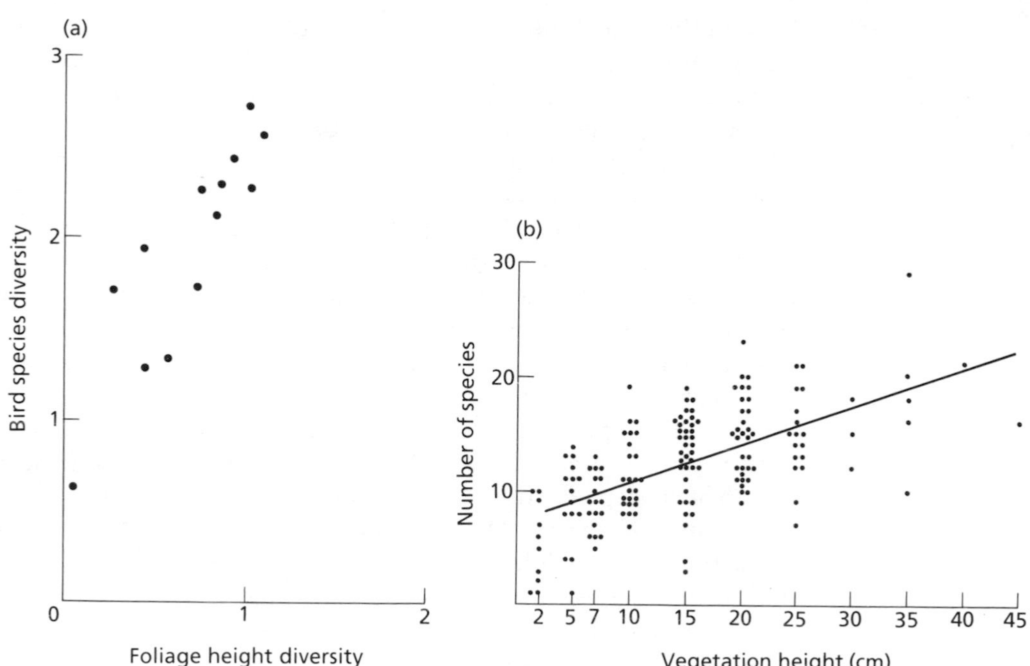

Fig. 8.11 Relationship between animal species diversity and vegetation structure. (a) Bird species diversity in relation to foliage height diversity at sites in eastern U.S.A. and Panama. Bird species diversity was measured by the Shannon Index, H:

$$H = -\sum_{i=1}^{S} P_i \ln(P_i)$$

where P_i is the proportion of the total abundance contributed by species i. There are S species in all. From MacArthur & MacArthur (1961).
(b) Diversity of leaf-hoppers (Auchenorhyncha) at English grassland sites, in relation to tallness of the vegetation. The line is the best fit linear regression. From Morris (1971).

encouraging particular plant species or ensuring that a forest has patches at different stages of regeneration after disturbance.

Creating new 'natural' areas

Conservation strictly means preserving existing ecosystems and their species. But sometimes we get an opportunity to create a new habitat for wild species, on land previously used for something else. Here I concentrate on abandoned farmland. Other types of land, including abandoned railway lines and industrial sites, mine waste, urban areas, each present their own opportunities and challenges, which I have not time to deal with. At present North America and much of Western Europe are producing more food than the inhabitants can eat, and one outcome is that spare farmland is being 'set aside'. Some of this set-aside land can be available for 'nature'. For farmland to be abandoned is not new. In many parts of Europe there are remains of terracing,

Farmland has been abandoned in the past

field outlines and other signs that present-day grassland, heath, shrubland or forest was formerly cultivated. Sometimes a decline in population was responsible, for example caused by the Black Death (bubonic plague) epidemic in Europe in AD 1348. Other abandonment followed changes in farming practice or migrations for other reasons. Much farmland in eastern U.S.A. was abandoned as the more fertile lands of the mid-West were opened up in the 19th and early 20th centuries. There are arid lands, e.g. in the eastern Mediterranean and the U.S. southwest, that carry signs of farming systems which operated centuries ago, where now there is only desert vegetation; and there are tropical areas, now covered by forest or savanna, where once cultivation must have been carried on, as shown by remains of towns or palaces (e.g. from the Maya culture of Central America), or by soil whose high fertility cannot be entirely natural (e.g. at sites along the lower Amazon).

When food production exceeds demand, this offers two sorts of option for wild species. Do we want to promote wild species within the agricultural landscape, for example by more and wider hedges, individual trees, strips and patches of rough grassland, ponds and marshy areas? Or should we concentrate on creation of new nature reserves? The best answer may well be 'both', one alternative in some areas, the other elsewhere. Here I concentrate on the second, the creation of new nature areas. A key question is, should we just abandon farmland and allow natural succession to take its course, or should we take active steps to promote development of the desired community, for example by planting and by introduction of animals? I consider this question first in relation to creation of new forests, then for new grassland. Natural regeneration of forest has already been considered in Chapter 5, but that was after forests had been felled, the timber removed and then the site immediately abandoned. We are now concerned with land that has been under cultivation for some years, where the prospects for natural regeneration are likely to be different.

New forests

Shenandoah National Park

Shenandoah National Park, in Virginia, U.S.A., is an example of creation of a forest. The Park, which is about 100 km from end to end and several kilometres wide, consists of a ridge which was presumably once almost entirely covered by mixed deciduous (hardwood) forest. When the Park was officially opened in 1935 there was no undisturbed forest left, with the possible exception of a few small patches. At that time some of the Park area was being cultivated, mostly as small family farms, and some was used for summer grazing of cattle and sheep belonging to lowland farms. The remainder was regrowth forest which had been clear-felled at least once and left to regenerate naturally. Lambert (1989) describes the history of the National Park in a non-scientific way, and the politics and negotiations behind its formation.

The inhabitants were offered financial inducements to move out, and those who refused were compulsorily evicted. The Great Smoky Mountains National Park, further southwest in the Appalachians, was formed about the same time (Campbell 1969). It also had much area that had been farmed or cut over, but unlike Shenandoah it also had substantial areas of forest that had never been felled, probably totalling about one-quarter of the Park area (Ambrose & Bratton 1990).

The policy in both these Parks was to let natural succession take place, with very little help from planting or other active management. I do not know whether this policy emerged from careful discussion at the time, by ecologically qualified people, of the best way to re-create forest, or whether it was thought obvious that 'nature's way is best'. According to Lambert (1989) there was no 'professional naturalist' on the Shenandoah National Park staff until 13 yrs after it was started. In western Europe the opposite assumption is usually made, that if we want to create a new forest we need to plant trees. Western Europe and eastern U.S.A. have broadly similar climates, and the natural vegetation in both is mostly deciduous forest, so why are the attitudes so different?

Old-field succession

Study of the succession in 'old fields', i.e. abandoned farmland, was already in progress in the 1930s in other parts of eastern U.S.A. One thorough and influential study was by Oosting (1942) in Duke Forest, central North Carolina. This 2000-ha area was acquired by Duke University in 1931 and had previously been used for farming by shifting cultivation. In 1931 it included fields still in cultivation and other fields abandoned at various times up to 100 yrs previously. The time of abandonment was known for some patches, and could be estimated in others by the ages of the oldest trees, so the different parts of the area provided a time sequence. During the first few years after abandonment of a cultivated field herbaceous species dominated, but within 5 yrs pines (*Pinus taeda*, *P. echinata* or *P. virginiana*) began to colonize, and within 10 yrs there was usually a fairly even-aged stand of small pines. As the pine stands grew up dicotyledonous woody species established beneath them. But the oldest stands, 90–110 yrs from abandonment of farming, were still dominated by pines; the dominant trees of old undisturbed forest in this area, oaks and hickories, were beginning to reach the canopy but were still sparse. So the forest was developing towards a structure and composition similar to what was there previously, but the succession would take well over a century to be completed. As far as I know, no detailed records have been kept of the re-establishment of the vegetation in Shenandoah National Park; a brief visit in 1990 made it clear to me that, more than 50 yrs after the Park was formed, although areas that were formerly farmland are now well covered by trees, the structure and species composition is very different from climax forest.

One key fact influencing the old-field succession is that many of the dominant tree species of the climax forest regenerate mainly vegetatively (see Chapter 5). Therefore they can regenerate well if a forest is clear-felled and then left. In many tropical forest regions the traditional system of shifting cultivation allows felled trees to regrow vegetatively, and the plot is abandoned when they become large enough to shade the crop seriously (Chapter 3). In North America the system of shifting cultivation, at least as practised by Europeans and their descendants, involved complete removal of trees, and recolonization by trees after cultivation was abandoned must therefore be by seed. So where did the seed come from? Oosting & Humphreys (1940) found that in the Duke Forest area during the herb stage of succession, i.e. a few years after cessation of farming, there were no viable seeds of any tree species in the soil. So all the tree seeds for the subsequent succession must have come from surrounding forest stands. McQuilkin (1940) measured the abundance of pine seedlings colonizing old fields in Virginia, North Carolina and South Carolina, along transects from the edge of the nearest pine stand which could provide seed. The results show that beyond about 100–200 m from the seed source, colonization by pine seedlings would be very slow.

Where do the seeds come from?

This research suggests that two features allowed the recolonization to take place in Duke Forest and similar areas: (1) even in climax oak–hickory forest there are a few pines (Oosting 1942) which can act as seed sources; (2) the fields used in shifting cultivation were small, often less than 1 ha. In a patchwork of such fields and regrowth forest most points on an abandoned field would be within 100 m of a forest edge. If the fields are much larger we should expect recolonization to take much longer, and this has happened in many places in eastern U.S.A. during the 20th century (e.g. see Gross in Jordan *et al.* 1987).

Natural tree colonization in Britain

In Europe traditionally less confidence has been placed in natural colonization of open land by trees, but it can happen. One example is provided by the Breckland area of eastern England. This area of freely drained sandy and calcareous soils in Norfolk and Suffolk is known, from pollen records, to have borne mixed deciduous forest until clearance for farming began about 4500 BP (Bennett 1983, 1986). From Medieval times parts were used for grazing by sheep and rabbits (Crompton & Sheail 1975). The rabbits were killed off in 1954 by myxomatosis, and the sheep had by then been withdrawn, so the grasslands and heaths were left without any grazing. Since then many of these areas have been colonized by trees, which in some parts now form dense forests (Marrs & Hicks 1986, Marrs *et al.* 1986). The species composition varies, however. For example, on Knettishall Heath birch is abundant, with some oak and Scots pine; whereas about 20 km away at Lakenheath Warren the developing forest is almost entirely Scots pine. There can be little doubt that the difference is due to what seed sources were available. Figure 8.12 shows the distribution of pines on Lakenheath 17 and 30 yrs after the rabbits disappeared. The two sources of pine seed were a large plantation abutting the north

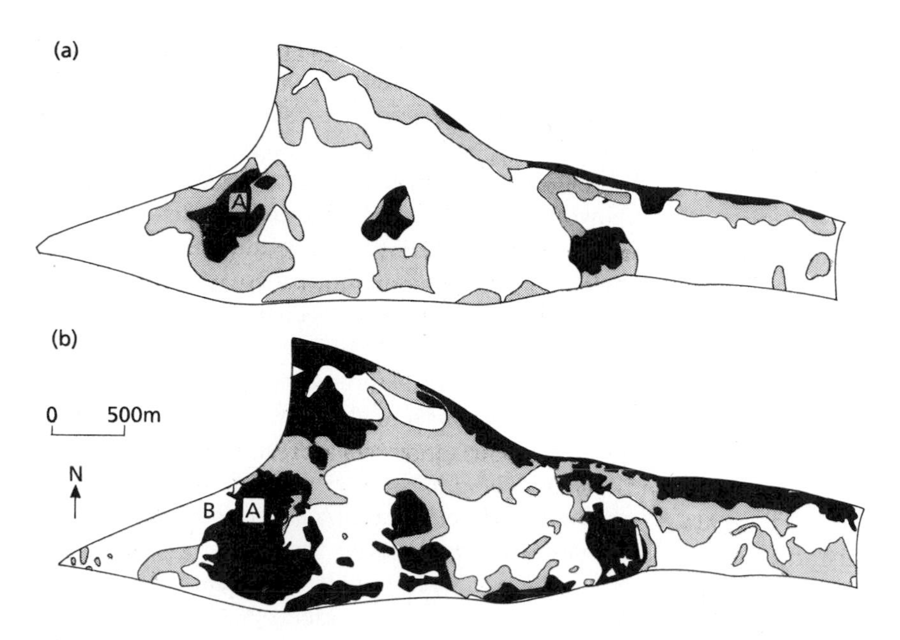

Fig. 8.12 Distribution of Scots pines in Lakenheath Warren, Suffolk, England, in 1971(a) and 1984(b). ■, canopy cover of pines more than 10%. □, pine cover less than 10%, or pine plus deciduous scrub. Remaining areas grassland or heath. A, position of house. B, area where pines were cleared in 1983. From Marrs *et al.* (1986).

edge and a few trees around the house (marked A). After 30 yrs free of grazing some parts less than 1 km from a seed source were still uncolonized by trees.

In upland Britain there are much larger expanses of open grassland or heathland, which once were covered by deciduous forest. If grazing animals are excluded there can be some sparse colonizing by hawthorn and rowan (e.g. in North Wales, Hill *et al.* 1992). These both have fleshy seeds which are dispersed by birds. The dominant species of the former forests in these areas, oak, has not been observed to colonize naturally even after several decades of sheep exclusion; poor dispersal of its large, heavy seed is presumably the reason. In areas as large as this the only way to re-create forest would be to plant trees.

How past farming systems may have affected tree colonization

The most important difference between western Europe and eastern U.S.A. influencing forest colonization is probably past farming systems (see Chapter 3). In eastern North America shifting cultivation was in use when the first Europeans arrived, and was continued into the 20th century. It provided a mosaic landscape within which recolonization of farmland by forest was continually taking place. In Europe the infield–outfield system predominated, probably for several millennia. The outfields were in some places forests, but elsewhere were heathland or grassland. The cultivated areas were much larger than under shifting cultivation, rarely moved, and were often a long way from the nearest forest. There was little to favour tree species that were good at colonizing abandoned farmland.

Recolonization of woods by herb species

The Forest chapter (Chapter 5) considered regeneration by trees after felling. But in this Conservation chapter we need to consider other species as well. Research in North America and in Britain has shown that natural recolonization by some herbaceous species is slow. Duffy & Meier (1992) found nine undisturbed, ancient forest sites in the southern Appalacian region of U.S.A., each of which had close by a patch of forest that had been felled and then abandoned 45–87 yrs previously. They determined the number of herbaceous species in 1-m^2 quadrats on the forest floor, and found that for every pair of sites the regrowth forest had fewer species than the ancient forest. The mean number of species per square metre was 10.9 in the ancient forest and 6.6 in the regrowth. Clearly the herbaceous flora of the regrowth had not yet returned to its former composition, even though ancient forest was nearby.

In Europe there are older secondary forests on former farmland. In the detailed study by Peterken & Game (1984) on woods in Lincolnshire that was mentioned earlier, they were able to classify 362 woods as 'ancient', i.e. there was forest on the site continuously since before AD 1600, or 'recent', i.e. on land that has been clear of forest for some period since 1600. Many of the ancient woods have probably been continuously forested since trees first colonized after the Ice Age. In Fig. 8.13 the straight line shows the relationship between number of plant species and wood area in ancient woods. The points are for individual recent woods, and it is clear that they usually have fewer species than ancient woods of the same area. Peterken & Game listed 62 woodland floor species that were found much more frequently in

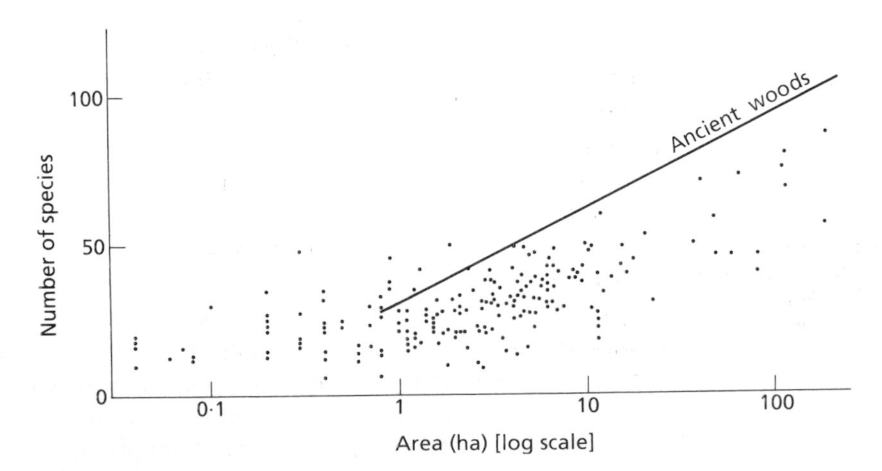

Fig. 8.13 Number of plant species in woods in Lincolnshire, England. Straight line: regression line for ancient woods, taken from Fig. 8.2(a). Points: individual recent woods. From Peterken & Game (1984).

the ancient than in the recent woods. They include species that most British people, whether trained ecologists or not, would recognize as typical woodland flowers, for example bluebell and wood anemone. Another list, of species about equally often found in ancient and recent woods, includes species often found in rough grassland, hedgerows, scrub or other open habitats. The species in the first list are evidently poor spreaders: they have failed to recolonize many of the recent woods, even though some of these woods are more than 300 years old. One of these ancient woodland species is dog's mercury, whose poor ability to spread is shown in Fig. 8.7.

This research shows that if we rely on natural colonization it can take centuries for the full species complement of forest to re-establish, and there is no clear proof that it ever would. So if we want to re-create an ancient forest as nearly as possible, artificial introduction of plant species will probably be necessary.

New grassland

Within much of Europe abandoned arable land will revert fairly quickly to some form of grassland, heathland or scrub. Therefore in Europe it has not often been thought necessary to create 'rough grassland'. In contrast, where it is desired to re-create prairie on former farmland in the American mid-west it has commonly been assumed that introduction of species will be necessary. The attitudes in the two continents are thus the opposite of those for forest.

Grassland on old arable land: natural succession in Britain

Although a vegetation cover dominated by perennials will develop within a few years on abandoned arable land in most parts of Europe, its species composition may take much longer to stabilize and conform to 'old grassland'. This is illustrated by work of Wells *et al.* (1976) on the Porton Ranges in Wiltshire, southern England. This chalkland area is part of a large uninhabited region reserved for military use. Much of the 28 km^2 of the Porton Ranges is now grassland but was cultivated in the past. When Wells *et al.* made their survey, the time since cultivation was abandoned, determined from old maps, ranged from less than 50 yrs to more than 130 yrs. It is likely that there was always a patchwork of arable and grassland, which would be important for providing seed of grassland species for recolonization. Wells *et al.* found that present-day grasslands of the same age sometimes differed from each other substantially in species composition; nevertheless they were able to list species that were characteristic of grasslands less than 50 yrs old and others characteristic of grasslands more than 130 yrs old. The '50-yr grassland group' includes species commonly found in roadside verges and others that occur in dune grassland. In other words, they occur in somewhat disturbed habitats, although they are not arable weeds. The '130-yr group' are species that are characteristic of old, long-undisturbed chalk grassland. So with chalk grassland, in spite of the small size of the individual plants, re-creation of a fairly natural vegetation by unassisted succession can take at least a century.

Re-creating prairie in U.S.A.

The first attempt to re-create North American prairie on farmland was the Curtis Prairie, a 24-ha site started in 1935 at the University of Wisconsin at Madison. Other attempts have been started since then in various parts of the mid-West, the largest being 250 ha inside a proton accelerator near Chicago. Some lessons learnt from these attempted re-creations are summarized by Kline & Howell (in Jordan *et al.* 1987) and Howell & Jordan (1991). When the Curtis Prairie was started it was assumed that, since the farmland had once been prairie, it would be sufficient to sow or plant in prairie species. These, it was assumed, would be more successful than arable weeds or other non-prairie species, which would in due course disappear. In practice prairie restorations have run up against several problems.

1 Some prairie species that were originally native to the area have proved consistently difficult to re-establish. Changing the technique of introduction—using whole plants or whole turfs rather than seed—may help but has not always solved the problem.

2 Some weedy species are very persistent. Attention needs to be given to removing these at the start, by repeated cultivation or herbicides. Later selective removal can be very time-consuming.

3 Some non-native species, especially introductions from Europe, tend to be very persistent, coexisting with the native prairie species.

4 Woody species often invade. This is partly because all the restorations are fairly small patches, so tree seed sources may be nearby. But it also indicates the importance of fire. There were fires in the prairies in pre-Columbian times, started naturally or by people, and they are probably essential to maintaining prairie vegetation, at least in the form first seen by white explorers.

5 The soil may have changed during the years of cultivation—for example the available nutrient status may have increased but the crumb structure been degraded (see Chapter 3)—and it may be necessary to reverse these changes. As mentioned in Chapter 3, in one prairie re-creation within 5 yrs after planting prairie species the crumb structure was back to a state very similar to old prairie.

These attempts at prairie restoration show how difficult it can be to re-create a natural ecosystem, even with active intervention. The areas are much smaller than, for example, Shenandoah National Park, and can receive much more active and detailed attention. Nevertheless Jordan *et al.* (1987) wrote of the oldest of them, Curtis Prairie, 50 yrs after it was started: 'none of the restored communities . . . was a finished replica of a natural model'.

Animals in re-created ecosystems

So far this discussion of creating new 'natural' areas has ignored animals. Can they be relied upon to return, if the right plants are there? Does it matter if some do not return? The answer to the first question is 'not always'. For one thing, the area may not be large enough to support a viable population. This chapter has also given examples of animals that fail to migrate across a few kilometres of unfavourable habitat. The Bay checkerspot butterfly is one. In Curtis Prairie ants occurred in a small part of the prairie that had not been ploughed, and have spread little from it during the following 50 yrs. Ant-hills can provide a habitat for some plant species not found elsewhere in grassland (King 1977). Large carnivores—wolves and mountain lions—were once present in Shenandoah, but have not returned since the Park was formed. The answer to the second question about animals ('does it matter?') is that some animals are keystone species, whose absence will have a great effect on the whole ecosystem. The absence of large carnivores from Shenandoah is one likely reason for the abundance of deer, which gives pleasure to visitors but is probably interfering with regeneration of the forests. The prairies that have been re-created are too small to support a viable population of bison. Apart from their effect by grazing, their wallows may have been a habitat for certain species.

Reintroduction of animals deserves more attention than it has so far received. There have been several successful introductions of rare butterflies to new sites in Britain (Thomas 1991). A dramatic success was the silver-studded blue (*Plebejus argus*): 90 adults were released at a site in Wales in 1942, and 41 yrs later there were estimated to be 60 000–90 000, though they had extended their area by only 2 km. Introducing mammals and birds presents more problems. It would be technically possible to try reintroducing wolves to Shenandoah National Park, but it is quite likely that the Park is too small to support a viable population. Estimated densities of grey wolf in North America range from 1 per 25 km^2 to 1 per 300 km^2 (Pimlott 1967), so the 800 km^2 of Shenandoah Park would support a maximum of about 30 adults, which might well be below the minimum viable population. The wolves would probably extend their hunting into neighbouring farmland to kill domestic animals, which would clearly cause severe displeasure among the human population.

Morton (in Jordan *et al.* 1987) describes an unsuccessful attempt to reintroduce two wren species to Barro Colorado Island. Although both species had formerly occurred there, and both were common in similar forest on the nearby mainland, neither survived for more than a few years after reintroduction. Morton attributes this to lack of particular habitats within the forest, disturbed forest for one and fast-flowing streams for the other. Species' requirements can be subtle.

What should we aim for?

If we decide to take an active role in re-creation of ecosystems, by introducing species or by other management, the question arises, what are we aiming for? One possible answer is, aim to re-create the natural ecosystem for the site. That is not the only possible answer. The aim of the 18th-century landscape gardeners and park creators in Europe was to learn from nature but to improve on it. Open landscape with spaced trees, grassy areas, vistas, lakes, perhaps the occasional statue or temple, were considered more attractive than the dense forest that would naturally cover the land. Today conservation at many sites in Europe aims to maintain ecosystems under some traditional management system of the past, for example woodland in which coppice is cut on a regular cycle, or hay meadows receiving no artificial fertilizer.

What would natural communities be like?

If we decide to aim for the natural state, the next question is, what would it have been? Western Europe has for so long been subject to farming, grazing by domestic animals, intensive exploitation of forests, that we have no remnants of most natural ecosystems to use as models. Pollen provides evidence on the composition of forests in the past, so should we aim to re-create the forests that existed just before the first Neolithic farmers arrived? In lowland Britain today ash, oak and beech are the main dominant tree species in many of the forest remnants. But pollen records show that in much of lowland Britain lime was formerly much commoner than it is now (Rackham 1976). It decreased partly because it is poor at spreading to new woods, partly because disturbance has favoured seedling establishment of oak, perhaps partly through many centuries of deliberate encouragement of oak by people. Today in woods left less disturbed this century than previously oak is regenerating poorly, while sycamore, an introduced species, regenerates well, ash and beech fairly well. Therefore there are at least three alternative ways to try to produce a 'natural' forest in lowland Britain (Peterken 1991).

1 Leave the present forest alone, to develop naturally. This would in many areas lead to abundance of sycamore, ash and/or beech but near-disappearance of oak.

2 Aim for an oak–ash forest, which many people regard as the traditional English wood. This would in many places require active removal of sycamore. The warden of my local National Nature Reserve ancient forest is already doing that.

3 Aim for a forest which, as far as we can guess, would have occurred 5000 years ago; planting-in of lime would probably be necessary.

I do not see a clear argument for choosing one of these alternatives rather than another. This illustrates the difficulty of deciding what is natural.

In North America it was once widely assumed that the land as first seen by European explorers was natural. That view does not seem tenable today. Farming was widely practised in the east, and in some parts of the southwest, before 1492. In forests and prairies fire had been used for management. Around 11 000 BP large mammals had suddenly become extinct, probably through human hunting (see earlier in this chapter) and this is likely to have affected the vegetation. In the tropics, too, vegetation is not all as natural as it once seemed. There has been widespread shifting cultivation and fire in tropical forests (see Chapters 3 and 5). Australia might seem a continent where people have had little effect over large areas. There are forests that can certainly never have been cut down, since the inhabitants before 1788 did not have the tools to do it. But the Aborigines did use fire to burn vegetation, they were good at killing animals, and may well have been responsible for the disappearance of many of the large mammal species long before the first arrival of Europeans.

This section has tried to show that stating that you wish to create a 'natural ecosystem' in an area is not a sufficiently precise aim. It is likely to save much wasted effort, and to increase the chances of success, if a more precise aim can be formulated. Examples of some possible aims are: (1) to re-create the system as it was at a stated time in the past, e.g. the mid-19th century or 5000 yrs ago; (2) to provide a grassland rich in plant species, or to provide a forest rich in bird species; (3) to provide a habitat into which certain named animal species can be successfully reintroduced; or (4) to provide an area where people can go to get away from the sights, sounds and stresses of urban life.

Conclusions

- Numbers of species of animals and plants on islands are consistently greater the larger the island. This is also true of habitat fragments on land, e.g. patches of forest in farmland. This relationship is due partly to greater habitat diversity in larger areas, and partly to a balance between invasions and extinctions.
- If part of a large ecosystem becomes isolated, some species are likely to die out. Reasons include: (1) an annual migration route has been blocked; (2) edge effects; or (3) the population has fallen below a minimum viable size.
- Small populations are at risk because of: (1) low genetic variability and inbreeding; and (2) the possibility that all individuals will die by chance. Models have predicted whether certain individual populations have a good chance of surviving.
- Some animal and plant species show limited ability to spread between suitable habitats. Habitat corridors have rarely been found to increase their spread.
- Disappearance of a single herbivore species (vertebrate or invertebrate) can have a major effect on a community. Past disappearance of large carnivores has had major effects, too, by removing control on herbivores.
- Coexistence and diversity of animal species can often be explained by niches. Diversity of plants on a landscape scale can, too, but on a smaller scale usually cannot.

• Structural complexity in vegetation can promote greater diversity in the resident animal community. Infertile soil often promotes species-richness in temperate grassland.
• When establishment of new forests or grassland on former farmland has been left to natural succession, the species composition has taken a century or more to approach that of long-established communities. But using much more active intervention to re-create prairie has not been conspicuously faster.

Further reading

Organization and integration within communities:
Kikkawa & Anderson (1986)

Diversity and what controls it:
Myers & Giller (1988)
Cockburn (1991)

Conservation:
Warren & Goldsmith (1983)
Soulé (1986)
Spellerberg *et al.* (1991)

Priorities and choices in conservation:
Usher (1986)

Rare species, minimum viable populations:
Soulé (1987)

Creating new 'natural' communities:
Jordan *et al.* (1987)

Acknowledgements

First I want to thank my dear wife Edna for her continued love and support.

Among those who contributed more directly to this book, I want to thank especially Susan Sternberg of Blackwell Scientific Publications. Her early enthusiasm helped to get the book off the ground, she played an important part in deciding its size and form, and guided me patiently through the stages of writing my first book.

Many other people, in Bristol and elsewhere, helped by making suggestions, by commenting on earlier drafts of chapters, and in other ways. In particular I want to thank John Beringer, Richard Campbell, Mike Martin, David Newbery, Elizabeth Newman, Henry Osmaston, David Paterson, John Porter, Sheila Ross, Tony Walsby and Neil Willey.

I am also very grateful to the inventors of the word processor. Before word processors existed, how did people ever manage to write books?

The following copyright-holders kindly gave permission for material to be used in the book.

Academic Press: Figs 3.6, 6.1.
American Association for the Advancement of Science: Fig. 2.5.
American Phytopathological Society: Fig. 6.9.
American Society of Agronomy: Fig. 3.5(a).
Biometrika Trust: Fig. 6.2.
Cambridge University Press: Figs 2.4, 2.5, 7.3.
Commonwealth Scientific and Industrial Research Organisation: Figs 3.4, 3.9.
Ecological Society of America: Figs 5.3, 6.7, 7.15, 8.6, 8.11.
Edward Arnold: Figs 4.3(a,c).
Elsevier Applied Science: Figs 8.2, 8.12.
Elsevier Sequoia: Fig. 3.3.
Fishing News Books: Fig. 4.10(b).
Genetical Society: Fig. 7.13.
Kluwer Academic Publishers: Figs 2.1, 3.1, 5.1.
Macmillan Magazines: Figs 6.8, 8.9.
Minister of Supply and Services, Canada: Fig. 4.9.
Dr D.M. Newbery: Fig. 5.4.
Professor M.L. Parry: Fig. 2.6.
Pergamon Press: Fig. 7.2.
Poyser Publishers: Fig. 7.8.
Regents of the University of Michigan: Fig. 8.4.
Dr S.M. Saulei: Table 5.5.

Society for Range Management: Fig. 6.6.
Springer-Verlag: Figs 3.2, 6.3.
United States Department of the Interior, National Park Service, Shenandoah National Park: photograph on front cover.
University of Wisconsin Press: Figs 4.3(b), 4.8.
John Wiley and Sons: Fig. 7.7.

Glossary

This glossary has several functions.

1 It gives the meaning of specialist terms and abbreviations used in the book.

2 If a species has been called by its English name in the text, the Latin (scientific) name is given here. A few of these may be out of date, if the name has been changed.

3 If a species was called only by its Latin name in the text, an English name is given here, or else some indication of what major group it belongs to.

Acer rubrum = red maple.

Aggregate (soil crumb) Formed by numerous soil mineral particles (e.g. individual clay particles) bound together, though with pores between them able to hold water.

Agropyron Grasses, some of them tussocky (bunchgrass). *Agropyron spicatum* has been renamed *Pseudoroegneria spicata.*

Agrostis = bentgrasses. *Agrostis vinealis* was formerly called *Agrostis canina.*

Alder = *Alnus.*

Alfalfa = lucerne = *Medicago sativa.*

Anchoveta, Peruvian = *Engraulis ringens.*

Anemone, wood = *Anemone nemorosa.*

Artemia = brine shrimp. In the Anostraca (fairy shrimp) group of Crustacea.

Ash = *Fraxinus.*

Auk, little = *Plotus alle.*

Autotroph An organism that does not obtain its energy from organic matter, but by photosynthesis (photoautotroph) or by oxidizing inorganic materials.

Balsa = *Ochroma lagopus.*

Bear = *Ursus*; grizzly bear = *Ursus arctos.*

Beech = *Fagus.*

Benthic Species that live within the bottom deposit of a lake or sea, or on the surface of the deposit.

Billion In this book means thousand million (10^9).

Birch = *Betula.*

Bluebell = *Hyacinthoides non-scripta.*

Blueberry = *Vaccinium angustifolium.*

Bluejay = *Cyanocitta cristata.*

Boar, wild = *Sus scrofa.*

Bobcat = *Lynx rufus.*

Boreal Cold temperate region; has cold winters, but summers warm enough for all soil to thaw. Boreal forest: native forest of boreal regions, in which conifers predominate.

BP = years before present.

Bq = bequerel. Amount of radioactive material that produces one disintegration per second.

Broccoli = *Brassica oleracea* var. *italica.*

Bromus tectorum = cheatgrass.

Brussels sprouts = *Brassica oleracea* var. *gemmifera.*

Business-as-usual Term used by IPCC (see Houghton *et al.* 1990) to mean a scenario in which no attempt is made to restrain use of fossil fuels and production of greenhouse gases.

C_3, C_4, CAM photosynthesis Three alternative carbon pathways involved in photosynthetic CO_2 fixation. In the C_3 pathway the initial step in CO_2 capture is its reaction with ribulose-1,5-bisphosphate, catalysed by the enzyme Rubisco. In the C_4 and CAM pathways the CO_2 reacts first with phosphoenol pyruvate, and the early steps result in it reaching ribulose bisphosphate at higher concentration than in the C_3 pathway. In CAM photosynthesis the CO_2 is taken in mostly at night, the stomata can remain closed by day, and hence transpirational water loss per unit C fixed is reduced.

CAI = current annual increment. New timber growth per year in a forest, usually expressed as $m^3 ha^{-1} yr^{-1}$.

Calluna vulgaris = heather (a low shrub).

CAM = Crassulacean acid metabolism. See under C_3.

Cardinal = *Cardinalis cardinalis.*

Carp, silver = *Hypophthalmichthys molitrix.*

Carya = hickory.

Casuarina Tropical and sub-tropical trees, sometimes known as she-oak.

CFC = chlorofluorocarbon.

Checkerspot Bay checkerspot butterfly = *Euphydryas editha bayensis.*

Cherry, black = *Prunus serotina.*

Chestnut blight (fungus) = *Cryphonectria parasitica*, formerly called *Endothia parasitica.*

Chipmunk = *Tamias striatus.*

Chironomids = non-biting midges. In the group Nematocera, in the Diptera.

Cholinesterase Enzyme which breaks down acetyl-choline, and is essential for functioning of nervous systems.

Cladoceran = water flea, a group in the Crustacea.

Clear-felling Felling all trees in a large area in one episode.

Cod Atlantic cod = *Gadus morhua*, Greenland cod = *Gadus ogac*, Pacific cod = *Gadus macrocephalus*, Polar cod = *Boreogadus saida.*

Conjugation Transfer of DNA between unicellular organisms through a temporary join.

Copepod Copepoda are a major group of Crustacea, mainly marine.

Coppice System of tree harvesting in which several or many shoots grow from the cut stump, these are later harvested and the stump is left for further cycles of growth and harvesting. The term also applies to the stems produced by this system.

CR = concentration ratio, e.g. see Table 7.5.

Cucumber = *Cucumis sativa.*

CV = coefficient of variation = standard deviation/mean.

Cyclodienes A group of chemically related compounds used (or formerly used) as insecticides. All have a several-ring molecular structure. Includes aldrin and dieldrin.

Daphnia In the Cladocera (water flea) group of the Crustacea.

Deer Red deer = *Cervus elaphus*, white-tailed deer = *Odocoileus virginianus.*

Desert pavement A layer of stones that forms naturally on the surface of soil in deserts when the finer material blows away.

Deterministic model A mathematical model that allows for no random variation, so a particular set of input values will always lead to exactly the same predicted outcome.

Diapause A period during the life of an insect, e.g. during the larval stage, when the metabolic rate is greatly reduced and the insect becomes better able to survive unfavourable conditions. Equivalent to dormancy, but applies only to insects.

Dilution plating A method for determining the number of bacteria or fungi in a sample. Involves making a series of dilute suspensions of the sample, spreading small drops on a nutrient agar medium, and counting the colonies that grow.

Dipterocarp A member of the family Dipterocarpaceae, trees occurring in the tropics, especially southeast Asia.

Dog's mercury = *Mercurialis perennis.*

Douglas fir = *Pseudotsuga menziesii.*

Eagle Bald eagle = *Haliaeetus leucocephalus*, golden eagle = *Aquila chrysaeetos.*

EC_{50} (EC = effective concentration) Concentration of chemical that causes 50% reduction in measured activity of a species, e.g. in growth rate or respiration rate.

Ectomycorrhiza Type of mycorrhiza characterized by sheath of fungal tissue round the outside of the infected root. Occurs in woody species.

Elephant African elephant = *Loxodonta africana*, Indian elephant = *Elephas maximus.*

Epizootic Sudden and temporary increase of a disease in an animal population.

Ericaceae The plant family which includes the heathers and blueberries.

Ericaceous Belonging to the plant family Ericaceae (see above).

Ericoid mycorrhiza Type of mycorrhiza found in members of the Ericaceae and a few related families. Characterized by coils of hyphae within cells of very fine roots.

Eucalypt Member of genus *Eucalyptus*, trees of the southern hemisphere, especially in Australia.

Euphausiid Marine crustaceans.

Eutrophic Applied to water of rivers, lakes and oceans: contains mineral nutrients in sufficient concentrations to support rapid algal growth.

Falcon, peregrine = *Falco peregrinus.*

FAO = Food and Agriculture Organization of the United Nations.

Farmyard manure Mixture of plant material (often straw) with faeces and urine from farm animals.

Festuca ovina = sheep's fescue (a grass).

Fir = *Abies*, balsam fir = *Abies balsamea.*

Fixation of nitrogen Incorporation of gaseous N_2 into an organic compound.

Flea Siphonaptera, rabbit flea = *Spilopsyllus cuniculi.*

Forb A herbaceous plant other than a grass.

Forest Vegetation in which trees form a continuous canopy.

Fossil fuel A fuel formed from partly decomposed plant material, i.e. coal, oil, natural gas or peat.

Fox, European = *Vulpes vulpes.*

Full-glacial period See Table 2.7.

G = giga = $\times 10^9$.

Gambusia affinis = mosquito-fish.

Gazelle, Thomson's = *Gazella thomsonii.*

gbh = girth at breast height, i.e. circumference of tree trunk at about 1.5 m above the ground.

GEO = genetically engineered organism.

Guillemot = *Uria aalge.*

Guppy = *Poecilia reticulata*, a fish.

ha = hectare = $10^4\,m^2$.

Hardwood In temperate regions: dicotyledonous tree species, and forests in which these dominate (in contrast to conifers which are called softwoods). In tropics: tree species that have dense, strong wood; they are usually slow-growing.

Hare, snowshoe = *Lepus americanus.*

Hawthorn = *Crataegus.*

Heather = *Erica* and *Calluna.*

Heavy metal Metal with relative density greater than 5; e.g. cadmium, copper, lead, nickel, zinc.

Hemlock = *Tsuga.* (A North American tree. A poisonous herb of Europe, called hemlock in Britain, is not mentioned in this book.)

Herring, Atlantic = *Clupea harengis.*

Heterotroph An organism that obtains its energy from organic material.
Hickory = *Carya.*
IPCC = Intergovernmental Panel on Climate Change. See Chapter 2.
J = joule.
Jay, blue = *Cyanocitta cristata.*
K, k = kilo = $\times 10^3$.
Kittiwake = *Rissa tridactyla.*
Ladybird = vedalia beetle = *Rodolia cardinalis.*
LAI = leaf area index = $\frac{\text{Total area of leaves}}{\text{Area of ground}}$.
Larch = *Larix.*
Late-glacial period See Table 2.7.
Latin America South America plus Central America including the Caribbean.
LC_{50} (LC = lethal concentration). Concentration of a chemical in water surrounding a population that is sufficient to kill 50% of them.
LD_{50} (LD = lethal dose). When this amount of a chemical is fed to each animal it kills 50% of the individuals.
Lettuce = *Lactuca sativa.*
Lime = *Tilia.* The small-leaved lime native in much of Europe is *Tilia cordata.*
Lion, mountain = *Felis concolor.*
Liriodendron tulipifera = tulip tree, yellow poplar.
Long-wave radiation. See Box 2.1.
Lynx, Canadian = *Felis lynx* = *Lynx canadensis.*
M = mega = $\times 10^6$.
Mackerel, Atlantic = *Scomber scombrus.*
MAI = mean annual increment = $\frac{\text{Volume of timber in a forest stand}}{\text{Age of stand}}$.
Maize = corn = *Zea mays.*
Maple = *Acer.* Red maple = *Acer rubrum*, sugar maple = *Acer saccharum.*
Mayfly = Ephemeroptera.
Merlin = *Falco columbarius.*
Mesquite = *Prosopis juliflora* (a shrub).
Micelles Flat, fine crystals of silicate, carrying negative charges, from which clay particles are constructed.
Minnow, fathead = *Pimephales promelas.*
Minuartia verna = vernal sandwort, a forb.
Moose = elk = *Alces alces.*
Mosquito = Culicidae.
Mouse Field mouse = *Apodemus sylvaticus*, white-footed mouse = *Peromyscus leucopus.*
Muskrat = *Ondatra zibethicus.*
Mycorrhiza Symbiotic association of plant root and fungus.
Nardus stricta = mat-grass.
Nutcracker = *Nucifraga*, Clark's nutcracker = *Nucifraga columbiana*, Eurasian nutcracker = *N. caryocatactes.*
Oak = *Quercus.*
Oligotrophic Applied to water of rivers, lakes and oceans: concentration of at least some nutrient elements low enough to severely limit plant growth.
Organochlorine Any organic compound that includes chlorine.
Organophosphorus Any organic compound that includes phosphorus.
Osprey = *Pandion haliaetus.*
Owl Northern spotted owl = *Strix occidentalis caurina*, short-eared owl = *Asio flammeus.*
Panda, giant = *Ailuropoda melanoleuca.*
Parasitoid An animal that is a parasite at one stage of its life cycle but free-living at another. Especially insects whose larvae are parasitic on another insect species but whose adult stage is free-living.

Parrot, Puerto Rican = *Amazona vittata.*
PCB = polychlorinated biphenyl. See Box 7.2, Fig. 7.12.
Peregrine falcon = *Falco peregrinus.*
Perennial ryegrass = *Lolium perenne.*
Photic Upper layer of water in an ocean or lake, in which the light intensity is great enough to allow photosynthesis.
Photorespiration Oxidation of ribulose-1,5-bisphosphate, which happens in light and is catalysed by Rubisco, the same enzyme that catalyses the reaction of this substance with CO_2. Photorespiration therefore competes against photosynthesis for substrate and reduces photosynthetic carbon fixation.
Photovoltaic cell A physical device which when illuminated generates an electric current.
Phytoplankton Planktonic algae and cyanobacteria; can include photosynthetic protozoa.
Pilchard, European = *Sardina pilchardus.*
Pine = *Pinus,* Monterey pine = *Pinus radiata,* piñon pine = *P. edulis,* Scots pine = *P. sylvestris.*
Pioneer A tree species whose seedlings can establish and grow only in a large area free of trees, not in undisturbed forest or small gaps.
Piscivorous Eats fish.
Plaice, European = *Pleuronectes platessa.*
Planktivorous Eats plankton.
Plankton Small organisms that live suspended in water. Some are unicellular, some multicellular, but all are small enough to be much affected by currents.
Plasmid DNA which is separate from the chromosome and which replicates separately. Found particularly in bacteria, which can transfer them between cells by conjugation.
Polygenic A character controlled by several or many genes.
Poplar = *Populus.*
Post-glacial period See Table 2.7.
Quasi-natural Used in Chapter 6 to include not only truly natural communities but also some substantially affected by people; but excluding crops, sown grassland and forestry plantations. My use of 'natural' in Chapter 8 was equivalent to this: see start of that chapter.
Quercus = oak, *Quercus prinus* = chestnut oak, *Q. rubra* = red oak.
***r*-species** A species adapted to colonize recently disturbed areas, where intensity of competition is low.
Rabbit, European = *Oryctolagus cuniculus.*
Razorbill = *Alca torda.*
Recruitment Applied to fish: the number of fish which, in a particular year, have joined the catchable stock, i.e. have during the year grown large enough to be caught by the nets used.
Rhizobium Bacteria which form a symbiotic N-fixing association in nodules on roots of leguminous plants.
Rhizosphere The region of soil very close to an individual root. The abundance of microorganisms is higher than elsewhere in the soil, because of organic materials from the root and dying root cells.
Robinia pseudoacacia = black locust.
Rowan = *Sorbus aucuparia.*
Rubisco = ribulose-1,5-bisphosphate carboxylase/oxygenase, a key enzyme in photosynthesis (see under C_3).
Rust fungus = Uredinales, in the Basidiomycetes. Obligate plant parasites.
Ryegrass = *Lolium.* Perennial ryegrass = *Lolium perenne.*
Sagebrush = *Artemisia* (a shrub). The term is also applied to vegetation in which *Artemisia* is prominent.
Sardine, Peruvian = *Sardinops sagax.*
Savanna Tropical vegetation in which perennial grasses are prominent. Trees or

shrubs may be present, but they do not form a continuous canopy.

Serpentine A basic rock with high magnesium content.

Shag = *Phalacrocorax aristotelis.*

Sheep = *Ovis.* Bighorn sheep = *Ovis canadensis.*

Short-wave radiation See Box 2.1.

Shrew = *Sorex.* Common shrew = *Sorex araneus.*

Simulation model Mathematical model which aims to make quantitatively accurate predictions.

Sloth Shasta ground sloth = *Northrotheriops shastensis.*

Sole, rock = *Lepidopsetta bilineata.*

Sparrowhawk = *Accipiter nisus.*

Spruce = *Picea.* Sitka spruce = *Picea sitchensis.*

Starling, European = *Sturnus vulgaris.*

Stochastic model Mathematical model which includes the possibility of chance events or random variation.

Stock Applied to fish: the number or biomass of fish (or of one species of fish) in an area that are large enough to be caught by the nets used.

Stocking density Applied to grazing mammals: the number of animals per unit ground area (e.g. per hectare).

Strategic model Mathematical model which aims not to provide quantitatively accurate predictions, but rather to provide basic insights.

Survivorship How mortality and survival of individuals in a population is related to their age. See Box 4.2.

Sustainable Can be continued indefinitely. See Chapter 1.

Sycamore = *Acer pseudoplatanus.*

T = tera = $\times 10^{12}$.

Tamarin, golden-handed = *Saguinus midas.*

Teak = *Tectona grandis.*

Thermocline A layer in a lake or sea where the temperature changes abruptly, usually (in summer) from warmer water above to cooler below.

Ton In this book means metric ton (= 1000 kg). Approximately equal to the old-fashioned ton.

Trout, rainbow = *Salmo gairdneri.*

Tuna Figure 4.2 referred particularly to skipjack tuna (*Katsuwonus pelamis*) and yellowfin tuna (*Thunnus albacares*).

UN = United Nations.

UNEP = United Nations Environment Programme.

Ungulate A hooved herbivorous mammal.

VAM = VA-mycorrhiza = vesicular–arbuscular mycorrhiza. The fungi form two characteristic types of structure, vesicles and arbuscules, inside the root but have little effect on the root's external appearance. The principal type of mycorrhiza in herbaceous plants; also occurs in some woody species.

Virulence How much a disease-causing organism affects the attacked species; often assessed by what percentage of infected individuals die and how quickly they die.

Vole, bank = *Clethrionomys glareolus.*

Warbler = *Dendroica.*

Weevil = Cuculionidae, in the Coleoptera.

Wildebeest = *Connochaetes taurinus.*

Willow = *Salix.*

Wolf = *Canis*; grey wolf of Europe and North America = *Canis lupus.*

Woodland Vegetation in which trees are present, but are far enough apart to form an open canopy.

Woodlouse Isopoda, in the Crustacea.

Zebra = *Equus burchelli.*

μ = micro = $\times 10^{-6}$.

References

Some of the titles have been abbreviated. These are indicated by [ ].

Abbott, I. & Loneragan, O. (1986). [Ecology of Jarrah (*Eucalyptus marginata*) in Western Australia]. Department of Conservation and Land Management, Perth.

Abuzinadah, R.A. & Read, D.J. (1986). [Utilization of peptides and proteins by ectomycorrhizal fungi.] *New Phytologist*, 103, 481–493.

Abuzinadah, R.A., Finlay, R.D. & Read, D.J. (1986). [Utilization of proteins by mycorrhizal plants of *Pinus contorta*.] *New Phytologist*, 103, 495–506.

Adams, J.M., Faure, H., Faure-Denard, L., McGlade, J.M. & Woodward, F.I. (1990). Increases in terrestrial carbon storage from the Last Glacial Maximum to the present. *Nature*, 348, 711–714.

Alabaster, J.S., Garland, J.H.N., Hart, I.C. & Solbe, J.F.deL.G. (1972). An approach to the problem of pollution and fisheries. *Symposia of the Zoological Society of London*, 29, 87–114.

Alderdice, D.F. & Forrester, C.R. (1971). Effects of salinity, temperature, and dissolved oxygen on early development of the Pacific cod. *Journal of the Fisheries Research Board of Canada*, 28, 883–902.

Alverson, W.S., Waller, D.M. & Solhem, S.L. (1988). Forests too deer: edge effects in northern Wisconsin. *Conservation Biology*, 2, 348–358.

Ambrose, J.P. & Bratton, S.P. (1990). Trends in landscape heterogeneity along the borders of the Great Smoky Mountains National Park. *Conservation Biology*, 4, 135–143.

Anderson, A.B. (1990). *Alternatives to Deforestation*. Columbia University Press, New York.

Anderson, R.M. (ed.) (1982). *Population Dynamics of Infectious Diseases*. Chapman & Hall, London.

Anderson, R.M. & May, R.M. (1979). Population biology of infectious diseases. *Nature*, 280, 361–367.

Anderson, R.M. & May, R.M. (1986). The invasion, persistence and spread of infectious diseases within animal and plant communities. *Philosophical Transactions of the Royal Society B*, 314, 533–570.

Anderson, R.M., Jackson, H.C., May, R.M. & Smith, A.M. (1981). Population dynamics of fox rabies in Europe. *Nature*, 289, 765–771.

Andersson, G., Berggren, H., Cronberg, G. & Gelin, C. (1978). Effects of planktivorous and benthivorous fish on organisms and water chemistry in eutrophic lakes. *Hydrobiologia*, 59, 9–15.

Andren, H. & Angelstam, P. (1988). Elevated predation rates as an edge effect in habitat islands: experimental evidence. *Ecology*, 69, 544–547.

Arno, S.F. (1980). Forest fire history in the northern Rockies. *Journal of Forestry*, 78, 460–465.

Ashton, D.H. & Willis, E.J. (1982). Antagonisms in the regeneration of *Eucalyptus regnans* in the mature forest. In: *The Plant Community as a Working Mechanism* (E.I. Newman, ed.), pp. 113–128. Blackwell Scientific Publications, Oxford.

Augspurger, C.K. (1983). Seed dispersal of the tropical tree, *Platypodium elegans*, and the escape of its seedlings from fungal pathogens. *Journal of Ecology*, 71, 759–771.

Augspurger, C.K. & Kelly, C.K. (1984). Pathogen mortality of tropical tree seedlings: experimental studies of the effects of dispersal distance, seedling density, and light conditions. *Oecologia*, 61, 211–217.

Bach, C.E. (1980). [Effects of plant density and diversity on the population dynamics of the striped cucumber beetle]. *Ecology*, 61, 1515–1530.

Bailey, K.M. & Houde, E.D. (1989). Predation on eggs and larvae of marine fishes and the recruitment problem. *Advances in Marine Biology*, 25, 1–83.

Baker, A.J.M. (1987). Metal tolerance. *New Phytologist* (Supplement), 106, 93–111.

Baker, K.F. & Cook, R.J. (1974). *Biological Control of Plant Pathogens*. Freeman, San Francisco, CA.

Barber, D.A. (1964). Influence of soil organic matter on the entry of caesium-137 into plants. *Nature*, 204, 1326–1327.

Barnes, R.S.K. & Hughes, R.N. (1988). *Introduction to Marine Ecology*, 2nd edition. Blackwell Scientific Publications, Oxford.

Barnes, R.S.K. & Mann, K.H. (1991). *Fundamentals of Aquatic Ecology*. Blackwell Scientific Publications, Oxford.

Barr, B.M. & Braden, K.E. (1988). *The Disappearing Russian Forest*. Rowman & Littlefield, Totowa, NJ.

Barrett, J.W. (1980). *Regional Silviculture of the United States*. Wiley, New York.

Battiston, G.A. *et al.* (1991). Transfer of Chernobyl fallout radionuclides from feed to growing rabbits: cesium-137 balance. *Science of the Total Environment*, 105, 1–12.

Bayliss-Smith, T.P. (1982). *The Ecology of Agricultural Systems*. Cambridge University Press, Cambridge.

Beckett, P.H.T. & Davis, R.D. (1977). Upper critical levels of toxic elements in plants. *New Phytologist*, 79, 95–106.

Beckett, P.H.T., Warr, E. & Davis, R.D. (1983). Copper and zinc in soils treated with sewage sludge: their 'extractability' to reagents compared with their 'availability' to plants. *Plant and Soil*, 70, 3–14.

Beddington, J.R., Free, C.A. & Lawton, J.H. (1978). Characteristics of successful natural enemies in models of biological control of insect pests. *Nature*, 273, 513–519.

Begon, M., Harper, J.L. & Townsend, C.R. (1990). *Ecology*, 2nd edition. Blackwell Scientific Publications, Oxford.

Bennett, K.D. (1983). [Devensian Late-glacial and Flandrian vegetational history of Hockham Mere, Norfolk, England]. *New Phytologist*, 95, 457–487.

Bennett, K.D. (1986). Competitive interactions among forest tree populations in Norfolk, England, during the last 10000 years. *New Phytologist*, 103, 603–620.

Berg, B., Ekbolm, G., Söderstrom, B. & Staaf, H. (1991). Reduction of decomposition rates of Scots pine needle litter due to heavy-metal pollution. *Water, Air, and Soil Pollution*, 59, 165–177.

Berger, J. (1990). Persistence of different sized populations: an empirical assessment of rapid extinctions in bighorn sheep. *Conservation Biology*, 4, 91–98.

Berglund, S., Davis, R.D. & L'Hermite, P. (eds) (1984). *Utilisation of Sewage Sludge on Land*. Reidel, Dordrecht.

Beringer, J.E. & Bale, M.J. (1988). The survival and persistence of genetically-engineered micro-organisms. In: *The Release of Genetically-engineered Micro-organisms* (M. Sussman, C.H. Collins, F.A. Skinner & D.E. Stewart-Tull, eds), pp. 28–46. Academic Press, London.

Berner, R.A. & Lasaga, A.C. (1989). Modeling the geochemical carbon cycle. *Scientific American*, 260(3), 54–61.

Betts, W.B. (ed.) (1991). *Biodegradation*. Springer-Verlag, London.

Beveridge, T.J. & Doyle, R.J. (eds) (1989). *Metal Ions and Bacteria*. Wiley, New York.

Beverton, R.J.H. & Holt, S.J. (1956). The theory of fishing. In: *Sea Fisheries: Their Investigation in the United Kingdom* (M. Graham, ed.), pp. 372–441. Arnold, London.

Billing, E. (1981). Hawthorn as a source of fireblight. In: *Pests, Pathogens and Vegetation* (J.M. Thresh, ed.), pp. 121–130. Pitman, Boston.

Billington, H.L. (1991). Effect of population size on genetic variation in a dioecious conifer. *Conservation Biology*, 5, 115–119.

Binkley, D. (1983). Ecosystem production in Douglas-fir plantations: interaction of red alder and site fertility. *Forest Ecology and Management*, 5, 215–227.

Birks, H.J.B. (1989). Holocene isochrone maps and patterns of tree-spreading in the British Isles. *Journal of Biogeography*, 16, 503–540.

Black, J.L. & Kenney, P.A. (1984). Factors affecting diet selection by sheep. II. Height and density of pasture. *Australian Journal of Agricultural Research*, 35, 565–578.

Blais, J.R. (1983). Trends in the frequency, extent, and severity of spruce budworm outbreaks in eastern Canada. *Canadian Journal of Forest Research*, 13, 539–547.

Bockman, O.C., Kaarstad, O., Lie, O.H. & Richards, I. (1990). *Agriculture and Fertilizers*. Norsk Hydro, Oslo.

Boddy, R.M., Urquiaga, S., Reis, V. & Döbereiner, J. (1991). Biological nitrogen fixation associated with sugar cane. *Plant and Soil*, 137, 111–117.

Bolan, N.S., Robson, A.D. & Barrow, N.J. (1987). [Effects of phosphorus application and mycorrhizal inoculation on root characteristics of subterranean clover and ryegrass]. *Plant and Soil*, 104, 294–298.

Bormann, F.H. & Likens, G.E. (1979). *Pattern and Process in a Forested Ecosystem*. Springer-Verlag, New York.

Bowen, G.D. & Nambiar, E.K.S. (1984). *Nutrition of Plantation Forests*. Academic Press, London.

Box, E.O., Holben, B.N. & Kalb, V. (1989). Accuracy of the AVHRR vegetation index as a predictor of biomass, primary productivity and net CO_2 flux. *Vegetatio*, 80, 71–88.

Bradley, R., Burt, A.J. & Read, D.J. (1981). Mycorrhizal infection and resistance to heavy metal toxicity in *Calluna vulgaris*. *Nature*, 292, 335–337.

Bradley, R., Burt, A.J. & Read, D.J. (1982). [The role of mycorrhizal infection in heavy metal resistance]. *New Phytologist*, 91, 197–209.

Brady, N.C. (1990). *The Nature and Properties of Soils*, 10th edition. Macmillan, New York.

Brafield, A.E. & Llewellyn, M.J. (1982). *Animal Energetics*. Blackie, Glasgow.

Brasier, C.M. (1983). The future of Dutch elm disease in Europe. In: *Research on Dutch Elm Disease in Europe* (D.A. Burdekin, ed.), pp. 96–104. Forestry Commission Bulletin 60, Her Majesty's Stationery Office, London.

Brenchley, W.E. & Warington, K. (1958). *The Park Grass Plots at Rothamsted 1856–1949*. Rothamsted Experimental Station, Harpenden.

Briffa, K.R. *et al.* (1990). A 1400-year tree-ring record of summer temperatures in Fennoscandia. *Nature*, 346, 434–439.

Brown, M.T. & Wilkins, D.A. (1985). Zinc tolerance of mycorrhizal *Betula*. *New Phytologist*, 99, 101–106.

Brown, N.D. & Whitmore, T.C. (1992). Do dipterocarp seedlings really partition tropical rain forest gaps? *Philosophical Transactions of the Royal Society B*, 335, 369–378.

Brunner, W., Sutherland, F.H. & Focht, D.D. (1985). Enhanced biodegradation of polychlorinated biphenyls in soil by analog enrichment and bacterial inoculation. *Journal of Environmental Quality*, 14, 324–328.

Buck, K.W. (1988). Control of plant pathogens with viruses and related agents. *Philosophical Transactions of the Royal Society B*, 318, 295–317.

Buffington, L.C. & Herbel, C.H. (1965). Vegetational changes on a semidesert grassland range from 1858 to 1963. *Ecological Monographs*, 35, 139–164.

Bunzl, K. & Kracke, W. (1989). Seasonal variation of soil-to-plant transfer of K and fallout ^{134,137}Cs in peatland vegetation. *Health Physics*, 57, 593–600.

Burdon, J.J. & Chilvers, G.A. (1977). Preliminary studies on a native eucalypt forest invaded by exotic pines. *Oecologia*, 31, 1–12.

Burdon, J.J. & Chilvers, G.A. (1982). Host density as a factor in plant disease ecology. *Annual Review of Phytopathology*, 20, 143–166.

Burdon, J.J. & Whitbread, R. (1979). Rates of increase of barley mildew in mixed stands of barley and wheat. *Journal of Applied Ecology*, 16, 253–258.

Burgman, M.A., Akcakaya, H.R. & Leow, S.S. (1988). The use of extinction models for species conservation. *Biological Conservation*, 43, 9–25.

Butler, G.C. (ed.) (1978). *Principles of Ecotoxicology*. Wiley, Chichester.

Caldwell, M.M., Richards, J.H., Johnson, D.A., Nowak, R.S. & Dzurec, R.S. (1981). Coping with herbivory: photosynthetic capacity and resource allocation in two semiarid *Agropyron* bunchgrasses. Oecologia, 50, 14–24.

Campbell, C.C. (1969). *Birth of a National Park in the Great Smoky Mountains*. University of Tennessee Press, Knoxville, TN.

Campbell, C.L. & Madden, C.V. (1990). *Introduction to Plant Disease Epidemiology*. Wiley, New York.

Campbell, R. (1989). *Biological Control of Microbial Plant Pathogens*. Cambridge University Press, Cambridge.

Cannell, M.G.R. (1980). Productivity of closely-spaced young poplar on agricultural soils in Britain. *Forestry*, 53, 1–21.

Cannell, M.G.R., Milne, R., Sheppard, L.J. & Unsworth, M.H. (1987). Radiation interception and productivity of willow. *Journal of Applied Ecology*, 24, 261–278.

Chapin, F.S. & Kedrowski, R.A. (1983). Seasonal changes in nitrogen and phosphorus and autumn retranslocation in evergreen and deciduous taiga trees. *Ecology*, 64, 376–391.

Chapman, J.D. (1989). *Geography and Energy*. Longman, Harlow.

Chapman, J.L. & Reiss, M.J. (1992). *Ecology: Principles and Applications*. Cambridge University Press, Cambridge.

Clark, J.S. (1990). Fire and climate change during the last 750 yr in northwestern Minnesota. *Ecological Monographs*, 60, 135–159.

Clarke, C.M.H. (1971). Liberations and dispersal of red deer in northern South Island districts. *New Zealand Journal of Forest Science*, 1, 194–207.

Cobbett, W. (1830). *Rural Rides*. Dent, London.

Cockburn, A. (1991). *An Introduction to Evolutionary Ecology*. Blackwell Scientific Publications, Oxford.

Collier, B.D., Cox, G.W., Johnson, A.W. & Miller, P.C. (1973). *Dynamic Ecology*. Prentice-Hall, Englewood Cliffs, NJ.

Common, M. (1988). *Environmental and Resource Economics*. Longman, London.

Connell, J.H. (1979). Tropical rain forests and coral reefs as open non-equilibrium systems. In: *Population Dynamics* (R.M. Anderson, B.D. Turner & L.R. Taylor, eds), pp. 141–163. Blackwell Scientific Publications, Oxford.

Connor, E.F. & McCoy, E.D. (1979). The statistics and biology of the species-area relationship. *American Naturalist*, 113, 791–833.

Cook, R.J. & Baker, K.F. (1983). *The Nature and Practice of Biological Control of Plant Pathogens*. American Phytopathological Society, St Paul, MN.

Cooke, G.W. (1976). A review of the effects of agriculture on the chemical composition and quality of surface and underground waters. In: *Agriculture and Water Quality*. Technical Bulletin 32 of the Ministry of Agriculture, Fisheries and Food, pp. 5–57. Her Majesty's Stationery Office, London.

Coppock, D.L., Ellis, J.E. & Swift, D.M. (1986). Livestock feeding ecology and resource utilization in a nomadic pastoral ecosystem. *Journal of Applied Ecology*, 23, 573–583.

Coughenour, M.B. *et al.* (1985). Energy extraction and use in a nomadic pastoral system. *Science*, 230, 619–625.

Coughtrey, P.J., Kirton, J.A. & Mitchell, N.G. (1989). Caesium transfer and cycling in upland pastures. *Science of the Total Environment*, 85, 149–158.

Crawley, M.J. (1983). *Herbivory. The Dynamics of Animal–Plant Interactions*. Blackwell Scientific Publications, Oxford.

Crawley, M.J. (1986). The population biology of invaders. *Philosophical Transactions of the Royal Society B*, 314, 711–730.

Crompton, G. & Sheail, J. (1975). [The historical ecology of Lakenheath Warren in Suffolk, England]. *Biological Conservation*, 8, 299–313.

Crosby, D.G., Tucker, R.K. & Aharonson, N. (1966). The detection of acute toxicity with *Daphnia magna*. *Food and Cosmetic Toxicology*, 4, 503–514.

Crowell, K.L. (1973). Experimental zoogeography: introductions of mice to small islands. *American Naturalist*, 107, 535–558.

Cullen, J.M. & Hasan, S. (1988). Pathogens for the control of weeds. *Philosophical Transactions of the Royal Society B*, 318, 213–222.

Curtis, R.O., Glendenen, G.W., Reukema, D.L. & DeMars, D.J. (1982). *Yield Tables for Managed Stands of Coast Douglas-fir*. USDA Forest Service General Technical Report PNW-135, Olympia WA.

Cushing, D.H. (1981). *Fisheries Biology: a Study in Population Dynamics*, 2nd edition. University of Wisconsin Press, Madison, WI.

Cushing, D.H. (1990). Plankton production and year-class strength in fish populations: an update of the match/mismatch hypothesis. *Advances in Marine Biology*, 26, 249–293.

Darwin, C. (1859). *The Origin of Species by Means of Natural Selection*. Murray, London.

da Silva, J.G., Serra, G.E., Moreira, J.R., Gonçalves, J.C. & Gollemberg, J. (1978). Energy balance for ethyl alcohol production from crops. *Science*, 201, 903–906.

Davidson, W.R., Hayes, F.A., Nettles, V.F. & Kellogg, F.E. (1981). *Diseases and Parasites of White-tailed Deer*. Miscellaneous Publication no. 7 of Tall Timbers Research Station, Tallahassee, FL.

Davis, M.B. (1976). Erosion rates and land-use history in southern Michigan. *Environmental Conservation*, 3, 139–148.

Davis, M.B. (1981). Quaternary history and the stability of forest communities. In: *Forest Succession* (D.C. West, H.H. Shugart & D.B. Botkin, eds), pp. 132–153. Springer-Verlag, New York.

Davis, M.B., Spear, R.W. & Shane, L.C.K. (1980). Holocene climate of New England. *Quaternary Research*, 14, 240–250.

Dawson, P., Weste, G. & Ashton, D. (1985). Regeneration of vegetation in the Brisbane Ranges after fire and infestation by *Phytophthora cinnamomi*. *Australian Journal of Botany*, 33, 15–26.

Debach, P. & Rosen, D. (1991). *Biological Control by Natural Enemies*, 2nd edition. Cambridge University Press, Cambridge.

Delcourt, H.R. & Delcourt, P.A. (1991). *Quaternary Ecology*. Chapman & Hall, London.

Dempster, J.P. (1969). Some effects of weed control on the numbers of the small cabbage white (*Pieris rapae*) on Brussels sprouts. *Journal of Applied Ecology*, 6, 339–345.

Dennis, B., Munholland, P.L. & Scott, J.M. (1991). Estimation of growth and extinction parameters for endangered species. *Ecological Monographs*, 61, 115–143.

Detwiler, R.P. & Hall, C.A.S. (1988). Tropical forests and the global carbon cycle. *Science*, 239, 42–47.

Diamond, J.M. (1969). Avifaunal equilibria and species turnover rates on the Channel Islands of California. *Proceedings of the National Academy of Sciences*, 64, 57–63.

Diamond, J.M. (1987). Human use of world resources. *Nature*, 328, 479–480.

Dixon, R.O.D. & Wheeler, C.T. (1986). *Nitrogen Fixation in Plants*, Blackie, Glasgow.

Doak, D. (1989). Spotted owls and old growth logging in the Pacific Northwest. *Conservation Biology*, 3, 389–396.

Duffy, D.C. & Meier, A.J. (1992). Do Appalachian herbaceous understories ever recover from clearcutting? *Conservation Biology*, 6, 196–201.

Dunbar, M.J. (ed) (1979). *Marine Production Mechanisms*. Cambridge University Press, Cambridge.

Dwyer, D.F., Rojo, F. & Timmis, K.N. (1988). Fate and behaviour in an activated sludge microcosm of a genetically-engineered micro-organism designed to degrade substituted aromatic compounds. In: *The Release of Genetically-engineered Micro-organisms* (M. Sussman, C.H. Collins, F.A. Skinner & D.E. Stewart-Tull, eds), pp. 77–88. Academic Press, London.

Dwyer, G., Levin, S.A. & Buttel, L. (1990). A simulation model of the population dynamics and evolution of myxomatosis. *Ecological Monographs*, 60, 423–447.

Eamus, D. & Jarvis, P.G. (1989). [The effects of increase in the atmospheric CO_2 concentration on trees and forests]. *Advances in Ecological Research*, 19, 1–55.

Eden, M.J. (1990). *Ecology and Land Management in Amazonia*. Bellhaven, London.

Ellison, L. (1960). Influence of grazing on plant succession of rangelands. *Botanical Review*, 26, 1–78.

Elton, C.S. (1958). *The Ecology of Invasions by Animals and Plants*. Methuen, London.

Engelmark, O. (1984). Forest fires in the Muddus National Park (northern Sweden) during the past 600 years. *Canadian Journal of Botany*, 62, 893–898.

Evans, D.G. & Miller, M.H. (1988). Vesicular–arbuscular mycorrhizas and the soil-disturbance-induced reduction of nutrient absorption in maize. *New Phytologist*, 110, 67–74.

Evans, J. (1984). *Silviculture of Broadleaved Woodland*. Forestry Commission Bulletin 62. Her Majesty's Stationery Office, London.

Evans, J. (1986). Nutrition experiments in broadleaved stands. *Quarterly Journal of Forestry*, 80, 85–104.

Evans, J. (1992). *Plantation Forestry in the Tropics*, 2nd edition. Clarendon Press, Oxford.

FAO (1979). *Eucalypts for Planting*. Food and Agriculture Organization of the United Nations, Rome.

FAO (1985). *The Forest Resources of the ECE Region*. Food and Agriculture Organization of the United Nations, Geneva.

FAO Yearbooks. Published by the Food and Agriculture Organization of the United Nations, Rome.

Fenner, F. (1983). Biological control, as exemplified by smallpox eradication and myxomatosis. *Proceedings of the Royal Society B*, 218, 259–285.

Fincham, J.R.S. & Ravetz, J.R. (1991). *Genetically Engineered Organisms*. Open University Press, Milton Keynes.

Fitter, A.H. (1986). Spatial and temporal patterns of root activity in a species-rich alluvial glassland. *Oecologia*, 69, 594–599.

Fogg, G.E. (1991). The phytoplanktonic ways of life. *New Phytologist*, 118, 191–232.

Fox, R.A. (1965). The role of biological eradication in root-disease control in replantings of *Hevea brasiliensis*. In: *Ecology of Soil-borne Plant Pathogens* (K.F. Baker & W.C. Snyder, eds), pp. 348–362. Murray, London.

Franklin, J.F. & Hemstrom, M.A. (1981). Aspects of succession in the coniferous forests of the Pacific Northwest. In: *Forest Succession* (D.C. West, H.H. Shugart & D.B. Botkin, eds), pp. 212–229. Springer-Verlag, New York.

Freedman, B. (1989). *Environmental Ecology*. Academic Press, San Diego.

Frissel, M.J. (ed.) (1978). *Cycling of Mineral Nutrients in Agricultural Ecosystems*. Elsevier, Amsterdam.

Frost, L.C. (1981). The study of *Ranunculus ophioglossifolius* and its successful conservation at the Badgeworth Nature Reserve, Gloucestershire. In: *The Biological Aspects of Rare Plant Conservation* (H. Synge, ed.), pp. 481–489. Wiley, Chichester.

Fulbright, D.W., Weidlich, W.H., Haufler, K.Z., Thomas, C.S. & Paul, C.P. (1983). Chestnut blight and recovering American chestnut trees in Michigan. *Canadian Journal of Botany*, 61, 3164–3171.

Furch, K. & Klinge, H. (1989). Chemical relationships between vegetation, soil and water in contrasting inundation areas of Amazonia. In: *Mineral Nutrients in Tropical Forest and Savanna Ecosystems* (J. Proctor, ed.), pp. 189–204. Blackwell Scientific Publications, Oxford.

Game, M. & Peterken, G.F. (1984). Nature reserve selection strategies in the woodlands of central Lincolnshire, England. *Biological Conservation*, 29, 157–181.

Garrod, D.J. (1988). North Atlantic cod: fisheries and management to 1986. In: *Fish Population Dynamics* (J.A. Gulland, ed.), pp. 185–218. Wiley, Chichester.

Gartside, D.W. & McNeilly, T. (1974). [The potential for evolution of heavy metal tolerance in plants. II. Copper tolerance]. *Heredity*, 32, 335–348.

Gasser, C.S. & Fraley, R.T. (1989). Genetically engineered plants for crop improvement. *Science*, 244, 1293–1299.
Gates, D.M. (1985). *Energy and Ecology*, Sinauer, Sunderland, MA.
Gates, J.E. & Gysel, L.W. (1978). Avian nest dispersion and fledgling success in field-forest ecotones. *Ecology*, 59, 871–883.
Getz, L.L., Cole, F.R. & Gates, D.L. (1978). Interstate roadsides as dispersal routes for *Microtus pennsylvanicus*. *Journal of Mammalogy*, 59, 208–212.
Gibbs, J.N. (1978). Development of Dutch elm disease epidemic in southern England, 1971–6. *Annals of Applied Biology*, 88, 219–228.
Giesy, J.P. & Graney, R.L. (1989). Recent developments in and intercomparisons of acute and chronic bioassays and indicators. *Hydrobiologia*, 188/189, 21–60.
Giller, K.E. & Day, J.M. (1985). [Nitrogen fixation in the rhizosphere.] In: *Ecological Interactions in Soil* (A.H. Fitter, D. Atkinson, D.J. Read & M.B. Usher, eds), pp. 127–147. Blackwell Scientific Publications, Oxford.
Gilligan, C.A. (1990). Comparison of disease progress curves. *New Phytologist*, 115, 223–242.
Golley, F.B. (ed.) (1983). *Tropical Forest Ecosystems*. Elsevier, Amsterdam.
Gomez-Pompa, A., Whitmore, T.C. & Hadley, M. (eds) (1991). *Rain Forest Regeneration and Management*. UNESCO/Parthenon, Paris.
Gordon, A.G. (1964). [Nutrition and growth of ash in the Lake District.] *Journal of Ecology*, 52, 169–187.
Gordon, I.J. & Lindsay, W.K. (1990). Could mammalian herbivores 'manage' their resources? *Oikos*, 59, 270–280.
Goudriaan, J., van Keulen, H. & van Laar, H.H. (1990). *The Greenhouse Effect and Primary Productivity in European Agro-ecosystems*. Pudoc, Wageningen.
Graham, R.W. (1986). Response of mammalian communities to environmental changes during the late Quaternary. In: *Community Ecology* (J. Diamond & T.J. Case, eds), pp. 300–313. Harper & Row, New York.
Grahame, J. (1987). *Plankton and Fisheries*. Arnold, London.
Grant, S.A., Torvell, L., Smith, H.K., Suckling, D.E., Forbes, T.D.A. & Hodgson, J. (1987). Comparative studies of diet selection by sheep and cattle: blanket bog and heather moor. *Journal of Ecology*, 75, 947–960.
Gray, J.S. (1989). Do bioassays adequately predict ecological effects of pollutants? *Hydrobiologia*, 188/189, 397–402.
Gray, N.F. (1989). *Biology of Wastewater Treatment*. Oxford University Press, Oxford.
Greaves, M.P. (1987). Side-effects testing: an alternative approach. In: *Pesticide Effects on Soil Microflora* (L. Somerville & M.P. Greaves, eds), pp. 183–190. Taylor & Francis, London.
Grime, J.P. (1973). Competitive exclusion in herbaceous vegetation. *Nature*, 242, 344–347.
Grubb, M.J. (1988). The potential for wind energy in Britain. *Energy Policy*, 16, 594–607.
Grubb, P.J. (1977). The maintenance of species-richness in plant communities: the importance of the regeneration niche. *Biological Reviews*, 52, 107–145.
Grue, C.E., Powell, G.V.N. & McChesney, M.J. (1982). Care of nestlings by wild female starlings exposed to an organophosphate pesticide. *Journal of Applied Ecology*, 19, 327–335.
Gudmundsson, F. (1951). The effects of the recent climatic changes on the bird life of Iceland. *Proceedings of the 10th International Ornithological Congress*, pp. 502–514.
Gulland, J.A. (ed.) (1988). *Fish Population Dynamics*, 2nd edition. Wiley, Chichester.
Hall, C.A.S., Cleveland, C.J. & Kaufmann, R. (1986). *Energy and Resource Quality*. Wiley, New York.
Hamilton, G.J. & Christie, J.M. (1971). *Forest Management Tables*. Forestry Commission. Her Majesty's Stationery Office, London.
Hardy, A. (1959). *The Open Sea: Its Natural History. Part II: Fish and Fisheries*. Collins, London.

Harley, J.L. & Smith, S.E. (1983). *Mycorrhizal Symbiosis*. Academic Press, London.
Harper, J.L. (1977). *Population Biology of Plants*. Academic Press, London.
Harper, J.L., Williams, J.T. & Sagar, G.R. (1965). [The heterogeneity of soil surfaces and its role in determining the establishment of plants from seed]. *Journal of Ecology*, 53, 273–286.
Harris, S. & Smith, G.C. (1987). The use of sociological data to explain the distribution and numbers of urban foxes in England and Wales. *Symposia of the Zoological Society of London*, 58, 313–328.
Harrison, S., Murphy, D.D. & Ehrlich, P.R. (1988). Distribution of the Bay checkerspot butterfly: evidence for a metapopulation model. *American Naturalist*, 132, 360–382.
Harte, J. (1985). *Consider a Spherical Cow*. Kaufmann, Los Altos, CA.
Hassell, M.P. (1978). *The Dynamics of Arthropod Predator–Prey Systems*. Princeton University Press, Princeton, NJ.
Hassell, M.P. (1980). Foraging strategies, population models and biological control: a case study. *Journal of Animal Ecology*, 49, 603–628.
Hayman, D.S. & Mosse, B. (1979). Improved growth of white clover in hill grasslands by mycorrhizal inoculation. *Annals of Applied Biology*, 93, 141–148.
Heinselman, M.L. (1973). Fire in the virgin forests of the Boundary Waters Canoe Area, Minnesota. *Quaternary Research*, 3, 329–382.
Heitefuss, R. (1989). *Crop and Plant Protection*. Wiley, New York.
Hellawell, J.M. (1986). *Biological Indicators of Fresh Water Pollution and Environmental Management*. Elsevier, London.
Henderson, M.T., Merriam, G. & Wegner, J. (1985). [Chipmunks in an agricultural mosaic]. *Biological Conservation*, 31, 95–109.
Hengeveld, R. (1989). *Dynamics of Biological Invasions*. Chapman & Hall, London.
Heske, F. (1938). *German Forestry*. Yale University Press, New Haven, CT.
Hickey, J.J. & Anderson, D.W. (1968). Chlorinated hydrocarbons and eggshell changes in raptorial and fish-eating birds. *Science*, 162, 271–273.
Hill, M.O., Evans, D.F. & Bell, S.A. (1992). Long-term effects of excluding sheep from hill pastures in North Wales. *Journal of Ecology*, 80, 1–13.
Hodgson, J., Forbes, T.D.A., Armstrong, R.H., Beattie, M.M. & Hunter, E.A. (1991). [The ingestive behaviour and herbage intake of sheep and cattle]. *Journal of Applied Ecology*, 28, 205–227.
Holdgate, M.W. (1991). Conservation in a world context. In: *The Scientific Management of Temperate Communities for Conservation* (I.F. Spellerberg, F.B. Goldsmith & M.G. Morris, eds), pp. 1–26. Blackwell Scientific Publications, Oxford.
Hopkin, S.P. (1989). *Ecophysiology of Metals in Terrestrial Invertebrates*. Elsevier, London.
Hopkin, S.P. (1990a). [Assimilation of zinc, cadmium, lead, copper and iron by terrestrial isopods]. *Journal of Applied Ecology*, 27, 460–474.
Hopkin, S.P. (1990b). Critical concentrations, pathways of detoxification and cellular ecotoxicology of metals in terrestrial arthropods. *Functional Ecology*, 4, 321–327.
Hopkins, M.S. & Graham, A.W. (1984). Viable soil seed banks in disturbed lowland tropical rainforest sites in North Queensland. *Australian Journal of Ecology*, 9, 71–79.
Hopkinson, C.S. & Day, J.W. (1980). Net energy analysis of alcohol production from sugarcane. *Science*, 207, 302–304.
Houghton, J.T., Jenkins, G.J. & Ephraums, J.J. (1990). *Climate Change. The IPCC Scientific Assessment*. Cambridge University Press, Cambridge.
Houghton, R.A. *et al.* (1983). [Changes in the carbon content of terrestrial biota and soils between 1860 and 1980]. *Ecological Monographs*, 53, 235–262.
Howell, D.G. & Murray, R.W. (1986). A budget for continental growth and denudation. *Science*, 233, 446–449.
Howell, E.A. & Jordan, W.R. (1991). Tallgrass prairie restoration in the North American midwest. In: *The Scientific Management of Temperate Communities*

for Conservation (I.F. Spellerberg, F.B. Goldsmith & M.G. Morris, eds), pp. 395–414. Blackwell Scientific Publications, Oxford.

Hubbell, S.P. & Foster, R.B. (1986a). [Commonness and rarity in a neotropical forest]. In: *Conservation Biology* (M.E. Soulé, ed.), pp. 205–231. Sinauer, Sunderland, MA.

Hubbell, S.P. & Foster, R.B. (1986b). Biology, chance, and history and the structure of tropical rain forest tree communities. In: Community Ecology (J. Diamond & T.J. Case, eds), pp. 314–329. Harper & Row, New York.

Huffaker, C.B. & Kennett, C.E. (1959). A ten-year study of vegetational changes associated with biological control of Klamath weed. *Journal of Range Management*, 12, 69–82.

Hughes, D.E. & McKenzie, P. (1975). The microbial degradation of oil in the sea. *Proceedings of the Royal Society B*, 189, 375–390.

Hulzebos, E.M., Adema, D.M.M., Dirven-van Breemen, E.M., Henzel, L. & van Gestel, C.A.M. (1991). QSARs in phytotoxicity. *Science of the Total Environment*, 109/110, 493–497.

Hunter, B.A., Johnson, M.S. & Thompson, D.J. (1987a,b,c). Ecotoxicology of copper and cadmium in a contaminated grassland ecosystem. *Journal of Applied Ecology*, 24. (a) I. Soil and vegetation contamination: pp. 573–586. (b) II. Invertebrates: pp. 587–599. (c) III. Small mammals: pp. 601–614.

Huntley, B. & Birks, H.J.B. (1983). *An Atlas of Past and Present Pollen Maps for Europe: 0 to 13 000 years ago*. Cambridge University Press, Cambridge.

Huntly, N. (1991). Herbivores and the dynamics of communities and ecosystems. *Annual Review of Ecology and Systematics*, 22, 477–503.

Hutchinson, G.E. (1959). Homage to Santa Rosalia, or why are there so many kinds of animals? *American Naturalist*, 93, 145–159.

Ingestad, T. (1982). Relative addition rate and external concentration; driving variables used in plant nutrition research. *Plant, Cell and Environment*, 5, 443–453.

Ingleby, K., Mason, P.A., Last, F.T. & Fleming, L.V. (1990). *Indentification of Ectomycorrhizas*. Her Majesty's Stationery Office, London.

IPCC1 (1990). Policymakers' summary from: *Climate Change, the IPCC Scientific Assessment* (J.T. Houghton, G.J. Jenkins & J.J. Ephraums, eds), pp. xi–xxxiii. Cambridge University Press, Cambridge.

Ives, J.D. & Messerli, B. (1989). *The Himalayan Dilemma*. Routledge, London.

Jackson, M.T., Ford-Lloyd, B.V. & Parry, M.L. (eds) (1990). *Climatic Change and Plant Genetic Resources*. Belhaven, London.

Janzen, D.H. (1972). Escape in space by *Sterculia apetala* seeds from the bug *Dysdercus fasciatus* in a Costa Rican deciduous forest. *Ecology*, 53, 350–361.

Jaramillo, V.J. & Detling, J.K. (1992). [Cattle grazing of simulated urine patches in North American grassland.] *Journal of Applied Ecology*, 29, 9–13.

Jasper, D.A., Abbott, L.K. & Robson, A.D. (1989). Acacias respond to phosphorus and to inoculation with VA mycorrhizal fungi in soils stockpiled during mineral sand mining. *Plant and Soil*, 115, 99–108.

Jasper, D.A., Robson, A.D. & Abbott, L.K. (1987). The effect of surface mining on the infectivity of vesicular–arbuscular mycorrhizal fungi. *Australian Journal of Botany*, 35, 641–652.

Jastrow, J.D. (1987). Changes in soil aggregation associated with tallgrass prairie restoration. *American Journal of Botany*, 74, 1656–1664.

Jedrzejewski, W., Jedrzejewska, B., Okarma, H. & Ruprecht, A.L. (1992). Wolf predation and snow cover as mortality factors in the ungulate community of the Bialowieza National Park, Poland. *Oecologia*, 90, 27–36.

Jenkinson, D.S. (1991). [The Rothamsted long-term experiments]. *Agronomy Journal*, 83, 2–10.

Jenkinson, D.S. & Rayner, J.H. (1977). The turnover of soil organic matter in some of the Rothamsted classical experiments. *Soil Science*, 123, 298–305.

Johns, A.D. (1985). Selective logging and wildlife conservation in tropical rainforest: problems and recommendations. *Biological Conservation*, 31, 355–375.

Johnsingh, A.J.T., Narenda Prasad, S. & Goyal, S.P. (1990). [Conservation status of the Chila-Motichur corridor for elephant movement in India]. *Biological Conservation*, 51, 125–138.

Johnson, W.C. & Webb, T. (1989). [The role of blue jays in the postglacial dispersal of trees in eastern North America]. *Journal of Biogeography*, 16, 561–571.

Jones, K.C., Sanders, G., Wild, S.R., Burnett, V. & Johnston, A.E. (1992). Evidence for a decline of PCBs and PAHs in rural vegetation and air in the United Kingdom. *Nature*, 356, 137–140.

Jordan, C.F. (1989). *An Amazonian Rainforest*. UNESCO/Parthenon, Paris.

Jordan, W.R., Gilpin, M.E. & Aber, J.D. (eds) (1987). *Restoration Ecology*. Cambridge University Press, Cambridge.

Jouzel, J. *et al.* (1987). Vostok ice core: a continuous isotope temperature record over the last climatic cycle (160 000 years). *Nature*, 329, 403–408.

Kaiser, K.L.E. & Esterby, S.R. (1991). [The toxicity of 267 chemicals to six species and the octanol/water partition coefficient]. *Science of the Total Environment*, 109/110, 499–514.

Kallio, M., Dykstra, D.P. & Binkley, C.S. (eds) (1987). *The Global Forest Sector*. Wiley, Chichester.

Keith, L.B., Cary, J.R., Rongstad, O.J. & Brittingham, M.C. (1984). Demography and ecology of a declining snowshoe hare population. *Wildlife Monographs*, no. 90.

Kennedy, D.N. & Swaine, M.D. (1992). Germination and growth of colonizing species in artificial gaps of different sizes in dipterocarp rain forest. *Philosophical Transactions of the Royal Society B*, 335, 357–366.

Kikkawa, J. & Anderson, D.J. (eds) (1986). *Community Ecology*. Blackwell Scientific Publications, Melbourne.

Kilbane, J.J., Chatterjee, D.K. & Chakrabarty, A.M. (1983). Detoxification of 2,4,5-trichlorophenoxyacetic acid from contaminated soil by *Pseudomonas cepacia*. *Applied and Environmental Microbiology*, 45, 1697–1700.

Killham, K. & Firestone, M.K. (1983). Vesicular–arbuscular mycorrhizal mediation of grass response to acidic and heavy metal depositions. *Plant and Soil*, 72, 39–48.

Kimball, B.A. (1983). Carbon dioxide and agricultural yield: an assemblage and analysis of 430 prior observations. *Agronomy Journal*, 75, 779–788.

King, T.J. (1977). [The plant ecology of ant-hills in calcareous grasslands]. *Journal of Ecology*, 65, 235–256.

Kjellberg, F. & Valdeyron, G. (1990). Species-specific pollination: a help or a limitation to range extension? In: *Biological Invasions in Europe and the Mediterranean Basin* (F. di Castri, A.J. Hansen & M. Debussche, eds), pp. 371–378. Kluwer, Dordrecht.

Klass, D.L. (1984). Methane from anaerobic fermentation. *Science*, 223, 1021–1028.

Knight, R.R. & Eberhardt, L.L. (1985). Population dynamics of Yellowstone grizzly bears. *Ecology*, 66, 323–334.

Knoop, W.T. & Walker, B.H. (1985). Interactions of woody and herbaceous vegetation in a southern African savanna. *Journal of Ecology*, 73, 235–253.

Kogan, M. (1986). *Ecological Theory and Integrated Pest Management Practice*. Wiley, New York.

König, E. & Gossow, H. (1979). Even-aged stands as habitat for deer in central Europe. In: *Ecology of Even-aged Forest Plantations* (E.D. Ford, D.C. Malcolm & J. Atherson, eds), pp. 429–451. Institute of Terrestrial Ecology, Cambridge.

Kornberg, H. & Williamson, M.H. (1986). Quantitative aspects of the ecology of biological invasions. *Philosophical Transactions of the Royal Society B*, 314, 501–742.

Koslow, J.A. (1992). Fecundity and the stock–recruitment relationship. *Canadian Journal of Fisheries and Aquatic Science*, 49, 210–217.

Krebs, C.J. (1985). *Ecology*, 3rd edition. Harper & Row, New York.

Krupa, S.V. & Kickert, R.N. (1989). [Impacts of ultraviolet radiation, carbon dioxide and ozone on vegetation]. *Environmental Pollution*, 61, 263–393.

Laevastu, T. & Favorite, F. (1988). *Fishing and Stock Fluctuations*. Fishing News Books, Farnham, Surrey.

Lamb, D. (1990). *Exploiting the Tropical Rain Forest. An Account of Pulpwood Logging in Papua New Guinea*. UNESCO/Parthenon, Paris.

Lamb, H.H. (1977). *Climate: Past, Present and Future*. Methuen, London.

Lambert, D. (1989). *The Undying Past of Shenandoah National Park*. Rinehart, Boulder, CO.

Lampert, W., Fleckner, W., Pott, E., Schober, U. & Störkel, K.-U. (1989). Herbicide effects on planktonic systems of different complexity. *Hydrobiologia*, 188/189, 415–424.

Lange, O.L., Nobel, P.S., Osmond, C.B. & Ziegler, H. (1983). *Physiological Plant Ecology. IV. Encyclopedia of Plant Physiology*, volume 12D. Springer-Verlag, Berlin.

Laskowski, R. (1991). Are the top carnivores endangered by heavy metal biomagnification? *Oikos*, 60, 387–390.

Law, R. & Blackford, J.C. (1992). [Self-assembling food webs: a global viewpoint of coexistence of species]. *Ecology*, 73, 567–578.

Lawlor, D.W. & Mitchell, R.A.C. (1991). The effects of increasing CO_2 on crop photosynthesis and productivity: a review of field studies. *Plant, Cell and Environment*, 14, 807–818.

Lawson, G.J., Callaghan, T.V. & Scott, R. (1984). Renewable energy from plants: bypassing fossilization. *Advances in Ecological Research*, 14, 57–114.

Lawton, J.H. & Brown K.C. (1986). The population and community ecology of invading insects. *Philosophical Transactions of the Royal Society B*, 314, 607–616.

Laycock, W.A. (1967). How heavy grazing and protection affect sagebrush-grass ranges. *Journal of Range Management*, 20, 206–213.

Lazenby, A. (1981). British grasslands; past, present and future. *Grass and Forage Science*, 36, 243–266.

Leonard, K.J. & Fry, W.E. (eds) (1989). *Plant Disease Epidemiology*. McGraw-Hill, New York.

Lieberman, D., Lieberman, M., Peralta, R. & Hartshorn, G.S. (1985). Mortality patterns and stand turnover rates in a wet tropical forest in Costa Rica. *Journal of Ecology*, 73, 915–924.

Lieth, H. & Whittaker, R.H. (1975). *Primary Productivity of the Biosphere*. Springer-Verlag, Berlin.

Likens, G.E., Bormann, F.H., Johnson, N.M., Fisher, D.W. & Pierce, R.S. (1970). Effects of forest cutting and herbicide treatment on nutrient budgets in the Hubbard Brook watershed-ecosystem. *Ecological Monographs*, 40, 23–47.

Likens, G.E., Bormann, F.H., Pierce, R.S., Eaton, J.S. & Johnson, N.M. (1977). *Biogeochemistry of a Forested Ecosystem*. Springer-Verlag, New York.

Lindow, S.E. & Panopoulos, N.J. (1988). Field tests of recombinant ice$^-$ *Pseudomonas syringae* for biological frost control in potato. In: *The Release of Genetically-engineered Micro-organisms* (M. Sussman, C.H. Collins, F.A. Skinner & D.E. Stewart-Tull, eds), pp. 121–138. Academic Press, London.

Lindow, S.E., Panopoulos, N.J. & McFarland, B.L. (1989). Genetic engineering of bacteria from managed and natural habitats. *Science*, 244, 1300–1307.

Longhurst, A.R. & Pauly, D. (1987). *Ecology of Tropical Oceans*. Academic Press, San Diego, CA.

Lönnroth, M., Johansson, T.B. & Steen, P. (1980). *Solar versus Nuclear. Choosing Energy Futures*. Pergamon, Oxford.

Loomis, R.S. (1984). Traditional agriculture in America. *Annual Review of Ecology and Systematics*, 15, 449–478.

Lovejoy, T.E. *et al.* (1986). Edge and other effects of isolation on Amazon forest fragments. In: *Conservation Biology* (M.E. Soule, ed.), pp. 257–285. Sinauer, Sunderland, MA.

Lynch, J.M. & Harper, S.H.T. (1985). The microbial upgrading of straw for agricultural use. *Philosophical Transactions of the Royal Society B*, 310, 221–226.

Maass, J.M., Jordan, C.F. & Sarukhan, J. (1988). [Soil erosion and nutrient losses in tropical agroecosystems]. *Journal of Applied Ecology*, 25, 595–607.

MacArthur, R.H. (1955). Fluctuations in animal populations, and a measure of community stability. *Ecology*, 36, 533–536.

MacArthur, R.H. (1958). Population ecology of some warblers of northeastern coniferous forests. *Ecology*, 39, 599–619.

MacArthur, R.H. & MacArthur, J.W. (1961). On bird species diversity. *Ecology*, 42, 594–598.

MacArthur, R.H. & Wilson, E.O. (1967). *The Theory of Island Biogeography*. Princeton University Press, Princeton, NJ.

Mack, R.N. (1981). [Invasion of *Bromus tectorum* into western North America]. *Agro-ecosystems*, 7, 145–165.

Mack, R.N. (1986). Alien plant invasion into the Intermountain West: a case history. In: *Ecology of Biological Invasions of North America and Hawaii* (H.A. Mooney & J.A. Drake, eds), pp. 191–213. Springer-Verlag, Berlin.

Macnair, M.R. (1983). The genetic control of copper tolerance in the yellow monkey flower, *Mimulus guttatus*. *Heredity*, 50, 283–293.

MacNicol, R.D. & Beckett, P.H.T. (1985). Critical tissue concentrations of potentially toxic elements. *Plant and Soil*, 85, 107–129.

Malajczuk, N. (1979). Biological suppression of *Phytophthora cinnamomi* in eucalypts and avocados in Australia. In: *Soil-borne Plant Pathogens* (B. Schippers & W. Gams, eds), pp. 635–652. Academic Press, London.

Malcolm, D.C., Evans, J. & Edwards, P.N. (eds) (1982). *Broadleaves in Britain: Future Management and Research*. Institute of Chartered Foresters, Edinburgh.

Mälkönen, E. (1974). Annual primary production and nutrient cycle in some Scots pine stands. *Communications Instituti Forestalis Fenniae*, 84.5, 1–87.

Manokaran, N. & LaFrankie, J.V. (1990). Stand structure of Pasoh Forest Reserve, a lowland rain forest in Peninsular Malaysia. *Journal of Tropical Forest Science*, 3, 14–24.

Marrs, R.H. & Hicks, M.J. (1986). [Study of vegetation change at Lakenheath Warren]. *Journal of Applied Ecology*, 23, 1029–1046.

Marrs, R.H., Hicks, M.J. & Fuller, R.M. (1986). Losses of lowland heath through succession at four sites in Breckland, East Anglia, England. *Biological Conservation*, 36, 19–38.

Marrs, R.H., Bravington, M. & Rawes, M. (1988). Long-term vegetation change in the *Juncus squarrosus* grassland at Moor House, northern England. *Vegetatio*, 76, 179–187.

Mason, C.F. (1991). *Biology of Freshwater Pollution*, 2nd edition. Longman, London.

Matson, P.A. & Ustin, S.L. (1991). The future of remote sensing in ecological studies. *Ecology*, 72, 1917–1945.

Mattson, D.J. & Reid, M.M. (1991). Conservation of the Yellowstone grizzly bear. *Conservation Biology*, 5, 364–372.

May, R.M. (1974). *Stability and Complexity in Model Ecosystems*. Princeton University Press, Princeton, NJ.

May, R.M. (ed.) (1981). *Theoretical Ecology*, 2nd edition. Blackwell Scientific Publications, Oxford.

May, R.M. & Anderson, R.M. (1983). Epidemiology and genetics in the coevolution of parasites and hosts. *Proceedings of the Royal Society B*, 219, 281–313.

May, R.M. & Hassell, M.P. (1988). Population dynamics and biological control. *Philosophical Transactions of the Royal Society B*, 318, 129–169.

McNaughton, S.J. (1985). Ecology of a grazing ecosystem: the Serengeti. *Ecological Monographs*, 55, 259–294.

McNeilly, T. (1987). Evolutionary lessons from degraded ecosystems. In: *Restoration Ecology* (W.R. Jordan, M.E. Gilpin & J.D. Aber, eds), pp. 271–286. Cambridge University Press, Cambridge.

McQuilkin, W.E. (1940). The natural establishment of pine in abandoned fields in the Piedmont Plateau region. *Ecology*, 21, 135–147.

Meharg, A.A. & Macnair, M.R. (1990). An altered phosphate uptake system in arsenate-tolerant *Holcus lanatus. New Phytologist*, 116, 29–35.

Meharg, A.A. & Macnair, M.R. (1991). Uptake, accumulation and translocation of arsenate in arsenate-tolerant and non-tolerant *Holcus lanatus. New Phytologist*, 117, 225–231.

Middleton, N. (1991). *Desertification*. Oxford University Press, Oxford.

Miles, J. (1974). Effects of experimental interference with stand structure on establishment of seedlings in Callunetum. *Journal of Ecology*, 62, 675–687.

Miller, H.G. & Miller, J.D. (1976). Effect of nitrogen supply on net primary production in Corsican pine. *Journal of Applied Ecology*, 13, 249–256.

Miller, H.G., Miller, J.D. & Cooper, J.M. (1980). Biomass and nutrient accumulation at different growth rates in thinned plantations of Corsican pine. *Forestry*, 53, 23–39.

Milliman, J.D. & Meade, R.H. (1983). World-wide delivery of river sediment to the oceans. *Journal of Geology*, 91, 1–21.

Monteith, J.L. (1981). Climatic variation and the growth of crops. *Quarterly Journal of the Royal Meteorological Society*, 107, 749–774.

Mooney, H.A. & Drake, J.A. (eds) (1986). *Ecology of Biological Invasions of North America and Hawaii*. Springer-Verlag, Berlin.

Moore, P.D. & Chapman, S.B. (1986). *Methods in Plant Ecology*, 2nd edition. Blackwell Scientific Publications, Oxford.

Morgan, P. & Watkinson, R.J. (1989). Microbiological methods for the cleanup of soil and ground water contaminated with halogenated organic compounds. *FEMS Microbiology Reviews*, 63, 277–300.

Morgan, R.P.C. (ed.) (1986). *Soil Erosion and its Control*. Van Nostrand, New York.

Morgan, R.P.C., Morgan, D.D.V. & Finney, H.J. (1984). A predictive model for the assessment of soil erosion risk. *Journal of Agricultural Engineering Research*, 30, 245–253.

Moriarty, F. (1988). *Ecotoxicology*, 2nd edition. Academic Press, London.

Morris, M.G. (1971). The management of grassland for the conservation of invertebrate animals. In: *The Scientific Management of Animal and Plant Communities for Conservation* (E. Duffey & A.S. Watt, eds), pp. 527–552. Blackwell Scientific Publications, Oxford.

Morton, A.J. (1977). Mineral nutrient pathways in a Molinietum in autumn and winter. *Journal of Ecology*, 65, 993–999.

Moss, B. (1988). *Ecology of Fresh Waters*, 2nd edition. Blackwell Scientific Publications, Oxford.

Moss, B. (1989). Water pollution and the management of ecosystems: a case study of science and scientist. In: *Towards a More Exact Ecology* (P.J. Grubb & J.B. Whittaker, eds), pp. 401–422. Blackwell Scientific Publications, Oxford.

Munawar, M., Dixon, G., Mayfield, C.I., Reynoldson, T. & Sadar, M.H. (eds) (1989). Environmental bioassay techniques and their application. *Hydrobiologia*, 188/189, 1–680.

Mundt, C.C. & Leonard, K.J. (1985). A modification of Gregory's model for describing plant disease gradients. *Phytopathology*, 75, 930–935.

Munro, J. (1967). The exploitation and conservation of resources by populations of insects. *Journal of Animal Ecology*, 36, 531–547.

Murphy, D.D. & Noon, B.N. (1992) [Reserve design for northern spotted owls]. *Ecological Applications*, 2, 3–17.

Murphy, D.D., Freas, K.E. & Weiss, S.B. (1990). An environment–metapopulation approach to population viability analysis for a threatened invertebrate. *Conservation Biology*, 4, 41–51.

Murray, F. & Wilson, S. (1990). Growth responses of barley exposed to SO_2. *New Phytologist*, 114, 537–541.

Myers, A.A. & Giller, P.S. (eds) (1988). *Analytical Biogeography*. Chapman & Hall, London.

Nair, P.K.R. (ed.) (1989). *Agroforestry Systems in the Tropics*. Kluwer, Dordrecht.

Newbery, D.M., Campbell, E.J.F., Lee, Y.F., Ridsdale, C.E. & Still, M.J. (1992). Primary lowland dipterocarp forest at Danum Valley, Sabah, Malaysia: structure, relative abundance and family composition. *Philosophical Transactions of the Royal Society B*, 335, 341–356.

Newhouse, J.R. (1990). Chestnut blight. *Scientific American*, 263(1), 74–79.

Newton, I. & Wyllie, I. (1992). Recovery of a sparrowhawk population in relation to declining pesticide contamination. *Journal of Applied Ecology*, 29, 476–484.

Newton, I., Bogan, J.A. & Haas, M.B. (1989). Organochlorines and mercury in the eggs of British peregrines. *Ibis*, 131, 355–376.

Ng, F.S.P. (1978). Strategies of establishment in Malayan forest trees. In: *Tropical Trees as Living Systems* (P.B. Tomlinson & M.H. Zimmerman, eds), pp. 129–162. Cambridge University Press, Cambridge.

Nobel, P.S. (1991a). *Physicochemical and Environmental Plant Physiology*. Academic Press, San Diego, CA.

Nobel, P.S. (1991b). Achievable productivities of certain CAM plants: basis for high values compared with C_3 and C_4 plants. *New Phytologist*, 119, 183–205.

Nowak, R.S. & Caldwell, M.M. (1984). A test of compensatory photosynthesis in the field: implications for herbivory tolerance. *Oecologia*, 61, 311–318.

Noy-Meir, I. (1975). Stability of grazing systems: an application of predator–prey graphs. *Journal of Ecology*, 63, 459–481.

Nye, P.H. & Greenland, D.J. (1960). *The Soil under Shifting Cultivation*. Commonwealth Agricultural Bureaux, Farnham Royal.

Nye, P.H. & Tinker, P.B. (1977). *Solute Movement in the Soil–Root System*. Blackwell Scientific Publications, Oxford.

O'Brien, S.J. & Knight, J.A. (1987). The future of the giant panda. *Nature*, 325, 758–759.

Odell, R.T., Walker, W.M., Boone, L.V. & Oldham, M.G. (1982). *The Morrow Plots*. University of Illinois Agricultural Experiment Station Bulletin 775. Urbana-Champaign.

Olsson, G. (1988). Nutrient use and productivity for different cropping systems in south Sweden during the 18th century. In: *The Cultural Landscape—Past, Present and Future* (H.H. Birks, H.J.B. Birks, P.E. Kaland & D. Moe, eds), pp. 123–137. Cambridge University Press, Cambridge.

Oosting, H.J. (1942). An ecological analysis of the plant communities of Piedmont, North Carolina. *American Midland Naturalist*, 28, 1–126.

Oosting, H.J. & Humphreys, M.E. (1940). Buried viable seeds in a successional series of old field forest soils. *Bulletin of the Torrey Botanical Club*, 67, 253–273.

Owen-Smith, N. (1989). Megafauna extinctions: the conservation message from 11 000 years BP. *Conservation Biology*, 3, 405–412.

Parry, M.L. & Carter, T.R. (1990). An assessment of the effects of climatic change on agriculture. In: *Climatic Change and Plant Genetic Resources* (M.T. Jackson, B.V. Ford-Lloyd & M.L. Parry, eds), pp. 61–82. Belhaven, London.

Parsons, A.J., Leafe, E.L., Collett, B. & Stiles, W. (1983a). [Leaf and canopy photosynthesis of continuously-grazed swards]. *Journal of Applied Ecology*, 20, 117–126.

Parsons, A.J., Leafe, E.L., Collett, B., Penning, P.D. & Lewis, J. (1983b). [Photosynthesis, crop growth and animal intake of continuously-grazed swards]. *Journal of Applied Ecology*, 20, 127–139.

Patmore, J.A. (1983). *Recreation and Resources*. Blackwell, Oxford.

Payne, C.C. (1988). Pathogens for the control of insects: where next? *Philosophical Transactions of the Royal Society B*, 318, 225–248.

Pearman, G.I., Etheridge, D., de Silva, F. & Fraser, P.J. (1986). Evidence of changing concentrations of atmospheric CO_2, N_2O and CH_4 from air bubbles in Antarctic ice. *Nature*, 320, 248–250.

Peet, R.K. & Christensen, N.L. (1980). Succession: a population process. *Vegetatio*, 43, 131–140.

Pennington, W. (1974). *The History of British Vegetation*, 2nd edition. English Universities Press, London.

Peterken, G.F. (1981). *Woodland Conservation and Management*. Chapman & Hall, London.

Peterken, G.F. (1991). Ecological issues in the management of woodland nature reserves. In: *The Scientific Management of Temperate Communities for Conservation* (I.F. Spellerberg, F.B. Goldsmith & M.G. Morris, eds), pp. 245–272. Blackwell Scientific Publications, Oxford.

Peterken, G.F. & Game, M. (1981). Historical factors affecting the distribution of *Mercurialis perennis* in central Lincolnshire. *Journal of Ecology*, 69, 781–796.

Peterken, G.F. & Game, M. (1984). Historical factors affecting the number and distribution of vascular plant species in the woodlands of central Lincolnshire. *Journal of Ecology*, 72, 155–182.

Phillips, D.L. & Shure, D.J. (1990). Patch-size effects on early succession in southern Appalachian forests. *Ecology*, 71, 204–212.

Piedade, M.T.F., Junk, W.J. & Long, S.P. (1991). The productivity of the C_4 grass *Echinochloa polystachya* on the Amazon floodplain. *Ecology*, 72, 1456–1463.

Pigott, C.D. (1975). Natural regeneration of *Tilia cordata* in relation to forest-structure in the Forest of Bialowieza, Poland. *Philosphical Transactions of the Royal Society B*, 270, 151–179.

Pigott, C.D. (1989). [Estimated ages of *Tilia cordata* trees at the northern limits of its range]. *New Phytologist*, 112, 117–121.

Pigott, C.D. & Huntley, J.P. (1978). Factors controlling the distribution of *Tilia cordata* at the northern limits of its geographical range. I. Distribution in north-west England. *New Phytologist*, 81, 429–441.

Pigott, C.D. & Huntley, J.P. (1980). Factors controlling the distribution of *Tilia cordata* at the northern limits of its geographical range. II. History in north-west England. *New Phytologist*, 84, 145–164.

Pigott, C.D. & Huntley, J.P. (1981). Factors controlling the distribution of *Tilia cordata* at the northern limits of its geographical range. III. Nature and causes of seed sterility. *New Phytologist*, 87, 817–839.

Pimentel, D. & Pimentel, M. (1979). *Food, Energy and Society*. Arnold, London.

Pimlott, D.H. (1967). Wolf predation and ungulate populations. *American Zoologist*, 7, 267–278.

Pitcher, T.J. & Hart, P.J.B. (1982). *Fisheries Ecology*. Croom Helm, London.

Pollard, E. (1973). [Woodland relic hedges in Huntingdon and Peterborough.] *Journal of Ecology*, 61, 343–352.

Pollard, E. & Relton, J. (1970). [A study of small mammals in hedges and cultivated fields]. *Journal of Applied Ecology*, 7, 549–557.

Porter, R.D. & Wiemeyer, S.N. (1969). Dieldrin and DDT: effects on sparrow hawk eggshells and reproduction. *Science*, 165, 199–200.

Porter, S.C. (ed.) (1983). *Late-Quaternary Environments of the United States. Volume I. The Late Pleistocene*. Longman, London.

Potts, G.R. (1991). The environmental and ecological importance of cereal fields. In: *The Ecology of Temperate Cereal Fields* (L.G. Firbank, N. Carter, J.F. Darbyshire & G.R. Potts, eds), pp. 3–21. Blackwell Scientific Publications, Oxford.

Powell, C.L. (1982). Selection of efficient VA mycorrhizal fungi. *Plant and Soil*, 68, 3–9.

Pratt, D.J. & Gwynne, M.D. (1977). *Rangeland Management and Ecology in East Africa*. Hodder & Stoughton, London.

Price, C. (1989). *The Theory and Application of Forest Economics*. Blackwell, Oxford.

Price, R. & Mitchell, C.P. (1985). *Potential for Wood as Fuel in the United Kingdom*. Energy Technology Support Unit, Harwell.

Prince, R.C. (1992). Bioremediation of oil spills, with particular reference to the spill from the *Exxon Valdez*. In: *Microbial Control of Pollution* (J.C. Fry, G.M. Gadd,

R.A. Herbert, C.W. Jones & I.A. Watson-Craik, eds), pp. 19–34. Cambridge University Press, Cambridge.

Proctor, J. (1987). Nutrient cycling in primary and old secondary rainforests. *Applied Geography*, 7, 135–152.

Quinn, J.F. & Hastings, A. (1987). Extinction in subdivided habitats. *Conservation Biology*, 1, 198–208.

Rackham, O. (1976). *Trees and Woodland in the British Landscape*. Dent, London.

Rackham, O. (1986). *The History of the Countryside*. Dent, London.

Rasmussen, P.E. & Collins, H.P. (1991). Long-term impacts of tillage, fertilizer, and crop residue on soil organic matter in temperate semiarid regions. *Advances in Agronomy*, 45, 93–134.

Ratcliffe, D.A. (1970). Changes attributable to pesticides in egg breakage frequency and eggshell thickness in some British birds. *Journal of Applied Ecology*, 7, 67–115.

Ratcliffe, D.A. (1980). *The Peregrine Falcon*. Poyser, Calton, Staffordshire.

Rawes, M. (1981). Further results of excluding sheep from high-level grasslands in the North Pennines. *Journal of Ecology*, 69, 651–669.

Reinert, R.E. (1972). [Accumulation of dieldrin in an alga, *Daphnia* and guppy]. *Journal of the Fisheries Research Board of Canada*, 29, 1413–1418.

Ricker, W.E. (1954). Stock and recruitment. *Journal of the Fisheries Research Board of Canada*, 11, 559–623.

Ricklefs, R.E. (1990). *Ecology*, 3rd edition. Freeman, New York.

Risch, S.J., Andow, D. & Altieri, M.A. (1983). [Agroecosystem diversity and pest control]. *Environmental Entomology*, 12, 625–629.

Rishbeth, J. (1988). Biological control of air-borne pathogens. *Philosophical Transactions of the Royal Society B*, 318, 265–281.

Roberts, T.M. (1984). Long-term effects of sulphur dioxide on crops: an analysis of dose–response relations. *Philosophical Transactions of the Royal Society B*, 305, 299–316.

Robinson, N.J. & Jackson, P.J. (1986). 'Metallothionein-like' metal complexes in angiosperms; their structure and function. *Physiologia Plantarum*, 67, 499–506.

Romme, W.H. & Despain, D.G. (1989). The Yellowstone fires. *Scientific American*, 261(5), 21–29.

Rothschild, B.J. (1986). *Dynamics of Marine Fish Populations*. Harvard University Press, Cambridge, MA.

Royama, T. (1984). Population dynamics of the spruce budworm *Choristoneura fumiferana*. *Ecological Monographs*, 54, 429–462.

Ruiz, M. & Ruiz, J.P. (1986). Ecological history of transhumance in Spain. *Biological Conservation*, 37, 73–86.

Runkle, J.R. (1982). Patterns of disturbance in some old-growth mesic forests of eastern North America. *Ecology*, 63, 1533–1546.

Russell, W.M.S. (1967). *Man, Nature and History*. Aldus, London.

Ryden, J.C., Ball, P.R. & Garwood, E.A. (1984). Nitrate leaching from grassland. *Nature*, 311, 50–53.

Salomons, W., Bayne, B.L., Duursma, E.K. & Förstner, U. (eds). (1988). *Pollution of the North Sea*. Springer-Verlag, Berlin.

Sanford, R.L., Saldarriaga, J., Clark, K.E., Uhl, C. & Herrera, R. (1985). Amazon rainforest fires. *Science*, 227, 53–55.

Sarnelle, O. (1992). Nutrient enrichment and grazer effects on phytoplankton in lakes. *Ecology*, 73, 551–560.

Saulei, S.M. (1984). Natural regeneration following clear-fell logging operations in the Gogol Valley, Papua New Guinea. *Ambio*, 13, 351–354.

Saulei, S.M. (1985). The recovery of tropical lowland rainforest after clear-fell logging in the Gogol Valley, Papua New Guinea. Ph.D. thesis, Aberdeen.

Saulei, S.M. & Swaine, M.D. (1988). Rain forest seed dynamics during succession at Gogol, Papua New Guinea. *Journal of Ecology*, 76, 1133–1152.

Savill, P.S. & Evans, J. (1986). *Plantation Silviculture in Temperate Regions*. Clarendon Press, Oxford.

Sayer, J.A. & Whitmore, T.C. (1991). Tropical moist forests: destruction and species extinction. *Biological Conservation*, 55, 199–213.

Schaefer, M.B. (1954). Some aspects of the dynamics of populations important to the management of commercial marine fisheries. *Bulletin of the Inter-American Tropical Tuna Commission*, 1, 27–56.

Schafer, E.W. (1972). The acute oral toxicity of 369 pesticidal, pharmaceutical and other chemicals to wild birds. *Toxicology and Applied Pharmacology*, 21, 315–330.

Schindler, D.W. & Fee, E.J. (1974). Experimental Lakes Area: whole-lake experiments in eutrophication. *Journal of the Fisheries Research Board of Canada*, 31, 937–953.

Schneider, S.H. (1989). The greenhouse effect: science and policy. *Science*, 243, 771–781.

Schneider, S.H. (1992). Will sea levels rise or fall? *Nature*, 356, 11–12.

Scholefield, D., Lockyer, D.R., Whitehead, D.C. & Tyson, K.C. (1991). A model to predict transformations and losses of nitrogen in UK pastures grazed by beef cattle. *Plant and Soil*, 132, 165–177.

Shackleton, N.J., Hall, M.A., Line, J. & Shuxi, C. (1983). [Carbon isotope data confirm reduced CO_2 in the ice age atmosphere.] *Nature*, 306, 319–322.

Shapiro, J. & Wright, D.I. (1984). Lake restoration by biomanipulation: Round Lake, Minnesota, the first two years. *Freshwater Biology*, 14, 371–383.

Sheikh, K.H. & Rutter, A.J. (1969). [Root distribution of *Molinia caerulea* and *Erica tetralix* in relation to soil porosity.] *Journal of Ecology*, 57, 713–726.

Sheppard, S.C. & Evenden, W.G. (1988). The assumption of linearity in soil and plant concentration ratios: an experimental evaluation. *Journal of Environmental Radioactivity*, 7, 221–247.

Silvertown, J. (1980). [The dynamics of a grassland ecosystem: the Park Grass experiment]. *Journal of Applied Ecology*, 17, 491–504.

Simmons, I.G. (1989). *Changing the Face of the Earth*. Blackwell, Oxford.

Sinclair, A.R.E., Krebs, C.J., Smith, J.N.M. & Boutin, S. (1988). Population biology of snowshoe hares. III. Nutrition, plant secondary compounds and food limitation. *Journal of Animal Ecology*, 57, 787–806.

Skellam, J.G. (1951). Random dispersal in theoretical populations. *Biometrika*, 38, 196–218.

Smith, D.W. (1985). Biological control of excessive phytoplankton growth and the enhancement of aquaculture production. *Canadian Journal of Fisheries and Aquatic Sciences*, 42, 1940–1945.

Smith, G. C. & Harris, S. (1991). [Rabies in urban foxes in Britain: the use of a model to examine the pattern of spread and evaluate different control regimes]. *Philosophical Transactions of the Royal Society*, 334, 459–479.

Somerville, L. & Greaves, M.P. (1987). *Pesticide Effects on Soil Microflora*. Taylor & Francis, London.

Soulé, M.E. (ed.) (1986). *Conservation Biology*. Sinauer, Sunderland, MA.

Soulé, M.E. (ed.) (1987). *Viable Populations for Conservation*. Cambridge University Press, Cambridge.

Southward, A.J. (1980). The Western English Channel—an inconstant system? *Nature*, 285, 361–366.

Speight, M.R. & Wainhouse, D. (1989). *Ecology and Management of Forest Insects*. Clarendon Press, Oxford.

Spellerberg, I.F., Goldsmith, F.B. & Morris, M.G. (eds) (1991). *The Scientific Management of Temperate Communities for Conservation*. Blackwell Scientific Publications, Oxford.

Sprague, J.B. (1969). Measurement of pollutant toxicity to fish. I. Bioassay methods for acute toxicity. *Water Research*, 3, 793–821.

Spurr, S.H. & Barnes, B.V. (1980). *Forest Ecology*, 3rd edition. Wiley, New York.

Stage, A.R., Renner, D.L. & Chapman, R.C. (1988). *Selected Yield Tables for Plantations and Natural Stands in Inland Northwest Forests*. US Department of Agriculture Forest Service, Research Paper INT-394.

Stanhill, G. (1976). Trends and deviations in the yield of the English wheat crop during the last 750 years. *Agro-Ecosystems*, 3, 1–10.

Steidl, R.J., Griffin, C.R., Niles, L.J. & Clark, K.E. (1991). Reproductive success and eggshell thinning of a reestablished Peregrine falcon population. *Journal of Wildlife Management*, 55, 294–299.

Steven, M.D. & Clark, J.A. (eds) (1990). *Applications of Remote Sensing in Agriculture*. Butterworths, London.

Stevenson, F.J. (1986). *Cycles of Soil*. Wiley, New York.

Stewart, G.J. & Carlson, C.A. (1986). The biology of natural transformation. *Annual Review of Microbiology*, 40, 211–235.

Stuart, A.J. (1991). Mammalian extinctions in the late Pleistocene of northern Eurasia and North America. *Biological Reviews*, 66, 453–562.

Sussman, M., Collins, C.H., Skinner, F.A. & Stewart-Tull, D.E. (eds) (1988). *The Release of Genetically-engineered Micro-organisms*. Academic Press, London.

Swain, A.M. (1973). A history of fire and vegetation in northeastern Minnesota as recorded in lake sediments. *Quaternary Research*, 3, 383–396.

Swift, M.J., Frost, P.G.H., Campbell, B.M., Hatton, J.C. & Wilson, K.B. (1989). [Nitrogen cycling in farming systems derived from savanna]. In: *Ecology of Arable Land* (M. Clarholm & L. Bergstrom, eds), pp. 63–76. Kluwer, Dordrecht.

Sykes, R.B. & Richmond, M.H. (1970). Intergeneric transfer of a β-lactamase gene between *Ps. aeruginosa* and *E. coli*. *Nature*, 226, 952–954.

Taiz, L. & Zeiger, E. (eds) (1991). *Plant Physiology*. Benjamin/Cummings, Redwood City, CA.

Talbot, H.W., Yamamoto, D.K., Smith, M.W. & Seidler, R.J. (1980). [Antibiotic resistance and its transfer among *Klebsiella* strains in botanical environments]. *Applied and Environmental Microbiology*, 39, 97–104.

Tanner, E.V.J. (1985). Jamaican montane forests: nutrient capital and cost of growth. *Journal of Ecology*, 73, 553–568.

Temple, P.H. (1972). Measurements of runoff and soil erosion at an erosion plot scale with particular reference to Tanzania. *Geografiska Annaler*, 54-A, 203–220.

Thang, H.C. (1987). Forest management systems for tropical high forest, with special reference to Peninsular Malaysia. *Forest Ecology and Management*, 21, 3–20.

Thirgood, J.V. (1981). *Man and the Mediterranean Forest*. Academic Press, London.

Thomas, J.A. (1991). Rare species conservation: case studies of European butterflies. In: *The Scientific Management of Temperate Communities for Conservation* (I.F. Spellerberg, F.B. Goldsmith & M.G. Morris, eds), pp. 149–197. Blackwell Scientific Publications, Oxford.

Thorarinsson, K. (1990). Biological control of the cottony-cushion scale: experimental tests of the spatial density-dependence hypothesis. *Ecology*, 71, 635–644.

Tilghman, N.G. (1989). Impacts of white-tailed deer on forest regeneration in northwestern Pennsylvania. *Journal of Wildlife Management*, 53, 524–532.

Tinker, P.B. & Gildon, A. (1983). Mycorrhizal fungi and ion uptake. In: *Metals and Micronutrients* (D.A. Robb & W.S. Pierpoint, eds), pp. 21–32. Academic Press, London.

Tisdall, J.M. & Oades, J.M. (1979). Stabilization of soil aggregates by the root systems of ryegrass. *Australian Journal of Soil Research*, 17, 429–441.

Tisdall, J.M. & Oades, J.M. (1980). The effect of crop rotation on aggregation in a red-brown earth. *Australian Journal of Soil Research*, 18, 423–433.

Tisdall, J.M. & Oades, J.M. (1982). Organic matter and water-stable aggregates in soils. *Journal of Soil Science*, 33, 141–163.

Tomanek, G.W. & Albertson, F.W. (1957). Variations in cover, composition, production, and roots of vegetation on two prairies in western Kansas. *Ecological Monographs*, 27, 267–281.

Tomashow, L.S. & Weller, D.M. (1990). Role of antibiotics and siderophores in biocontrol of take-all disease of wheat. *Plant and Soil*, 129, 93–99.

Tomlinson, J.A. & Carter, A.L. (1970). [Studies in the seed transmission of cucumber mosaic virus in chickweed]. *Annals of Applied Biology*, 66, 381–386.

Tomlinson, J.A., Carter, A.L., Dale, W.T. & Simpson, C.J. (1970). Weed plants as sources of cucumber mosaic virus. *Annals of Applied Biology*, 66, 11–16.

UN Statistics Yearbooks. United Nations, New York.

UNEP (1991). *Environmental Data Report*, 3rd edition. United Nations Environment Programme. Blackwell, Oxford.

Usher, M.B. (ed.) (1986). *Wildlife Conservation Evaluation*. Chapman & Hall, London.

Van Alfen, N.K. (1982). Biology and potential for disease control of hypovirulence of *Endothia parasitica*. *Annual Review of Phytopathology*, 20, 349–362.

van Campo, E., Duplessey, J.C., Prell, W.L., Barratt, N. & Sabatier, R. (1990). [Comparison of terrestrial and marine temperature estimates for the past 135 kyr off southeast Africa.] *Nature*, 348, 209–212.

van den Bosch, F., Hengeveld, R. & Metz, J.A.J. (1992). Analysing the velocity of animal range expansion. *Journal of Biogeography*, 19, 135–150.

Vander Wall, S.B. & Balda, R.P. (1977). Coadaptations of the Clark's nutcracker and the piñon pine for efficient seed harvest and dispersal. *Ecological Monographs*, 47, 89–111.

van Doren, C.S., Priddle, G.B. & Lewis, J.E. (eds) (1979). *Land and Leisure*, 2nd edition. Methuen, London.

Van Elsas, J.D., Govaert, J.M. & van Veen, J.A. (1987). Transfer of plasmid pFT30 between Bacilli in soil as influenced by bacterial population dynamics and soil conditions. *Soil Biology and Biochemistry*, 19, 639–647.

Verstraete, M.M. & Schwartz, S.A. (1991). Desertification and global change. *Vegetatio*, 91, 3–13.

Vickery, P.J. (1972). Grazing and net primary production of a temperate grassland. *Journal of Applied Ecology*, 9, 307–314.

Virgil (29 BC). *Georgics*. Translation into English by C. Day Lewis. Oxford University Press, Oxford.

Vitousek, P.M. (1991). Can planted forests counteract increasing atmospheric carbon dioxide? *Journal of Environmental Quality*, 20, 348–354.

Vitousek, P.M., Ehrlich, P.R., Ehrlich, A.H. & Matson, P.A. (1986). Human appropriation of the products of photosynthesis. *BioScience*, 36, 368–373.

Vogtmann, H., Temperli, A.T., Künsch, U., Eichenberger, M., Ott, P. (1984). Accumulation of nitrates in leafy vegetables grown under contrasting agricultural systems. *Biological Agriculture and Horticulture*, 2, 51–68.

Vreman, K., Van der Struijs, T.D.B., Van den Hoek, J., Berende, P.L.M. & Goedhart, P.W. (1989). Transfer of Cs-137 from grass and wilted grass silage to milk of dairy cows. *Science of the Total Environment*, 85, 139–147.

Waage, J. (1989). The population ecology of pest–pesticide–natural enemy interactions. In: *Pesticides and Non-target Invertebrates* (P.C. Jepson, ed.), pp. 81–93. Intercept, Wimborne.

Waage, J.K. & Greathead, D.J. (1988). Biological control: challenges and opportunities. *Philosophical Transactions of the Royal Society B*, 318, 111–128.

Walker, B.H., Ludwig, D., Holling, C.S. & Peterman, R.M. (1981). Stability of semi-arid savanna grazing systems. *Journal of Ecology*, 69, 473–498.

Walker, C.H. (1990). Kinetic models to predict bioaccumulation of pollutants. *Functional Ecology*, 4, 295–301.

Walsh, J.J. (1981). A carbon budget for overfishing off Peru. *Nature*, 290, 300–304.

Ward, D.M., Atlas, R.M., Boehm, P.D. & Calder, J.A. (1980). Microbial biodegradation and chemical evolution of oil from the *Amoco* spill. *Ambio*, 9, 277–283.

Waring, R.H. & Schlesinger, W.H. (1985). *Forest Ecosystems: Concepts and Management*. Academic Press, Orlando, FL.

Warren, A. & Goldsmith, F.B. (eds) (1983). *Conservation in Perspective.* Wiley, Chichester.

Watt, A.S. (1947). Pattern and process in the plant community. *Journal of Ecology*, 35, 1–22.

Watt, A.S. (1960). Population changes in acidiphilous grass-heath in Breckland, 1936–57. *Journal of Ecology*, 48, 605–629.

Wayne, R.K. *et al.* (1991). Conservation genetics of the endangered Isle Royale gray wolf. *Conservation Biology*, 5, 41–51.

Webb, S.L. (1986). Potential role of passenger pigeons and other vertebrates in the rapid Holocene migrations of nut trees. *Quaternary Research*, 26, 367–375.

Wegner, J.F. & Merriam, G. (1979). Movements by birds and small mammals between a wood and adjoining farmland habitats. *Journal of Applied Ecology*, 16, 349–357.

Wellburn, A. (1988). *Air Pollution and Acid Rain.* Longman, London.

Wells, T.C.E., Sheail, J., Ball, D.F. & Ward, L.K. (1976). Ecological studies on the Porton Ranges: relationships between vegetation, soils and land-use history. *Journal of Ecology*, 64, 589–626.

West, L.T. *et al.* (1991). Cropping system effects on interrill soil loss in the Georgia Piedmont. *Soil Science Society of America Journal*, 55, 460–466.

Weste, G. (1986). Vegetation changes associated with invasion by *Phytophthora cinnamomi* of defined plots in the Brisbane Ranges, Victoria, 1975–1985. *Australian Journal of Botany*, 34, 633–648.

Weste, G. & Marks, G.C. (1987). The biology of *Phytophthora cinnamomi* in Australasian forests. *Annual Review of Phytopathology*, 25, 207–229.

Whitmore, T.C. (1984). *Tropical Rain Forests of the Far East*, 2nd edition. Clarendon Press, Oxford.

Whitney, G.G. (1986). Relation of Michigan's presettlement pine forests to substrate and disturbance history. *Ecology*, 67, 1548–1559.

Whitney, G.G. (1990). The history and status of the hemlock–hardwood forests of the Allegheny Plateau. *Journal of Ecology*, 78, 443–458.

Whittaker, R.H. (1956). Vegetation of the Great Smoky Mountains. *Ecological Monographs*, 26, 1–80.

Whittaker, R.H. (1966). Forest dimensions and production in the Great Smoky Mountains. *Ecology*, 47, 103–121.

Whittaker, R.H. (1975). *Communities and Ecosystems*, 2nd edition. Collier-Macmillan, London.

Whittaker, R.H., Likens, G.E., Bormann, F.H., Eaton, J.S. & Siccama, T.G. (1979). The Hubbard Brook ecosystem study: forest nutrient cycling and element behavior. *Ecology*, 60, 203–220.

Wild, A. (ed.) (1988). *Russell's Soil Conditions and Plant Growth*, 11th edition. Longman, Harlow.

Wild, S.R., Berrow, M.L. & Jones, K.C. (1991). The persistence of polynuclear aromatic hydrocarbons (PAHs) in sewage sludge amended agricultural soils. *Environmental Pollution*, 72, 141–157.

Wilkins, D.A. (1978). The measurement of tolerance to edaphic factors by means of root growth. *New Phytologist*, 80, 623–633.

Williams, E.D. (1978). *Botanical Composition of the Park Grass Plots at Rothamsted.* Rothamsted Experimental Station, Harpenden.

Williams, M. (1989). *Americans and their Forests.* Cambridge University Press, Cambridge.

Williams, R.J.P. (ed.) (1978). *Phosphorus in the Environment: its Chemistry and Biochemistry.* Excerpta Medica, Amsterdam.

Williamson, H. & Brown K.C. (1986). The analysis and modelling of British invasions. *Philosophical Transactions of the Royal Society B*, 314, 505–522.

Wilson, J.T. & Wilson, B.H. (1985). Biotransformation of trichloroethylene in soil. *Applied and Environmental Microbiology*, 49, 242–243.

Wilson, R.A., Smith G. & Thomas M.R. (1982). Fascioliasis. In: *The Population*

Dynamics of Infectious Diseases (R.M. Anderson, ed.), pp. 262–319. Chapman & Hall, London.

Winner, R.W., Boesel, M.W. & Farrell, M.P. (1980). Insect community structure as an index of heavy-metal pollution in lotic systems. *Canadian Journal of Fisheries and Aquatic Science*, 37, 647–655.

Winteringham, F.P.W. (1985). *Environment and Chemicals in Agriculture*. Elsevier, London.

Wong, M., Wright, S.J., Hubbell, S.P. & Foster, R.B. (1990). [The consequences of outbreak defoliation in a tropical tree.] *Journal of Ecology*, 78, 579–588.

Wood, H.A. & Granados, R.R. (1991). Genetically engineered Baculoviruses as agents for pest control. *Annual Review of Microbiology*, 45, 69–87.

Wood, R.K.S. & Way, M.J. (eds) (1988). Biological control of pests, pathogens and weeds: developments and prospects. *Philosophical Transactions of the Royal Society B*, 318, 109–376.

Woodwell, G.M., Wurster, C.F. & Isaacson, P.A. (1967). DDT residues in an East Coast estuary: a case of biological concentration of a persistent insecticide. *Science*, 156, 821–823.

Woodwell, G.M. *et al.* (1978). Biota and the world carbon budget. *Science*, 199, 141–146.

Wratten, S.D. & Powell, W. (1991). Cereal aphids and their natural enemies. In: *The Ecology of Temperate Cereal Fields* (L.G. Firbank, N. Carter, J.F. Darbyshire & G.R. Potts, eds), pp. 233–257. Blackwell Scientific Publications, Oxford.

Yeaton, R.I., Travis, J. & Gilinsky, E. (1977). Competition and spacing in plant communities: the Arizona upland association. *Journal of Ecology*, 65, 587–595.

Zavitkovski, J. (1979). [Energy production in irrigated, intensively cultured plantations of *Populus* and jack pine.] *Forest Science*, 25, 383–392.

Zentmeyer, G.A. (1985). Origin and distribution of *Phytophthora cinnamomi*. In: *Ecology and Management of Soilborne Plant Pathogens* (C.A. Parker, A.D. Rovira, K.J. Moore & P.T.W. Wong, eds), pp. 71–72. American Phytopathological Society, St Paul, MN.

Index